W9-COB-375

INTRODUCTION TO
AGRICULTURAL ECONOMICS

INTRODUCTION TO AGRICULTURAL ECONOMICS

Third Edition

John B. Penson, Jr.
Texas A&M University

Oral Capps, Jr.
Texas A&M University

C. Parr Rosson III
Texas A&M University

Prentice
Hall

Upper Saddle River, New Jersey

Library of Congress Cataloging-in-Publication Data
Penson, John B.
   Introduction to agricultural economics / John B. Penson, Jr., Oral
Capps, Jr., C. Parr Rosson III.—3rd ed.
     p. cm.
   Includes bibliographical references and index.
   ISBN 0-13-019572-3
   1. Agriculture—Economic aspects.   I. Capps, Oral.   II. Rosson,
C. Parr. III. Title.
HD1415.P374 2002
338.1—dc21   00-069881

Executive Editor: Debbie Yarnell
Associate Editor: Kimberly Yehle
Production Editor: Lori Dalberg, Carlisle Publishers Services
Production Liaison: Eileen O'Sullivan
Director of Manufacturing and Production: Bruce Johnson
Managing Editor: Mary Carnis
Marketing Manager: Jimmy Stephens
Senior Design Coordinator: Miguel Ortiz
Cover Design: Miguel Ortiz
Composition: Carlisle Communications, Ltd.
Printing and Binding: R.R. Donnelley Harrisonburg

Prentice-Hall International (U.K.) Limited, *London*
Prentice-Hall of Australia Pty. Limited, *Sydney*
Prentice-Hall Canada, Inc., *Toronto*
Prentice-Hall Hispanoamericana, S.A., *Mexico*
Prentice-Hall of India Private Limited, *New Delhi*
Prentice-Hall of Japan, Inc., *Tokyo*
Prentice-Hall Singapore, Pte. Ltd.
Editora Prentice-Hall do Brasil, Ltda., *Rio de Janeiro*

10 9 8 7 6 5 4 3 2
ISBN 0-13-019572-3

CONTENTS

CHAPTER 4   *CONSUMER EQUILIBRIUM AND MARKET DEMAND 74*

CHAPTER 9   ECONOMICS OF PRODUCT SUBSTITUTION   197

CHAPTER 10   MARKET EQUILIBRIUM AND PRODUCT PRICE: PERFECT COMPETITION   212

CHAPTER 11   *MARKET EQUILIBRIUM AND PRODUCT PRICE: IMPERFECT COMPETITION* 231

PART IV

GOVERNMENT IN THE FOOD AND FIBER INDUSTRY 267

CHAPTER 12   GOVERNMENT INTERVENTION IN AGRICULTURE 269

CHAPTER 15   *CONSEQUENCES OF BUSINESS FLUCTUATIONS   351*

CHAPTER 18   *MACROECONOMIC POLICY AND AGRICULTURE 422*

CHAPTER 22   AGRICULTURAL TRADE POLICY   525

CHAPTER 23   EMERGING ISSUES IN AGRICULTURAL TRADE: THE FORMATION OF PREFERENTIAL TRADING ARRANGEMENTS   552

PREFACE

The purpose of this book is to provide beginning students in agriculture with a systematic introduction to the basic concepts and issues in economics as they relate to a major segment of the U.S. economy—namely, the food and fiber industry. This requires that the student understand the microeconomic and macroeconomic forces influencing the decisions of producers and consumers of food and fiber products, which include (1) farmers and ranchers, (2) the agribusinesses that supply them with production inputs and credit, (3) the agribusinesses that process food products and manufacture fiber products, and (4) the agribusinesses that provide marketing and related services at the wholesale and retail levels to both domestic consumers and overseas markets.

We begin the book by answering the question raised by the title of Chapter 1, "What is agricultural economics?" We do this by first defining the field of economics and then developing a definition of agricultural economics based on the role that agricultural economists play at both the micro and macro levels. Chapter 2 discusses the changing structure of agriculture during the post–World War II period and the structure of those sectors in the economy that supply farmers and ranchers with inputs and process their output.

Part II helps students understand the economic decisions made by consumers of food and fiber products. The topics covered here include the forces influencing consumer behavior (Chapter 3) and the concept market demand for a particular product (Chapter 4). Part II concludes with the elasticity of demand (Chapter 5). The specification of various elasticity measures is supplemented by empirical examples and their relevance to decision making in the food and fiber industry, including the potential magnitude of consumer response and the implication for producer revenue.

Part III turns to the supply side of the market. Chapter 6 describes some basic fundamentals of assessing current business performance. Issues related to resource use and production response by businesses in the short run are explained in Chapter 7. This is followed by a discussion of the economic forces underlying the firm's input use (Chapters 8) and the firm's choice of commodities to produce (Chapter 9). Chapter 10 introduces the market supply curve, followed by the determination of market clearing prices and quantities under perfect and imperfect competition in Chapter 11. Empirical examples illustrate the magnitude and applicability of the relationships covered in the chapter.

Part IV addresses the role of government in the food and fiber industry. Chapter 12 outlines the general nature of government involvement in farm economic issues, consumer issues, resource issues, and international trade issues. Resource issues are also addressed in other chapters as well. Chapter 13 explores the relationship between market equilibrium and farm program policy, introducing the variety of approaches taken to support prices and incomes of farmers over the last 50 years. This includes the current farm commodity legislation set to expire after 2002.

Part V focuses on the macroeconomics of agriculture. Chapter 14 outlines the general linkage between product markets and national output, Chapter 15 illustrates the consequences of business fluctuations in the economy, Chapter 16 documents the importance of monetary policy to the economic performance of the economy, and Chapter 17 does the same thing for fiscal policy. Part V establishes the relationship between events in the general economy and their impacts on agriculture and other sectors in the U.S. food and fiber industry. Chapter 18 concludes this part of the book with a brief walk through the last several decades, illustrating graphically how changes in the macroeconomy affected the economic performance of agriculture.

Part VI focuses on international agricultural trade issues. Chapter 19 examines the growth and instability of agricultural trade, including the relative dependence on exports and imports. Chapter 20 focuses on the foreign exchange market, the international monetary system, and the effects of foreign exchange rates on U.S. agricultural trade. Chapter 21 explores the rationale for why nations trade and who gains from trade. Chapter 22 addresses agricultural trade policy, the rationale for restricting free trade, and the trade policy decision-making mechanisms and institutions. Finally, Chapter 23 examines the formation of preferential trading arrangements and their potential impact.

Each chapter concludes with a summary of the chapter's purpose and the major points covered. In addition, the key terms used in the chapters are defined. Most chapters also include a number of exercises the readers can use to test their understanding of key issues covered in the chapter. There is also a list of references and further readings.

This textbook differs substantially from the traditional introductory agricultural economics textbook in several ways. First, the book explicitly goes beyond the farm gate to address the entirety of the food and fiber industry, which accounts for about 17% of U.S. national output. Second, the book places an unusually strong emphasis on the macroeconomics of agriculture (five chapters) and international trade (five chapters). The experience of the 1980s and 1990s certainly has shown that farmers and ranchers, agribusinesses, financial institutions, and consumers of food and fiber products are significantly affected by macroeconomic policies and trade agreements.

Supplementary materials to complement this textbook—an instructor's manual containing chapter outlines as well as a test bank of questions and problems (with solutions) that have been classroom tested over an extended time period, and transparency masters that help present key concepts from the text—are available. A slide show of key features of each chapter in Powerpoint will also be made available.

We wish to thank the many students who have given us comments and suggestions during the developmental phases of this edition.

John B. Penson, Jr.
Oral Capps, Jr.
C. Parr Rosson III

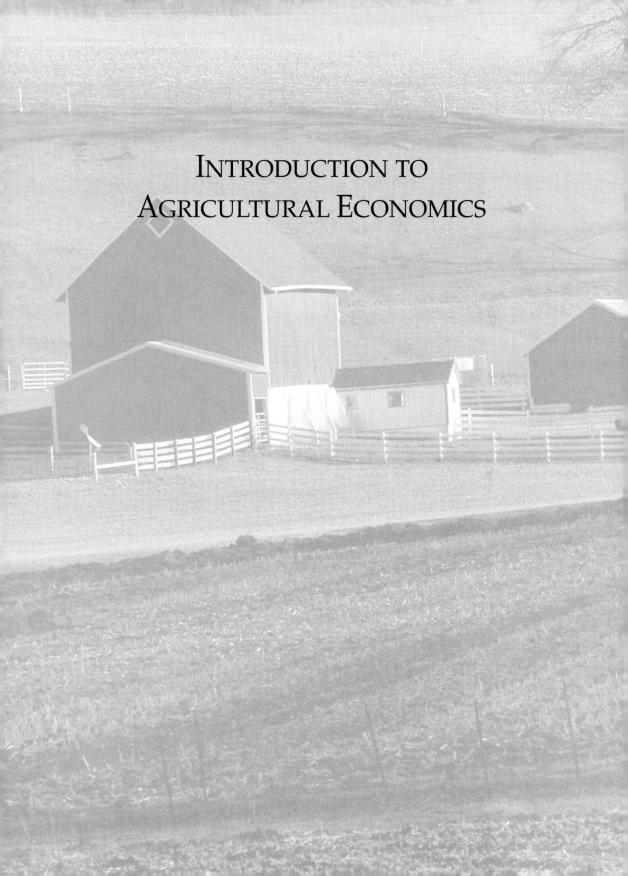

INTRODUCTION TO
AGRICULTURAL ECONOMICS

P A R T

I

INTRODUCTION

1

WHAT IS AGRICULTURAL ECONOMICS?

*While agriculture prospers all other arts alike
are vigorous and strong; but where the land is
forced to remain desert, the spring that feeds
the other arts is dried up.*

Xenophon
(c. 440–355 B.C.)

TOPICS OF DISCUSSION

The roots of agricultural economics can be traced back to ancient Egypt and to the first agricultural economist, Joseph. Joseph interpreted the dreams of the Pharaoh of Egypt and correctly predicted seven years of feast and seven years of famine.

What is agricultural economics? If you were to say "Agricultural economics is the application of economic principles to agriculture," you would be technically correct—but in a narrow context. This definition does not recognize the economic, social, and environmental issues addressed by the agricultural economist profession. To perceive agricultural economics as being limited only to the economics of farming and ranching operations would be incorrect. These operations annually account for only 2% of the nation's output. Actually, the scope of agricultural economics goes well beyond the farm gate to encompass a broader range of food- and fiber-related activity, which annually accounts for approximately 19% of the nation's output.

Before we define agricultural economics further, let us first examine the scope of economics and the role that agricultural economists play in today's economy. This examination will allow us to propose a more definitive answer to the question raised by the chapter title. A more in-depth assessment of the nation's food and fiber industry will be presented in Chapter 2.

SCOPE OF ECONOMICS

Two frequently used clichés describe the economic problem: "You can't have your cake and eat it too," and "There's no such thing as a free lunch." Because we—individually or collectively—cannot have everything we desire, we must make

"And so, extrapolating from the best figures available, we see that the current trends, unless dramatically reversed, will inevitably lead to a situation in which the sky will fall."
Drawn by Lorenz; © 1972 The New Yorker Magazine, Inc.

choices. Consumers, for example, must make expenditure decisions with a budget in mind. Their objective is to maximize the satisfaction they derive from allocating their time between work and leisure, and from allocating their available income to consumption and saving, given the current prices and interest rates. Producers must make production, marketing, and investment decisions with a budget in mind. Their objective is to maximize the profit of the firm, given its current resources and current relative prices. After considering the costs and benefits involved, society must also make choices on how to allocate its scarce resources among different government programs most efficiently.

Scarce Resources

The term *scarcity* refers to the finite quantity of resources that are available to meet society's needs. Because nature does not freely provide enough of these resources, there is only a limited quantity available. **Scarce resources** can be broken down into the following categories: (1) natural and biological resources, (2) human resources, and (3) manufactured resources.

Natural and Biological Resources.   Land and mineral deposits are examples of scarce **natural resources.** The quality of these natural resources in the United States differs greatly from region to region. Some land is incapable of growing anything in its natural state, and other lands are extremely fertile. Still other areas are rich in coal deposits, or oil and natural gas reserves. In recent years, our society has also become aware of the increasing scarcity of fresh water. Whereas energy-related natural resources have represented critical scarce resources in recent decades, water could become *the* critical scarce natural resource as we approach the end of this century. In addition to natural resources, scarce resources also include biological resources such as livestock, wildlife, and different genetic varieties of crops.

Human Resources.   **Human resources** are the services provided by laborers and management to the production of goods and services, and are also considered scarce. Laborers, for example, provide services that, combined with scarce non-human resources, produce economic goods.[1] Steel workers provide the labor input to producing steel. Farm laborers provide the labor input to producing crops and livestock. Labor is considered scarce even when the country's labor force is not fully employed (i.e., when the unemployment rate is 10%). Laborers supply services in response to the going wage rate and to the returns that they derive from

---

[1]Goods and services produced from scarce resources are also scarce and are referred to as economic goods. Economic goods are in contrast to free goods, in which the quantity desired is available at a price of zero. Air has long been a free good, but pollution (a negative good), which makes the air unfit to breathe, is changing this notion in some areas of the country.

leisure. Agribusinesses may not be able to hire all the labor services they desire at the wage they wish to pay.

Management, another form of human resource, provides entrepreneurial services, which may entail the formation of a new firm, the renovation or expansion of an existing firm, the taking of financial risks, and the supervision of the use of the firm's existing resources so that its objectives can be met. Without entrepreneurship, large-scale agribusinesses would cease operating efficiently.

Manufactured Resources.   The third category of scarce resources is **manufactured resources,** or capital. Manufactured resources are machines, equipment, and structures. A product that has not been used up in the year it was made is also considered a manufactured resource. For example, inventories of corn raised but not fed to livestock or sold to agribusinesses represent a manufactured resource.

Scarcity is a relative concept. Nations with high per capita incomes and wealth face the problem of scarcity like nations with low per capita incomes and wealth. The difference lies in the degree to which resource scarcity exists and the forms that it takes. For example, some nations have more abundant water supplies and mineral deposits, and richer soils than others. Box 1.1 lists some scarce resources for the United States.

Making Choices

Resource scarcity forces consumers and producers to make choices. These choices have a time dimension. The choices consumers make today will have an effect on how they will live in the future. The choices businesses make today will have an effect on the future profitability of their firms. Your decision to go to college rather than get a job today was probably based in part on your desire to increase your future earning power or eventual wealth, knowing what your earning potential would be if you did not attend college.

The choices one makes also have an associated **opportunity cost.** The opportunity cost of going to college now is the income you are currently foregoing by not getting a job at this time. The opportunity cost of a consumer taking $1,000 out of his or her savings account to buy a new stereo system is the interest income this money would have earned if left in the bank. An agribusiness firm considering the purchase of a new computer system must also consider the income it could receive by using this money for another purpose. The bottom line expressed in economic terms is whether the economic benefits exceed the costs, including foregone income.

Sometimes the choices we make are constrained not only by resource scarcity but also by noneconomic considerations. These forces may be political, legal, or moral. For example, some states have "blue laws" that prohibit the sale of specific commodities on Sundays. A variety of regulations at the federal and state level that govern the production of food and fiber products, including environmental and

---

BOX 1.1   *Scarce Resources*

Human Resources

141.1 million people in U.S. civilian labor force
6.3 million people unemployed

Manufactured Resources

121.4 million net tons of steel making capacity
3.9 million miles of highways

Natural Resources

3.5 million square miles of land surface
352 million acres of forest land
954 million acres of land in farms
507 billion short tons of proven coal resources

Source: *1999 Statistical abstract of the United States.*

---

food safety concerns, exist. For example, specific chemicals are banned from use in producing and processing food products because of their potential health hazard. The Big Green movement in California in 1990 sought to ban the use of all agricultural chemicals that were shown to pose health hazards to laboratory animals.

Most resources are best suited for a particular use. For example, the instructor of this course is probably better qualified to teach this course than to perform open-heart surgery. By focusing the use of our resources on a specific task, we are engaging in **specialization.** With a given set of human and nonhuman resources, specialization of effort generally results in a higher total output. Individuals should do what they do comparatively better than others, given their endowment of resources. Some individuals might specialize in fields such as professional athletics or law. Others might specialize in agricultural economics. States and nations may find it to their advantage to specialize in the production of coffee, rice, or computers and import other commodities for which their endowment of natural, human, and manufactured resources is ill suited. As illustrated in Figure 1.1, Kansas has a surplus of wheat production but a shortage of orange production, while Florida has a surplus of orange production and a shortage of wheat production. Both states have a shortage of potato production, while Idaho has plenty to spare. Specialization in production provides the basis for trade because these products are processed and shipped to their point of consumption by consumers.

FIGURE 1.1   Specialization and resource allocation.

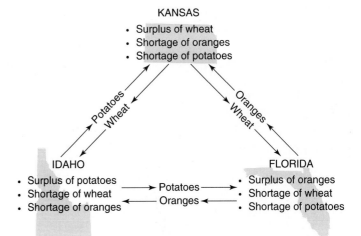

Society as a whole must make choices that might alter the allocation of resources from that which individuals collectively might have desired. For example, all nations normally allocate some resources to military uses. Society as a whole must decide how best to allocate its resources between the production of civilian goods and services versus production of military goods, popularly referred to as the choice of "guns versus butter."

DEFINITION OF ECONOMICS

With the foregoing concepts of resource scarcity and choice in mind, we may now define the nature and scope of the field of economics as follows:

> **Economics** is a social science that deals with how consumers, producers, and societies choose among the alternative uses of scarce resources in the process of producing, exchanging, and consuming goods and services.

Microeconomics versus Macroeconomics

As with most disciplines, the field of economics can be divided into several branches. **Microeconomics** and **macroeconomics** are two major branches of economics. Microeconomics focuses on the economic actions of individuals or specific groups of individuals. For example, microeconomists are concerned with the economic behavior of consumers or producers of eggs, and with the determination of the price of eggs. Microeconomics ignores the interrelationships among markets and assumes that other events taking place in the economy outside the market will remain constant.

Macroeconomics focuses on broad aggregates, such as the growth of the nation's gross domestic product (GDP), the gaps between the economy's potential

GDP and its current GDP, and trade-offs between unemployment and inflation. For example, macroeconomists are concerned with identifying the monetary and fiscal policies that would reduce inflation, promote growth of the nation's economy, and improve the nation's trade balance (exports minus imports). Macroeconomics explicitly accounts for the interrelationships between the nation's labor, product, and money markets and the economic decisions of foreign governments and individuals.

Despite the differences between micro- and macroeconomics, there is no conflict between these two branches. After all, the economy in the aggregate is certainly affected by the events taking place in individual markets.

A word of caution: we must be careful when generalizing the aggregate or macroeconomic consequences of an individual or microeconomic event. If not, we run the risk of committing a **fallacy of composition,** meaning that which is true in an individual situation is not necessarily true in the aggregate. For example, suppose Walt Wheatman adopts a new technology that doubles his wheat production. If the other 300,000 wheat farmers in the United States and other wheat producers worldwide do not follow suit, Walt's income will rise sharply. It would be wrong for Walt or others to conclude, however, that all wheat farmers would achieve income gains if they adopted this new technology also. If other wheat producers did respond, supply would expand significantly, and wheat prices would fall dramatically.

Positive versus Normative Economics

The study of economics can also be divided between **positive economics** and **normative economics.** Positive economics focuses on what-is and what-would-happen-if questions and policy issues. No value judgments or prescriptions are made. Instead, the economic behavior of producers and consumers is explained or predicted. For example, policymakers may be interested in knowing how consumers and producers would respond to a tax cut. Or a canning plant may be interested in knowing whether a profit could be made if a particular vegetable were canned.

Normative economics focuses on determining "what should be." For example, policymakers might inquire as to which of several alternative policies *should be* adopted to maximize the economic welfare of producers and consumers. At the micro level, a canning plant might be interested in knowing what vegetables it *should be* canning to maximize profit.[2]

Alternative Economic Systems

An economic system can be defined as the institutional means by which resources are used to satisfy human desires, in which the term *institutional* refers to the laws, habits, ethics, and customs of the nation's citizens. **Capitalism** is a free

---

[2]For a more in-depth discussion of positive and normative economics, see Friedman M: *Essays in positive economics,* Chicago, 1974, University of Chicago Press.

market economic system in which individuals own resources and have the right to employ their time and resources however they choose, with minimal legal constraints from government. Prices signal the value of resources and economic goods. Capitalism differs sharply from **socialism** in centrally planned economies, because resources are generally collectively owned and the government decides how human and nonhuman resources are to be utilized across the various sectors of the economy. Prices are largely set by government and administered to consumers and farmers.

The United States has what is commonly referred to as a **mixed economic system;** that is, markets are not entirely free to determine price in some markets but are in others. Government's intervention in farming, for example, is well known. Loan guarantees to large firms and deposit guarantees to savings and loan depositors are forms of government intervention in the private sector. Government also controls numerous aspects of transportation, communications, education, and finance. Welfare programs are also indicative of a mixed economic system.

DEFINITION OF AGRICULTURAL ECONOMICS

Because agricultural economics involves the application of economics to agriculture, we may define this field of study as follows:

> **Agricultural economics** is an applied social science that deals with how producers, consumers, and societies use scarce resources in the production, processing, marketing, and consumption of food and fiber products.

WHAT DOES AN AGRICULTURAL ECONOMIST DO?

The application of economics to agriculture in a complex market economy such as that of the United States has a long and rich history. We can summarize this activity by discussing the activities of agricultural economists at the microeconomic level and at the macroeconomic level.

Role at Microeconomic Level

Agricultural economists at the micro level are concerned with issues related to resource use in the production, processing, distribution, and consumption of products in the **food and fiber system**. Production economists examine resource demand by businesses and their supply response. Market economists focus on the flow of food and fiber through market channels to its final destination and the determination of prices at each stage. Financial economists are concerned with issues related to the financing of businesses and the supply of capital to these firms. Resource economists focus on the use and preservation of the nation's natural resources. Other economists are interested in the formation of government programs for specific commodities that will support the incomes of farmers and provide food and fiber products to low-income consumers.

Role at Macroeconomic Level

Agricultural economists involved at the macro level are interested in how agriculture and agribusinesses affect domestic and world economies and how the events taking place in other sectors affect these firms and vice versa. For example, agricultural economists employed by the Federal Reserve System must evaluate how changes in monetary policy affect the price of food. Macroeconomists with a research interest may use computer-based models to analyze the direct and indirect effects that specific monetary or fiscal policy proposals would have on the farm business sector. Macroeconomists employed by multinational food companies examine foreign trade relationships for food and fiber products. Others address issues in the area of international development.

Marginal Analysis

Economists frequently are concerned with what happens at the margin. A microeconomist may focus on how the addition of another input by a business, or the purchase of another product by a consumer, will change the economic well-being of the business and the consumer. A macroeconomist, on the other hand, may focus on how a change in the tax rate on personal income may change the nation's output, interest rates, and inflation, and the federal budget deficit. The key word in this example is *change*. Or, more specifically, how a change in price, quantity, etc. will affect other prices and quantities in the economy, and how this might change the economic well-being of consumers, businesses, and the economy as a whole. Many of the chapters to follow include a discussion of marginal analysis as a means of making economic decisions at the firm, household, or economy level.

Key agencies that agricultural economists deal with include the U.S. Department of Agriculture (USDA) and the American Farm Bureau Federation (AFBF). The current U.S. secretary of agriculture is Dan Glickman, and the current president of AFBF is Bob Stallman.

WHAT LIES AHEAD?

Chapter 2 gives an overview of the structure of the nation's food and fiber system and the important role it plays in the United States' general economy. The remaining parts of the book can be summarized as follows:

- Part II focuses on understanding consumer behavior in the marketplace, particularly in explaining the demand for food and fiber products. Chapter 3 presents the theory of consumer behavior. Chapter 4 describes the conditions for consumer equilibrium and determination of market demand. Chapter 5 discusses the measurement and interpretation of demand elasticities.
- Part III changes the focus from the behavior of consumers to the behavior of producers of food and fiber products, and to the determination of market equilibrium in which both consumers and producers desire to be given existing

economic conditions. Chapter 6 outlines key concepts in assessing business performance. Chapter 7 describes how to measure the costs of production and revenue. Chapter 8 describes the economics of input substitution. Chapter 9 describes the economics of product substitution. Chapter 10 describes output and price under conditions of perfect competition. Finally, Chapter 11 describes output and price under conditions of imperfect competition.

- Part IV examines the political setting in which producers and consumers of food and fiber products in the United States find themselves. Chapter 12 focuses on the rationale for government intervention and considers farm economic issues, consumer issues, resource and environmental issues, and international issues. Chapter 13 outlines the development and application of income and price supports in the United States and contrasts these subsidies with those received by farmers in other countries.

- Part V switches attention to the macroeconomy—what makes it tick and the important links between the food and fiber system and the rest of the economy. Chapter 14 discusses product markets and national output. Chapter 15 focuses briefly on measures of macroeconomic performance. Chapter 16 focuses on the creation of money in the economy and the monetary policy tools used by the nation's central bank. Chapter 17 discusses the federal budget deficit and the nation's fiscal policy tools used by Congress and the president. Chapter 18 examines what macroeconomic activity means for farmers, the price of food, and other important developments in the nation's food and fiber system.

SUMMARY

The purpose of this chapter was to define the field of agricultural economics as a subset of the general field of economics. The major points made in this chapter may be summarized as follows:

1. Scarce resources are human and nonhuman resources that exist in a finite quantity. Scarce resources can be subdivided into three groups: (1) natural and biological resources, (2) human resources, and (3) manufactured resources.

2. Resource scarcity forces both consumers and farmers to make choices.

3. Most resources are best suited to a particular use. Specialization of effort may lead to a higher total output.

4. The field of economics can be divided into microeconomics and macroeconomics. Microeconomics focuses on the actions of individuals—specifically with the economic behavior of consumers and farmers. Microeconomic analysis is largely partial equilibrium in nature; events outside the market in question are assumed to be constant. Macroeconomics focuses on broad aggregates, including the nation's aggregate performance as measured by gross domestic product (GDP), unemployment, and inflation. Macroeconomic

analysis is normally general equilibrium in nature; events in all markets are allowed to vary.

5. Positive economic analysis focuses on what-is and what-would-happen-if questions and policy issues. Normative economic analysis focuses on what-should-be policy issues.

6. Capitalism, or free market economics, and socialism, or centrally planned economics, represent economic systems at the opposite ends of the spectrum. The U.S. economy represents a mixed economic system. Some markets are free to determine price, and other market prices are regulated.

DEFINITION OF KEY TERMS

Agricultural economics: an applied social science that deals with how producers, consumers, and societies use scarce resources in the production, processing, marketing, and consumption of food and fiber products.

Capitalism: a free market economic system in which individuals own resources and have the right to employ their time and resources however they choose, with minimal legal constraints from government.

Economics: a social science that studies how consumers, producers, and societies choose among the alternative uses of scarce resources in the process of producing, exchanging, and consuming goods and services.

Fallacy of composition: economic reasoning that is true for one individual but not for society as a whole.

Farm business sector: the sector of the food and fiber system that represents an aggregation of firms that produce raw agricultural products (e.g., farms and ranches).

Food and fiber system: an economic system consisting of business entities that are involved in one way or another with the supply of food and fiber products to consumers.

Human resources: the services provided by laborers and management to the production of goods and services.

Macroeconomics: branch of economics that focuses on the broad aggregates, such as the growth of gross domestic product, the money supply, the stability of prices, and the level of employment.

Manufactured resources: resources such as plows, tractors, tools, buildings, and other improvements to land that are manufactured by human beings; often referred to as capital.

Microeconomics: branch of economics that focuses on the economic actions of individuals or specific groups of individuals.

Mixed economic system: an economic system in which markets are not entirely free to determine price in some markets but are in others. Government controls in selected markets and welfare programs are indicative of a mixed economic system.

Natural resources: resources such as land and mineral deposits, which are available without additional effort on the part of the owners.

Normative economics: a branch of economics that focuses on determining what-should-be issues and questions. Unlike positive economics, it assigns specific values with specific goals or objectives.

Opportunity cost: the economic sacrifice of not doing something else or foregoing another opportunity.

Positive economics: a branch of economics that focuses on what-is and what-would-happen-if questions and issues; does not involve value judgments or policy prescriptions to reach a particular objective.

Scarce resources: a finite quantity of resources that are available to meet society's needs.

Socialism: an economic system in which resources are generally collectively owned and government decides through central planning how human and non-human resources are to be utilized in different sectors of the economy; prices are largely set by the government and administered to consumers and producers.

Specialization: the separation of productive activities between persons or geographical areas in such a manner that none of these persons or regions is completely self-sufficient.

REFERENCES

Friedman M: *Essays in positive economics,* Chicago, 1974, University of Chicago Press.
Statistical abstract of the United States, Washington, D.C., 1999, U.S. Government Printing Office.

A REVIEW OF GRAPHICAL ANALYSIS

In many of the chapters to follow, you must understand the construction and interpretation of graphs. We begin with the construction of a graph from the numbers in a table documenting the relationship between two variables.

Constructing a Graph

Two variables can be related in different ways. For example, there is a direct relationship between yields and fertilizer usage (at least over some relevant range). The higher the application of fertilizer, the higher the yield. In more general terms, the increase in one variable may be associated with an increase in another variable. Two variables can also be inversely related. As the price of gasoline increases, individuals will find ways to reduce their consumption of this product. Here, an increase in one

TABLE 1.1   Two Related Variables: Price and Quantity

Price per Pair	Quantity Sold during the Week	Location on Graph
$9	20	A
8	30	B
7	40	C
6	50	D
5	60	E
4	70	F

variable is associated with a decrease in another variable. Finally, you will encounter instances later in this book in which the relationship between two variables is mixed. For example, consider the relationship between yields and rainfall. Yields will increase sharply as we move from a situation of no rainfall to some normal amount. Beyond this level of rainfall, however, yields may actually begin to decline as a result of farmers not being able to get into the fields at the proper time, low-lying areas being washed out, and so on. This scenario suggests that there is a maximum point beyond which this curve begins decreasing instead of increasing.

To illustrate how to graph two related variables, let us assume that a local farm input supply dealer has noted a relationship between the price he charges for work gloves and the number of pairs of work gloves sold during the week (see Table 1.1). The data in Table 1.1 should suggest to you that there is an inverse relationship between the price of a pair of work gloves and the number of pairs sold. Price is decreasing at the same time the quantity is increasing.

The price-quantity relationship in Table 1.1 can be viewed as coordinates on a graph. In economics, it is customary to put the dollar values (price in this instance) on the vertical, or Y, axis and quantity on the horizontal, or X, axis. Figure 1.2 shows the location of these price-quantity coordinates on a graph. Point A, for example, represents the observation that 20 pairs of gloves will be sold if the price per pair is $9. Assume for the moment that the sales of work gloves are perfectly divisible; that is, we can sell one-fourth or one-eighth of a work glove. This division allows us to have a quantity relationship at every possible price between the $4 to $9 range cited in Table 1.1 and also allows us to connect points A through F with a solid line. This line is normally referred to as a linear curve by economists, although this line does not curve at all. The term *nonlinear curve* is used to distinguish between a true curve and a straight line (linear curve).

Slope of a Linear Curve

An important feature of a curve to an economist is its slope, or the ratio of the change in the vertical axis to the change in the horizontal axis (rise over run). To illustrate the calculation of the slope for a linear curve, let us return to the price-quantity relationship observed for work gloves in Table 1.1. The slope of the linear

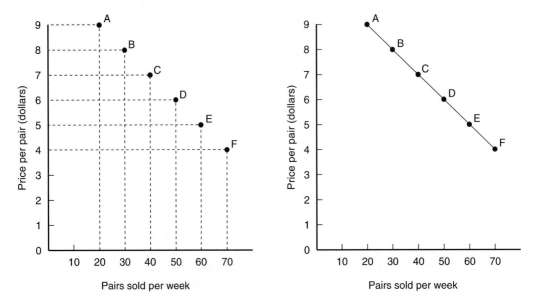

FIGURE 1.2   Graphing relationship between price and quantity.

curve is found by dividing the change in the values on the *Y*, or vertical, axis by the corresponding change in the values on the *X*, or horizontal, axis. As we move from point *A* to point *B* on this curve, the price per pair of work gloves falls by $1 because we moved from $9 a pair to $8 a pair. The corresponding change in quantity of gloves sold per week was 10 pairs, or 30 pairs minus 20 pairs. The slope of this curve therefore would be:

$$\text{slope} = \text{change in price} \div \text{change in quantity}$$
$$= -\$1.00 \div 10 \text{ pairs}$$
$$= -\$0.10 \text{ per pair}$$

Thus, the slope of this linear curve at *all* points along this curve is $0.10. A specific property of a linear curve (which you should prove to yourself by examining other points along the curve) is that its slope is the same between any two points (i.e., its slope is constant).

Because economists often discuss basic demand and supply relationships in the terms of the slopes of these curves, you must understand the difference between a positive slope, a negative slope, a zero slope, and an infinite slope. Each of these slopes is illustrated in Figure 1.3.

Positive- and negative-sloped curves can take on a variety of slopes. Figure 1.4, for example, shows that a positive-sloped linear curve (also called a ray) starting from the origin at a 45-degree angle has a slope of 1.0. At all points along this line, the quantity of *Y* is exactly equal to the quantity of *X* (verify this for 8 units of

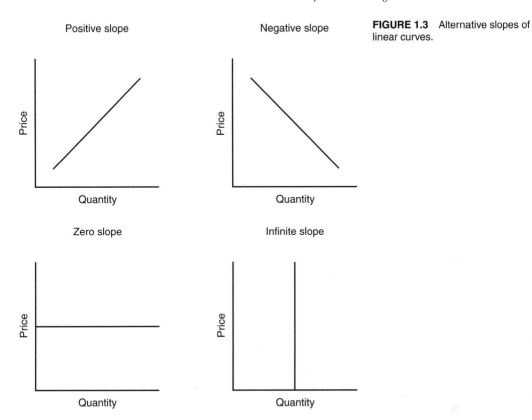

Positive slope

Negative slope

FIGURE 1.3   Alternative slopes of linear curves.

Zero slope

Infinite slope

each good). A line with a positive slope of 0.5 will be flatter than the line having a positive slope of 1.0. Here, four units of *Y* are associated with only eight units of *X*. Finally, lines with a slope of greater than 1.0 will be steeper than either of the first two curves.

Figure 1.4 suggests the following conclusion: the greater (smaller) a positive slope, the steeper (flatter) a linear curve will be. The opposite is true for negative sloped linear curves.

Slope of a Nonlinear Curve

Although the slope of a linear curve is constant over the entire range of the curve, the slope of a nonlinear curve is not. A nonlinear curve, in fact, can exhibit a positive, negative, and zero slope. Consider the nonlinear curve presented in Figure 1.5, which shows a time path for a business cycle over a period of time.

The slope at specific points along a nonlinear curve is calculated by computing the slope of a linear curve tangent to the nonlinear curve at these points. The slope at point *A* in Figure 1.5 is positive, indicating a positive growth in the economy at that point. The slope at point *B* is zero, indicating no change in the nation's

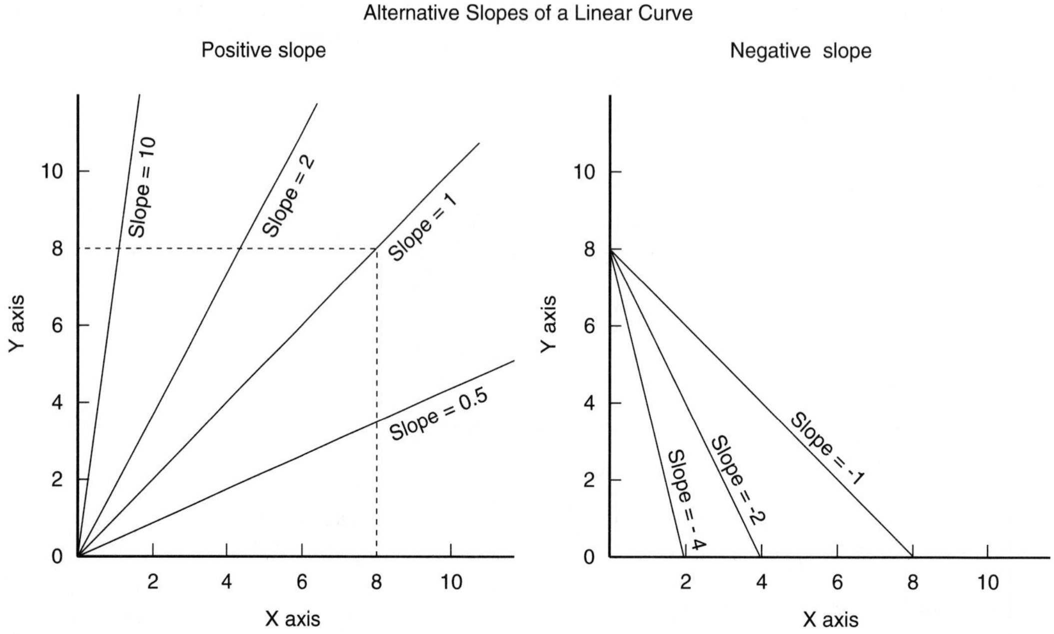

Alternative Slopes of a Linear Curve

Positive slope

Slope = 10
Slope = 2
Slope = 1
Slope = 0.5

Y axis

X axis

Negative slope

Slope = -1
Slope = -2
Slope = -4

Y axis

X axis

FIGURE 1.4   A ray from the origin (zero units of *Y* and zero units of *X*) with a 45-degree angle will have a slope of one. Linear curves with a positive (negative) slope of less than one will be flatter (steeper), while positive (negative) sloped curves with a slope greater than one will be steeper (flatter).

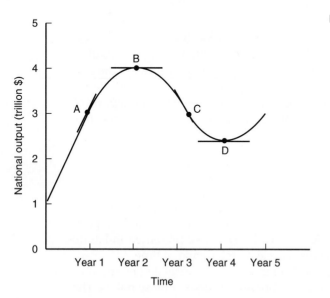

FIGURE 1.5   Slope of a nonlinear curve.

National output (trillion $)

Year 1   Year 2   Year 3   Year 4   Year 5

Time

output during the period. The slope of point *C* is negative, indicating a negative growth in the economy, or a recession. At some point in time, the economy will bottom out (point *D*) and begin a period of positive economic growth.

EXERCISES

1. The Stinson family owns a farm in Amarillo, Texas. Three alternatives exist for how to use the farm:

   a. Grow wheat. Wheat yield would be 70 bushels per acre. The price of wheat is $3.50/bu and production expenses are $140/ac.

   b. Grow barley. Barley yield would be 50 bushels per acre. The price of barley is $2.50/bu and production expenses are $150/ac.

   c. Lease out the acres. The Stinson's neighbor, Auld McDonald will pay $80/ac for leasing, but the Stinsons would still have expenses of $35/ac.

   Based on this information, answer the following:

   - Which alternative should the Stinsons undertake? Why?

   - Given your answer above, what is the Stinson's opportunity cost per acre?

   - What is the total economic cost per acre for your answer?

2. Which type of economics describes a study whose objective is to determine what the minimum income levels of farmers should be?

3. What branch of economics is concerned with national unemployment? Which deals with the effects of food safety on consumer demand for shrimp?

2

THE U.S. FOOD
AND FIBER INDUSTRY

Increased linkage to world markets and to macroeconomic policies, technology, resource considerations, and domestic farm policies will be major factors shaping U.S. agriculture over the rest of this century.

Former ERS Administrator John Lee, 1983

TOPICS OF DISCUSSION

Agriculture is important in the development of an economy. The efficient use of resources in the early stages of development can free labor and capital for use in other sectors of the economy. When the manufacturing sectors of the economy develop, they begin to supply farmers with machinery, improved seeds, fertilizer, and other manufactured resources, and to supply nonfarm products to the rest of the economy. The existence of technological advances embodied in these manufactured resources can further enhance the productivity of agriculture.

Farmers and ranchers today are highly integrated partners in the U.S. economy. Their links to other sectors in the domestic economy include not only the markets to which they sell their output but also the financial markets from which they borrow funds, the labor markets from which they hire labor and seek off-farm employment, and the manufactured input markets from which they purchase chemicals, fertilizers, and equipment.

INDICES

Before embarking on the description of and trends associated with the food and fiber industry, it is necessary to first understand the notion of how economists report measures of economic activity. Trends in output, productivity, and prices are commonly reported through the aid of *index* values. An index is nothing more than a percentage comparison from a fixed point of reference or benchmark. By comparing output and prices of apples in various years, for example, economists can describe changes in apple prices and apple production relative to the benchmark, or base period.

To illustrate, consider Table 2.1. Apple production from 1985 to 1997 ranged from 4,222 million pounds to 5,832 million pounds. Apple prices during the same period ranged from $0.68 per pound to $0.91 per pound. Let 1990 be the base period or benchmark. To calculate the output index associated with apple production for any given period, simply divide the apple production in that period by the production level in the base period. So, to arrive at the output index of 0.76 for 1985, we divide 4,222 million pounds (production in 1985) by 5,515 million pounds (production in the base period). In the same way, to obtain the output index of 1.06 for 1997, we divide 5,832 million pounds (production in 1997) by 5,515 million

TABLE 2.1   Output and Price Indices for Apples, 1985 to 1997

Year	Apple Production (Million Pounds)	Output Index	Price of Apples ($/Pound)	Price Index
1985	4,222	0.76	$0.68	0.94
1990	5,515	1.00	0.72	1.00
1997	5,832	1.06	0.91	1.26

Source: *Food Consumption Prices, and Expenditures,* 1998.

pounds. Similar calculations are made and exhibited in Table 2.1 to obtain the price indices of apples.

The index of the base period is *always* either 1 or 100, and the choice of the base period is arbitrary. If the base period changes, the corresponding set of indices also changes. Indices are unitless measures. Interpretation of the indices is quite important. For example, in Table 2.1, the price index of 0.94 in 1985 is interpreted as follows: relative to 1990 (the base period), apple prices in 1985 were lower by 6%. By the same token, apple prices in 1997 were higher by 26% relative to 1990. So, in short, indices provide a straightforward way to make comparisons.

Indices commonly used in practice include the Consumer Price Index, the Wholesale Price Index, and the Index of Prices Received (or Paid) by Farmers. Perhaps the most visible price index today is the Consumer Price Index (CPI). Price indices are useful in that they are measures of inflation. Inflation or its opposite, deflation, affects prices, interest rates, expenditures, and disposable income.

Nominal values refer to economic measures for which no adjustments to inflation (or deflation) have been made. Suppose that your annual income in 1980 was $25,000 and that, at present, it is $50,000. Your *nominal* income today is twice that of 1980. Are you twice as well off today as in 1980? The answer is no, because prices of goods and services that you buy have also changed. Most prices of goods and services have increased since 1980, although they have not necessarily doubled. To arrive at the actual change in purchasing power of your income from 1980 to present, some adjustments for price changes need to be made.

Real values refer to economic measures for which adjustments to inflation have been made. To derive the real value, economists use the following:

$$Real\ Value = \frac{Nominal\ Value}{Price\ Index} \qquad (2.1)$$

To illustrate, consider Table 2.2. Total expenditures for food away from home are exhibited for the years 1980, 1985, 1990, and 1998. Nominal expenditures ranged from $120.3 million in 1980 to $354.4 million in 1998. The corresponding CPI ranged from 0.868 in 1980 to 1.611 in 1998. Using the formula (2.1), the associated

TABLE 2.2   Nominal and Real Total Food Away from Home Expenditures

Year	Nominal Expenditures (Million Dollars)	CPI (Base Period 1982–84 = 1.00)	Real Expenditures (Million Dollars)
1980	120.3	0.868	138.6
1985	170.5	1.056	161.5
1990	252.7	1.324	190.8
1998	354.4	1.611	219.9

Source: *Food Review*, December 1999.

real expenditures ranged from $138.6 million to $219.9 million expenditures in 1998. In this way, adjustments for inflation are made. The $219.9 million figure is in terms of dollars related to base period of the CPI, 1982–84.

WHAT IS THE FOOD AND FIBER INDUSTRY?

The term *agriculture* means different things to different people. Some might think solely of farmers and ranchers when they use this term; others might think of agribusiness firms such as Tyson Foods and Kroger in their definition. In recent years, many agricultural economists have referred to the **food and fiber industry** when describing the agricultural scene. The food and fiber industry includes farms, ranches, and agribusinesses.

The food and fiber system—from the farmer to the consumer—is one of the largest sectors in the U.S. economy. This system produced output valued at $843.5 billion, or 16% of the nation's output, and employed 20.7 million full-time workers in 1990. The system includes all economic activities supporting farm production, such as machinery repair and fertilizer production, food processing and manufacturing, transportation, and wholesale and retail distribution of food and apparel products.

As Figure 2.1 suggests, the food and fiber system encompasses the activities of the farm input supply sector, the farm sector, the processing and manufacturing sector, and the wholesale and retail trade sector. In addition to farms and ranches captured in the farm sector, the food and fiber system includes such firms as:

- John Deere, DeKalb Seed, Ralston-Purina, and other firms that supply goods and services to farmers and ranchers,
- Swift, Green Giant, and other firms that utilize raw agricultural products in fiber manufacturing and food processing operations, and
- Fleming, Kroger, and other firms that distribute finished food and fiber products at the wholesale and retail level.

Employment

The food and fiber industry is the nation's largest employer, with more than 22 million people working in food- and fiber-related jobs, as shown in Table 2.3. One out of every six jobs in the nation's businesses are tied in one way or another to the food and fiber industry. Employment in specific sectors of the U.S. food and fiber industry grew during the past two decades, while other sectors did not. Employment in farming operations has been declining, while employment in the wholesale and retail trade and restaurant sectors has been growing.

Employment in specific sectors of the U.S. food and fiber industry is important to the growth of rural America. The food and fiber industry is a major employer in nonmetropolitan areas of the country. Farms and farm-related businesses

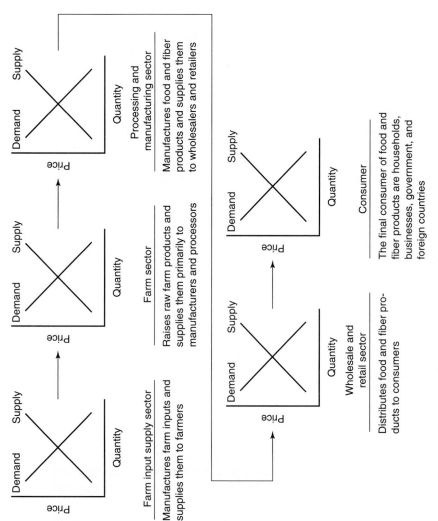

FIGURE 2.1 The food and fiber industry consists of the farm input supply sector, the farm sector, the processing and manufacturing sector, and the wholesale and retail sectors. These sectors are linked together by a series of markets in which farmers purchase production inputs and sell their raw products to processors and manufacturers, which, in turn, create food and fiber products that are distributed to customers in wholesale and retail markets.

TABLE 2.3 Contribution of the Food and Fiber System to the U.S. Economy, 1996

Industry	Value Added to GDP ($ Billion)	Share of Food and Fiber System's Contribution to GDP (Percent)	Share of GDP (Percent)	Number of Workers (Thousands)	Share of Food and Fiber System Employment (Percent)	Share of Total U.S. Employment (Percent)
Farming	71.3	7.1	0.9	1,637	7.2	1.2
Total inputs	295.4	29.6	3.9	4,343	19.1	3.2
Mining	13.4	1.3	0.2	60	0.3	—
Forestry, fishing, and agricultural services	8.7	0.9	0.1	315	1.3	0.2
Manufacturing	94.4	9.5	1.2	1,186	5.2	0.9
Services	178.9	17.9	2.3	2,782	12.3	2.1
Total manufacturing and distribution:	631.0	63.3	8.3	16,716	73.7	12.5
Manufacturing—						
Food processing	108.0	10.8	1.4	1,316	5.8	1.0
Textiles	48.2	4.8	0.6	1,352	6.0	1.0
Leather	0.3	—	—	7	—	—
Tobacco	18.4	1.8	0.2	45	0.2	—
Distribution—						
Transportation	33.8	3.4	0.4	602	2.7	0.4
Wholesaling and retailing	283.1	28.4	3.7	6,519	28.7	4.9
Foodservice	139.2	14.0	1.8	6,874	30.3	5.1
Total food and fiber system	997.7	100.0	13.1	22,694	100.0	16.9

— = less than 0.1 percent

Source: U.S. Department of Agriculture.

employ one-third or more of the local labor force in 800 counties across the United States. These businesses account for 30% (or approximately 6 million) of jobs in nonmetropolitan areas.

Output

All sectors except farming experienced growth in output during the same time period, with some sectors expanding more sharply than others. Among the fastest growing sectors are the food processing and manufacturing sectors. Table 2.3 shows that the food and fiber system contributed $997.7 billion or 13.1 percent to the nation's gross domestic product (GDP) in 1996. The largest single contributor was the wholesaling and retailing industry (3.7%).

CHANGING COMPLEXION OF FARMING

The nature of farming and ranching operations in the nation's farm sector has changed dramatically during the post–World War II period. The focus of this section is on the collective structure and performance of farmland ranches and on the changing complexion of farming activities in the United States. We can assess these attributes by examining recent trends in the physical structure, productivity, profitability, and financial structure of farms in general.

Physical Structure

An examination of the changing physical structure of the farm sector must necessarily focus on things such as the number and size of farms and ranches, their ownership and control, and the ease of entry into the farm sector.

Number and Size of Farms. As illustrated in Figure 2.2, a trend toward fewer but larger farms has been occurring. The number of farms has declined from 5.6 million in 1950 to about 2.1 million by 1999 (U.S. Department of Agriculture). This situation represents a 50% decline in the number of farms in the last 40 years.

Where did the farmland go when these farms left the sector? Three million acres of farmland are removed from the sector each year for nonagricultural uses. About 1 million of these acres are considered to be prime farmland. This figure may appear to be insignificant when we consider that there are more than 400 million acres of cropland and another 100 million plus acres that could be converted to cropland. However, we may need more than 450 million acres to meet demand by the year 2030, given existing technology. If the loss of prime farmland continues at 1 million acres per year, we would lose nearly 50 million acres by 2030. It is estimated that yields would have to increase by 1.1% per year to avoid being short of the cropland needed by the year 2030, given existing technology.

Trends in the Number and Average Size of Farms

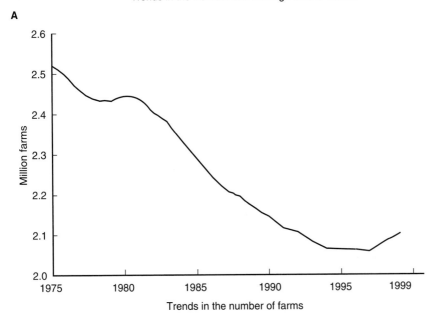

Trends in the number of farms

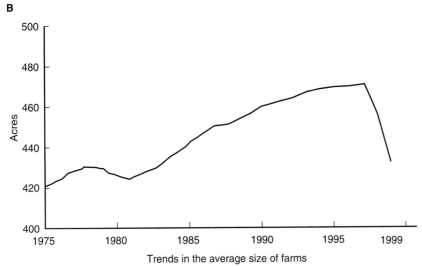

Trends in the average size of farms

FIGURE 2.2 Two significant trends in the changing structure of the farm sector are the declining number of farms and the rising average size of farms. *A,* The number of farms in this country has declined from nearly 6 million farms in 1950 to roughly 2.1 million farms in 1999. *B,* The average size of farms has roughly doubled during this period. This situation suggests that the land sold by farmers leaving the sector during this period has largely remained in agriculture. (*Source:* U.S. Department of Agriculture.)

The majority of the land sold by farmers leaving the sector, however, is purchased by other farmers who wish to expand their operations. The average farm size during the post-World War II period has doubled, from about 200 acres to more than 400 acres per farm, as shown in Figure 2.2, *B*. National averages can, however, give a somewhat misleading picture of the physical structure of the farm sector. For example, consider the 400-acres-average-size statistic mentioned above. Roughly 250,000 farms in the United States have 50 or fewer acres, while approximately 150,000 farms have 1,000 acres or more. Therefore, the 433-acre national average does not give an accurate representation of many of today's farming operations.

It is interesting to note that the share of total farm receipts earned by the 50,000 largest farms has been increasing during the last 30 years. These farms accounted for approximately 50% of total sales in 1997 compared with 30% in 1967 and only 23% in 1960. This statistic is amazing when you stop to consider that these 50,000 farms represent only 2% of the total number of farms in this country. These farms have average assets totaling more than $1 million. They also receive the majority of all direct government payments. The data suggests that a large share of the resources and production is concentrated in a relatively small number of large farming operations.

Capital versus Labor.   Although the amount of land farmed in the United States has remained relatively stable over time, a large expansion in the use of manufactured resources, such as machinery and chemicals, has occurred. Embodied in these resources are new technologies that have enabled farmers to expand their production substantially over time. For example, there has been a substantial increase in tractor horsepower on farms during the post-World War II period. A surprising statistic to some is the fact that almost twice as much capital is invested in buildings and machinery per worker in agriculture as for all U.S. businesses in general. Fertilizer use has increased more than fivefold during this same period, although expenditures on agricultural chemicals plateaued in the early 1980s. Farming and ranching operations can definitely be considered a highly capital-intensive business activity.

In sharp contrast to the increased use of capital in the farm sector is the declining relative contribution of labor. The role of labor relative to capital declined from 28% in 1950 to 20% by 1994, as shown in Figure 2.3. Labor accounted for almost 40% of the value of all resources used in farming activities in 1950 but fell below 20% by the end of the 1990s.

This shift in the relative use of capital and labor reflects the changing relative productivity of these resources and their relative prices. Improvements in farm machinery and variable production inputs such as hybrid seed, fertilizer, and agricultural chemicals led to an increase in production per unit of labor input and have fostered the growth of large-scale, specialized farms.[1]

---

[1]Specialization in the production of farm commodities has become increasingly apparent during the post–World War II period. Such specialization has come about primarily as the result of the development of capital-intensive production technologies and the proliferation of government programs that have reduced the need for farm diversification as a method of reducing exposure to risk.

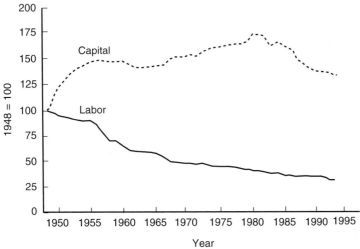

FIGURE 2.3   There has been a rapid expansion of the use of manufactured resources (capital) in the farm sector during the post–World War II period. The new technologies embodied in these resources have enabled farmers to raise their output and to cut back on their use of human resources (labor). The contribution of farmland has been relatively stable during this period. (*Source: Agricultural Productivity in the United States*, USDA, 1998.)

A farmer can cover more acreage today than ever before. In the 1990s, farmers, who represented only 3% of the nation's total civilian labor force, were producing enough output to feed the nation's population. Yet enough output remained to account for about one-fifth of the nation's total export revenue.

Advances in machinery technology and achievements in farm chemicals have affected regional production patterns and once-conventional cropping practices. For example, farmers used to practice crop rotation and diversification to conserve their soil and control pests. Farm chemicals were increasingly used by farmers in the 1970s and 1980s to grow one crop exclusively year after year. Disease-control techniques have also allowed livestock farms to specialize in one particular type of livestock and to utilize confinement production practices. Many groups concerned with the environment and food safety have urged the adoption of limited input (i.e., low chemical) production practices in the 1990s.

Hired farm labor continues to represent a significant production resource in areas where fruit and vegetable production are extensive, although some substitution of capital for labor (i.e., greater use of manufactured resources and less reliance on human resources) has taken place here also.

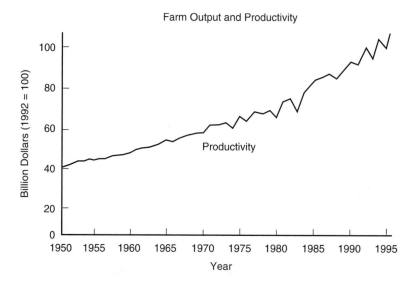

Farm Output and Productivity

FIGURE 2.4   The productivity, or level of farm output per unit of farm input, has increased sharply during the post–World War II period. Yields per crop acre, a measure of productivity, have risen 94% during this period. One reason for this rising productivity has been the technological advances embodied in farm inputs. The trend in productivity during the 1980s shows the adverse effects of the 1983 and 1988 droughts on this general upward trend. (*Source: Economic Report of the President,* February 2000.)

Productivity

Earlier we stated that the total quantity of resources used in producing raw agricultural products has remained relatively stable during the post–World War II period. **Productivity,** or the level of output per unit of input, in the farm sector has increased rather dramatically during the post–World War II period. Crop production per acre in the 1990s has doubled since the 1950s. This statistic reflects both the productivity of the land and the changing productivities and amounts of other inputs used with land to produce crops. Year-to-year variations in crop yields reflect unusual weather patterns in addition to other factors.

Figure 2.4 shows the annual trends in output and input use in agriculture during the 1950 to 1996 period. The otherwise upward trend in farm productivity during the 1980s was influenced by the droughts in 1983 and 1988.

Recent years have seen the development of several biotechnologies that enhance the productivity of farming and ranching operations, such as bovine somatotropin, or BST. This hormone, which is administered to the dairy cow by daily injections, increases milk production per cow. As the article "Those Terrifying Cows" in Box 2.1 suggests, consumers have not been convinced that this biotechnology in the nation's milk supply is safe. This article takes a pro-biotech

BOX 2.1   *Those Terrifying Cows*

No modern advance is more vulnerable to damaging public assault today than agricultural biotechnology. It promises to produce a more bountiful, cheaper food supply. But for years the promise has had to confront demagogic scaremongering about the science itself, which in turn frightens consumers, which in turn causes not very-courageous supermarket executives to repudiate the new technology. Consider the story of BST.

Since 1985, many dairy farmers have been hoping that the Food and Drug Administration would approve the use of a genetically engineered version of the growth hormone that cows secrete naturally. This hormone is bovine somatotropin, or BST, and four U.S. companies have been trying to bring it to market. When given to cows, they produce up to 25% more milk on the same amount of feed.

The FDA determined as far back as 1985 that milk from cows treated with BST is safe for human consumption. And the productivity burst from wide use of BST holds out the hope of lower prices for one of man's most basic foods. This sort of extraordinary innovation is, of course, one way to make life easier for people living on tight budgets, but the opposition won't stand for it.

In April, Wisconsin's Governor Tommy Thompson signed legislation to bar the use of BST there until June 1, 1991. Minnesota has enacted a similar ban. A moratorium, Governor Thompson explained, would provide "sufficient time to allow for additional farmer and consumer education on the use of" BST. Unfortunately, it is also providing time for some miseducation.

At the forefront of the anti-BST movement is free-media specialist Jeremy Rifkin, filing petitions with the FDA and raising the possibility of time-consuming litigation. Ben & Jerry's ice cream, serving a market that can afford to pay extra to feel better about living well, has already announced that it won't attach its name to dairy products made of milk from BST-treated cows. Last month, Consumers Union, which feels obliged to launch a left-wing policy offensive periodically, objected to BST. Five major grocery chains have announced their unwillingness to handle BST milk so long as the controversy lasts.

Opponents of BST make two arguments: that cheaper milk will force smaller dairy farms to close, and that BST renders cows more susceptible to infection and thus endangers the safety of the country's milk supply. Both claims are false.

Competition dooms not the smallest producers but the least efficient producers. In the dairy business, the most efficient producers are also often among the smallest. As an official from one of the companies that makes BST told a congressional subcommittee in 1986: "The average size of the highest yielding herds (top 20%) in the state of New York is approximately 60. The average size of the lowest yielding herds (bottom 20%) is also 60—exactly the same." The effect of BST is to cut *all* dairy farmers' fodder costs—the saving is proportionally the same for a big farm as for a small one.

Worries over the safety of the milk supply are a more serious issue. Happily, a panel of doctors and scientists convened by the National Institutes of Health unanimously concluded on December 7 that milk from BST cows is safe for human consumption. The panel confirmed the observation made by the deputy director of the FDA's Center for Veterinary Medicine in December 1989 that "BST is one of the safest products we've ever looked at."

So if BST is so great, why would anyone want to keep it off the market? BST is the first big agricultural biotechnology breakthrough. If the fear-of-science movement can kill so safe and beneficial a product, they—and the biotechnology companies—will know that they can stop other products.

For biotech's opponents, this is D-Day: If BST gets past them, the way is clear for innovation after innovation. There can be little doubt that this would be tough on the least efficient farmers, who would prefer the status quo. For the more politicized opponents, the game is undoubtedly about power. Once the large corporations bringing BST to market have proved that genetic engineering is a safe technology, government regulators will spend less time with the subject. That in turn diminishes whatever leverage the "public-interest" groups have been able to accumulate over this area of the country's economic life.

In 1950, before artificial insemination and new feeding methods became common on American dairy farms, the average cow yielded some 5,314 pounds of milk a year. Today, the average cow produces more than 14,300. That is the achievement of scientific farming. Better achievements are still to come—if the public and policy makers are willing to stand up to the scaremongers.

Source: *Wall Street Journal,* January 7, 1991.

stance, concluding that better achievements will be forthcoming if the public and policymakers are "willing to stand up to the scaremongers." How do you stand on this issue?

Profitability

Gross income from farming activities and related production expenses were relatively stagnant from the mid-1950s to the early 1970s. Beginning in 1972, however, gross income from farming activities, which consists of cash receipts from marketing crops and livestock and government payments to farmers, has risen erratically, while production expenses have risen steadily.

These trends have led to volatile shifts in net farm income during the 1970s, 1980s, and 1990s. Net farm income is defined as:

$$\begin{array}{ccc} \text{net farm} \\ \text{income} \end{array} = \begin{array}{ccc} \text{gross farm} \\ \text{income} \end{array} - \begin{array}{ccc} \text{production} \\ \text{expenses} \end{array} \qquad (2.2)$$

and gross farm income is defined as:

$$\begin{array}{ccc} \text{gross farm} \\ \text{income} \end{array} = \begin{array}{ccc} \text{cash receipts from} \\ \text{farm marketings} \end{array} + \begin{array}{ccc} \text{government} \\ \text{payments} \end{array} + \begin{array}{ccc} \text{other income from} \\ \text{farm sources} \end{array} \qquad (2.3)$$

As depicted in Figure 2.5, *A*, production expenses began to increase persistently after 1973 due in part to the rising prices of petroleum and rising interest rates on farm debt. The combination of erratic gross farm income and rapidly rising production expenses led to a relatively flat net farm income. Inflation further eroded net farm income during the 1980s. In 1983, net farm income adjusted for the effects of inflation was the lowest in terms of **purchasing power** since the depression years of the 1930s.

The off-farm income earned by farm families has increased steadily during the last several decades. In fact, off-farm income typically raises the average farm family income above the national median family income. Off-farm income in 1983 accounted for almost 70% of total farm family income. Again, national totals can be somewhat misleading. Much of this off-farm income is earned by farm operating families on smaller farms, where one or more family members have a full-time job in the city.

Financial Structure

The overall financial structure of farms can be assessed by examining the major components of the **balance sheet** for the farm sector. This balance sheet, which is published each year by the U.S. Department of Agriculture (*Economic Indicators of the Farm Sector*), indicates the value of real estate assets (e.g., farmland and buildings) and non–real estate assets (e.g., machinery, trucks, and crop and livestock inventories) on farms and the financial assets (e.g., checking account balances and savings accounts) and liabilities of the nation's farms. Equity, the "balancing" entry in the balance sheet, is given by:

$$\text{equity} = \begin{array}{ccc} \text{value of real} \\ \text{estate assets} \end{array} + \begin{array}{ccc} \text{value of non-real} \\ \text{estate assets} \end{array} + \begin{array}{ccc} \text{financial} \\ \text{assets} \end{array} - \text{liabilities} \qquad (2.4)$$

Trends in these balance sheet items in recent years are presented in Figure 2.6. Examining Figure 2.6, *A* and *B*, we see that the change in the value of farm real

Income from Farming Activities

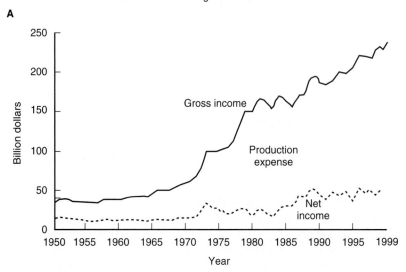

A

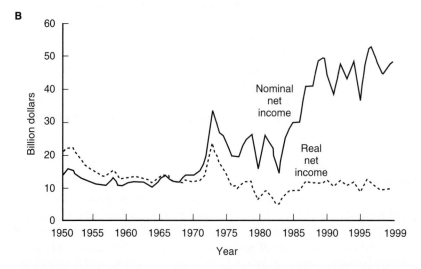

B

FIGURE 2.5 One of the key economic statistics watched by policymakers is farmers' real net income, which reflects what is happening to gross farm income and farm production expenses (net farm income is equal to gross farm income minus farm production expenses) and the effects that inflation has on the purchasing power of money. Some farm families have offset the relatively flat trend in real net income by working in off-farm enterprises. (*Source: Economic Report of the President,* August 2000.)

Balance Sheet of the Farm Sector

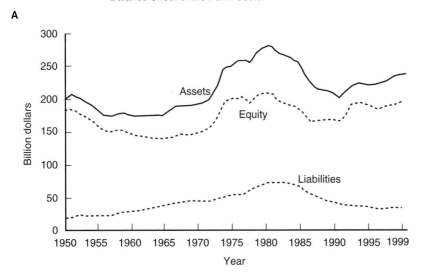

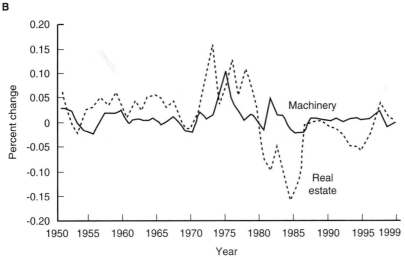

FIGURE 2.6   The primary factor explaining the rising equity (i.e., total assets minus total debt) during this period was the appreciation of farmland values reflected in *B*. The declines in farmland values and equity during the early 1980s reflected the declining economic conditions in the farm sector. (*Source:* U.S. Department of Agriculture.)

estate represents the major component of the change in the total value of all farm assets. We also see that farm indebtedness doubled in less than 20 years, but then began to decline after 1984 when many farmers lost their farms because of declining net incomes and equity.

OTHER SECTORS IN THE FOOD AND FIBER INDUSTRY

From the wide fluctuations in net farm income during the 1980s and 1990s, it is easy to see that farms are directly affected by the economic health of the rest of the economy, both domestic and foreign. In turn, the other sectors in the food and fiber industry are dependent on a healthy farm sector. Farm input suppliers, such as John Deere Company and the DuPont Chemical Company, are dependent on a healthy farm sector for a strong market for their products. Similarly, food processors and manufacturers are dependent on the farm sector for a steady stream of inputs to their production processes. Let us look more closely now at some of the characteristics of the other firms involved in the U.S. food and fiber industry.

Farm Input Suppliers

Farm input suppliers provide the nation's farmers and ranchers with the inputs they need to produce crops and livestock. There are six broad categories of farm input suppliers: (1) the feed manufacturing industry, (2) the fertilizer industry, (3) the agricultural chemical industry, (4) the farm machinery and equipment industry, (5) the hired farm labor market, and (6) farm lenders.

The relative importance of individual categories of farm production expenses is illustrated in Figure 2.7. This figure indicates that 10% of total farm production expenses go toward the purchase of fertilizer and agricultural chemicals. Feed, seed and livestock purchased for feeding operations constitute another 24% of total farm production expenses. Interest payments to banks and other farm lenders (7%); the consumption of capital reflected in the wear and tear on farm tractors and other machinery, equipment, and buildings during the year (10%); and wages paid to hired labor (10%) also represent significant input expenditure categories.

Of the 6,500 feed manufacturing firms in the United States, approximately one-fourth are farm cooperatives. Perhaps the most dramatic structural change in this industry has been the movement of some feed manufacturers into direct participation in production. For example, broiler production used to be prevalent on many farms operating on a small scale; feed manufacturers have helped foster large-scale, highly capitalized operations, which has resulted in a more standardized, less expensive product to consumers.

Conglomerate firms producing a wide variety of products are found in the fertilizer industry (e.g., petroleum producers, refiners, and chemical companies). This industry occasionally experiences either an excess supply of fertilizer products and falling prices, or input shortages and rising prices. The year-to-year fluc-

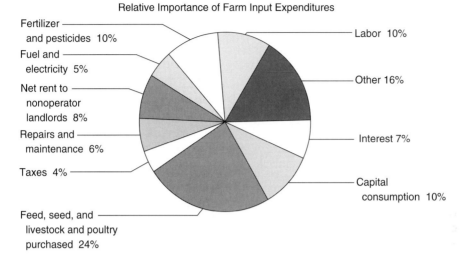

Relative Importance of Farm Input Expenditures

Fertilizer and pesticides 10%

Fuel and electricity 5%

Net rent to nonoperator landlords 8%

Repairs and maintenance 6%

Taxes 4%

Feed, seed, and livestock and poultry purchased 24%

Labor 10%

Other 16%

Interest 7%

Capital consumption 10%

FIGURE 2.7   Purchases of farm inputs in farm labor markets, farm financial markets, fertilizer and chemical markets, and farm machinery and equipment markets represent major linkages between farmers and farm input suppliers. (*Source:* U.S. Department of Agriculture.)

tuations in fertilizer sales are shown in Figure 2.8, *A*. This industry is fairly competitive, but less competitive than the feed manufacturing industry. For example, the top four firms producing nitrogen account for only one-fourth of the market. The top four firms producing potash account for slightly more than one-half of the market. A similar market share is noted for manufacturers of phosphate fertilizer. The four largest firms in this industry supply a little more than one-third of total fertilizer sales. The eight largest firms supply almost one-half of total fertilizer sales.

The agricultural chemical industry is also characterized by conglomerates that produce a wide variety of chemical products. A prime example is DuPont, which manufactures a variety of nonfarm-oriented products (e.g., paints) in addition to their line of agricultural chemical products. The concentration of firms in the agricultural chemical industry is apparently high. For example, the four leading brands of corn insecticide account for more than three-fourths of total sales.

The farm machinery and equipment industry comprises some large firms that offer a full line of machinery and a more diverse group of manufacturers that supply specialized equipment (e.g., irrigation equipment, grain-handling equipment, etc.). For example, about 80% of all two-wheel-drive tractor sales and nearly the same proportion of all combine sales are accounted for by the four top firms. The declining trend in machinery sales during the 1980s, which were characterized by high costs of capital and idled crop land, is shown in Figure 2.8, *B*.

A final input supplier to farmers is the nation's farm lenders. Because much of the farm production expenses and farm capital expenditures for new tractors

Purchases of Selected Farm Inputs

A

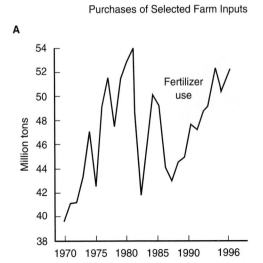

B

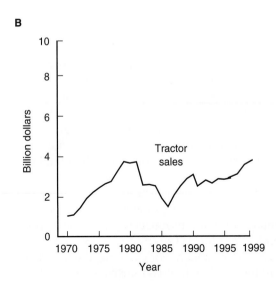

FIGURE 2.8 Purchases of special farm input categories declined during the 1950s as the sector experienced some ownership consolidation. The federal government also required farms to set aside up to one-third of their cropland to qualify for government payments. These conditions translated into reduced sales by farm input suppliers. (*Source:* U.S. Department of Agriculture.)

and other large ticket items are financed with borrowed funds, these lenders are important to the food and fiber industry. The trends in farm liabilities before and during the 1980s are presented in Figure 2.6, *A*. Farm indebtedness began to decline after 1984, when many farmers lost their farms, and the debt was written off or reduced to lower their exposure to financial risk. The major farm lending institutions to which farmers and ranchers must make annual principal and interest payments are identified in Figure 2.9. The Farmers Home Administration (FmHA) is the lender of last resort to farmers who cannot obtain credit else-

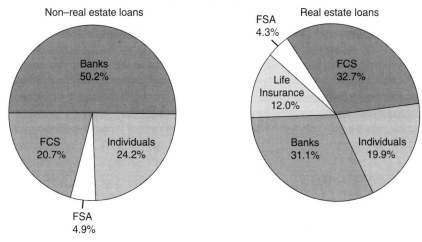

FIGURE 2.9   Among those lending institutions that provide loan funds to the nation's farmers are commercial banks, the Farm Credit System (FCS), and the Farm Service Agency (FSA). (*Source:* U.S. Department of Agriculture.)

where. Commercial banks and the Farm Credit System (FCS), a borrower-owned, cooperative lending institution with quasi-governmental backing, are the major farm lenders.

In summary, farms are confronted by sellers who have a great deal of market power in the markets in which farmers purchase most of their production inputs. This situation generally means that farmers are price takers rather than price setters in these markets.

Food Processors, Wholesalers, and Retailers

Farmers are linked to consumers by a complex system of food processing, wholesaling, retailing, and food service firms. The U.S. food marketing sector, the network of processors, wholesalers, retailers, and restaurateurs that market food from farmers to consumers, was responsible for $456 added to GDP in 1996 (see Table 2.3). Food processing contributed an estimated $108 billion, retailers and wholesalers added $283.1 billion, and eating and drinking places (food service) added $139.2 billion. (See Table 2.4.)

At critical stages in the flow of products from farmers to processors and eventually to consumers, various categories of middlemen participate. Middlemen firms perform one or more of the marketing functions (i.e., exchange, physical, or facilitating) as the product moves through the marketing channels to consumers. Middlemen firms can be classified as merchant middlemen firms, agent middlemen firms, speculative middlemen firms, processors and manufacturers, and facilitative organizations.

TABLE 2.4   Nominal Personal Consumption Expenditures on Food

Year	Personal Consumption Expenditures on Food (Billion Dollars)	
	Nominal	Real 1992 = 100
1970	$143.8	$477.2
1975	223.1	502.6
1980	356.0	558.7
1981	383.5	557.9
1982	963.4	565.1
1983	423.8	579.7
1984	447.4	589.9
1985	467.6	602.2
1986	492.0	614.0
1987	515.3	664.6
1988	553.5	690.7
1989	591.9	703.5
1990	636.9	722.4
1991	657.6	721.4
1992	669.3	725.6
1993	697.9	745.1
1994	728.2	764.9
1995	755.8	777.0
1996	786.0	786.0
1997	817.0	799.1
1998	853.4	820.6
1999	903.0	850.8

Source: *Economic report of the president*, February 2000.

Merchant Middlemen Firms.   **Merchant middlemen firms** buy and sell commodities for their own economic gain. A **retailer,** one type of merchant middleman, purchases products in final form for resale to the ultimate consumers of goods. A **wholesaler,** another type of merchant middleman, principally sells products in final form to retailers. A portion of their sales activity, however, may involve direct sales to the ultimate consumer of the product. Some wholesalers may purchase products directly from farms and ranches and assemble them for shipment to other wholesalers or processors. Other wholesalers may never come in direct contact with farmers, but instead function in large urban centers as intermediaries between rural wholesalers and urban retailers. Merchant middlemen, as their name implies, own title to the products they assemble and sell.

Agent Middlemen Firms.   **Agent middlemen firms** do *not* own title to the products they handle, but instead serve as an agent for firms that do. Instead of earning income on the difference between what they can buy and sell commodities for, agent middlemen receive fees or commissions for their efforts. Brokers and commission firms are the two major categories of agent middlemen. Brokers normally do not exert physical control over the product or have the power to negotiate a price. A grain broker working on a grain exchange, for example, is acting on buy and sell instructions for his or her client. A commission firm normally sees to the physical handling of the economy and supervises the closing of the sale. Thus, the firm's powers are much broader than those of a broker. Livestock commission firms, for example, take control of the livestock and seek to make the best deal they can for their clients.

Speculative Middlemen Firms.   **Speculative middlemen firms** differ from merchant middlemen firms in that they do less in merchandising and more in seeking out and bearing price risk to profit from fluctuations in commodity prices. They often buy and resell commodities over a relatively short period of time (days). It is common for a grain trader to buy and sell grain several times during the same day, if price fluctuations are perceived to be favorable.

Food Processors and Manufacturers.   **Food processors and manufacturers** may not only perform their obvious role of transforming raw farm products into final goods but may also perform some of the functions identified for merchant and agent middlemen to control their costs. For example, meat packers may act as their own purchasing agent by buying feed livestock directly from farms and ranches. They may also serve as their own wholesaler by dealing directly with retail stores. Processors and manufacturers often communicate directly to the ultimate consumer of the product by advertising and other promotional means.

Facilitative Organizations.   **Facilitative organizations** serve a variety of functions, including gathering and distributing market information and establishing and enforcing policies and standards for their trade. Grain exchanges, stockyards, and trade associations all represent facilitative organizations. Their income comes from fees paid for the use of their facilities or services.

Value Added Process

Figure 2.10 illustrates the value added to the product as it flows through the food and fiber industry to consumers. The majority of the value of farm output goes to assemblers and brokers; only a small proportion is retained for farm consumption or is sold directly to consumers at roadside stands and through other means. Brokers and assemblers then market these commodities to food manufacturers and processors or market them directly to wholesalers, brokers, and chain warehouses. Along the way, imports of these commodities to the United States add to the product flow, while sales to government industries and exports to other countries diminish the product flow. Wholesalers, brokers, and chain warehouses supply

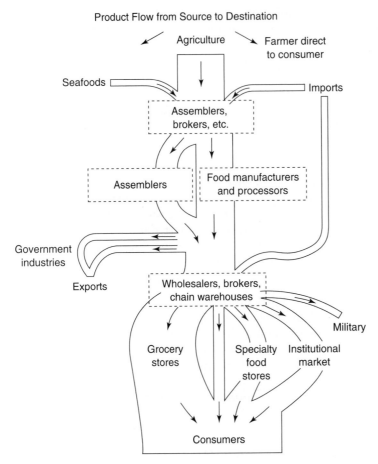

Product Flow from Source to Destination

FIGURE 2.10 This figure illustrates the channels through which food and fiber products flow. The width of these channels reflects their relative magnitude. Almost all raw agricultural products flow to brokers and other middlemen firms that either pass them on to wholesalers or sell them to manufacturers and processors.

processed food products to grocery stores, specialty food stores, institutions, and restaurants that sell directly to consumers and to the military.

We can examine the value added process at a more micro level by studying the transformation of wheat into bread. The wheat in a loaf of bread sold at the local grocery store originated with the shipment of wheat from farmers to a network of local and terminal elevators, where it was then assembled and sold to milling and baking firms for processing into bread. Finally, the finished bread was either sold directly to retail stores or purchased by chain stores for further distribution. Table 2.5 illustrates the value added at each stage.

Structure of Food Industry.   The food processing industry and the wholesale and retail trade industries are characterized by a relatively small number of firms that account for a substantial portion of total industry sales. The 50 largest food

TABLE 2.5   Value Added for a Loaf of Bread

Product	Type of Firm	Product Sold	Paid	Received	Value Added
Wheat	Farm	Wheat	—	$.08	$.08
Milling	Miller	Flour	$.08	$.50	$.42
Baking	Bakery	Bread in bulk	$.50	$.72	$.22
Marketing	Store	Distributed bread	$.72	$1.02	$.30
					$1.02

processing firms, for example, account for almost one-half of the processed food market. The top 50 food wholesaling firms that move food products through channels to the retail level account for more than 60% of the wholesale food market. There is less concentration in the food retailing sector (35%) and food service industry (20%).

Although aggregate concentration has increased, the number of food marketing companies has remained relatively constant. In the food processing sector, the number of firms amounts to slightly more than 16,000, down from 32,000 in 1963. In the wholesaling sector, the number of firms is about 36,000, down from 43,000 in 1963. The number of food retail stores and food service companies totals approximately 200,000, down from 220,000 in 1977. "Mom and Pop" grocery stores are being replaced by large supermarket and convenience store chains. The chain stores, by combining wholesaling, retailing, and (in some cases) processing operations, have been able to lower their operating costs and price the small independent stores out of business.

Employment in the food processing, wholesaling, and retailing industry has more than doubled during the past 25 years. Of the 16.7 million individuals employed in this industry in 1996, 8% were associated with processing operations, 39% worked in food wholesaling and retailing establishments, and 41% worked in eating and drinking establishments.

The Marketing Bill.   The marketing bill for food is defined as the portion of food expenditures associated with the activities of firms beyond the farm gate. As shown in Figure 2.11, the marketing bill represents 80% of every dollar spent on food. Labor expenses, packaging expenses, transportation costs, and advertising expenses of food processors, wholesalers, and retailers are among the major categories of expenditures made by these firms. For example, 39 cents of every dollar spent on food represents labor expenses incurred beyond the farm gate.

The share received by farmers and ranchers varies considerably by commodity. For example, wheat farmers receive only 5.5% of the value of a loaf of white bread. Stated another way, the farm value of a 30-slice loaf of white bread is *three* slices. Conversely, the farm value of a dozen eggs is 37.6%, or approximately five eggs.

What a Dollar Spent for Food Paid for in 1994

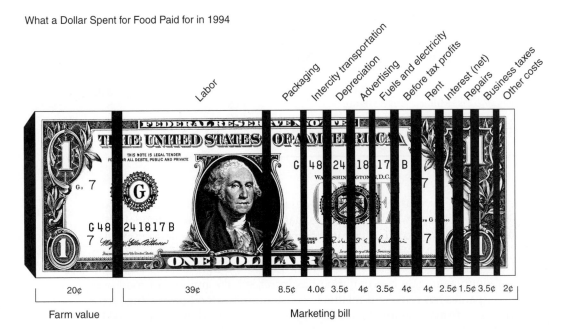

| 20¢ | 39¢ | 8.5¢ | 4.0¢ | 3.5¢ | 4¢ | 3.5¢ | 4¢ | 4¢ | 2.5¢ | 1.5¢ | 3.5¢ | 2¢ |

Farm value   Marketing bill

FIGURE 2.11   The marketing bill is the value of food expenditures contributed by firms beyond the farm gate. The farm value of total food expenditures is 20%. (*Source:* Calculated by ERS from government and private sources.) (*Notes:* Includes food eaten at home and away from home. Other costs include property taxes and insurance, accounting and professional services, promotion, bad debts, and many miscellaneous items.)

The Top Firms.   A national ranking of the top food processing, wholesaling, retailing, and restaurant firms in the country based upon a dollar volume of sales is presented in Table 2.6. Many of the names are easily recognizable as you drive down the main street of your hometown. Others, such as ConAgra, are large conglomerate food processing firms that process a wide range of products such as Armour meat products, Orville Redenbacher's popping corn, and Banquet frozen foods, which appear on food store shelves nationwide.

The 50 largest food processing firms account for three-fourths of total media advertisements and nearly all of network television advertisements for these products. In 1998, food processors spent about $22 billion on advertising in seven media (magazines, newspaper supplements, network television, spot television, network radio, billboards, and cable television). The food retailing and food service sectors contributed roughly 40 percent of this total.

Fiber Manufacturers

Earlier in this chapter, we said that the output of the nation's textile manufacturers has been expanding since the early 1970s. By 1996, these firms were adding $48.2 billion to GDP and employing 1.35 million workers (see Table 2.3).

TABLE 2.6 Rank of Top 10 Food Processing, Wholesaling, and Retailing Companies

Rank	Food Processing	Wholesaling	Retailing
1	Phillip Morris	Super Valu Stores	Kroger
2	PepsiCo	Fleming Co.	American Stores Co.
3	Coca-Cola	Wakefern Food Corp.	Safeway
4	ConAgra	Nash Finch Co.	Albertson's
5	IBP	Richfood Holdings	Ahold International
6	Anheuser Busch	C&S Wholesale Grocers	Winn-Dixie
7	Sara Lee	Assoc. Grocers, Inc.	Publix Supermarkets
8	H.J. Heinz	Roundy's, Inc.	Food Lion, Inc.
9	RJR Nabisco	Spartan Stores, Inc.	Great Atlantic & Pacific Tea Co.
10	CPC International	Certified Grocers of CA	H.E. Butt Grocery Co.
11			
12			
13			
14			
15			
16			
17			
18			
19			
20			

Source: U.S. Department of Agriculture.

Trends in the per capita consumption of fibers in the United States show that cotton consumption per capita was expanding relative to synthetic manufactured fibers such as polyester, which remained relatively stable. Wool fibers represented an extremely small share of the total production of textile manufacturers.

Shippers and Handlers

The transportation of food and fiber products—when commodities move from the nation's farms and ranches to processors and manufacturers and then on to wholesalers and retailers—is an extremely important dimension of the U.S. food and fiber industry. More than 400 million tons of food products flow annually over 200,000 miles of railroad tracks, 3 million miles of intercity highways, and 26,000 miles of improved waterways.

As you will recall from Table 2.3, employment in this sector was 602,000 workers. Food and fiber transporters contributed $33.8 billion to GDP in 1996.

The transportation of fresh fruits and vegetables is of particular importance because of the perishability of the product. Trucks are replacing railroads as the mode of shipping fresh fruits and vegetables. Piggybacks, or trucks containing fresh fruits and vegetables traveling on railroad flatbed cars, represent a small component of total fruit and vegetable shipments.

Importance of Export Markets

The volume of agricultural exports is important not only to the producers of the commodities marketed but also to the firms that move these commodities through marketing channels to foreign ports. These firms provide storage and transportation facilities and marketing services, which add value to commodities as they move overseas. The volume of agricultural exports is also important because it offsets, at least in part, imports of nonagricultural products and thus lowers the nation's trade deficit. The United States provided approximately 70% of the world's coarse grain exports and 65% of world soybean exports in the late 1990s.

As shown in Table 2.7, exports of wheat typically exceed the level of domestic use for feed, food, and seed. If the export demand for wheat is weak, unsold production will accumulate in the form of carryover stocks and will depress domestic wheat prices and the revenue wheat farmers receive from selling their production. The carryover stocks of grain from one year to the next in this figure show the magnitude of the capacity of the nation's grain storage facilities for wheat. Unusual variations may also reflect abnormal growing and harvesting conditions during the year.

World events influence domestic consumers and producers. Growth in wheat production by Argentina, Australia, Canada, and the European Community (EC) increases supplies on world markets, depresses wheat prices, and lowers revenue received by U.S. wheat producers. Domestic consumers of bread and other bakery products will benefit because wheat prices will decrease, causing wheat products

TABLE 2.7   Supply and Utilization of Wheat (Million Bushels)

	Total	Feed & Residual	Other Domestic Use	Exports	Stocks
1996/97	2,796	308	993	1,002	444
1997/98	3,020	251	1,007	1,040	722
1998/99	3,373	356	989	1,042	946
1999/00	3,342	290	1,012	1,090	950
2000/01	3,293	225	1,021	1,100	947

Source: *Agricultural Outlook*, August 2000.

to drop in price. Other world events, including World Trade Organization (WTO) negotiations, a drive for food security in traditional food importing nations, and trends in foreign exchange rates and world weather shocks will also influence world trade flows and market shares and affect U.S. consumers and producers.

SUMMARY

The purpose of this chapter was to acquaint you with the structure and performance of the farm sector during the post–World War II period and its role in the nation's food and fiber industry. The major points made in this chapter may be summarized as follows:

1. The U.S. food and fiber industry consists of different groups of business entities called sectors, which are in one way or another associated with the supply of food and fiber products to consumers. In addition to the **farm sector,** this industry consists of firms that supply manufactured inputs to farms and ranches, firms that process raw food and fiber products, and firms that distribute food and fiber products to consumers.

2. Among the physical structural changes taking place in the farm sector during the post–World War II period is the trend toward fewer but larger farms. We have also seen a tremendous expansion in the use of manufactured inputs, such as machinery and chemicals, and a decline in labor use. Rising capital requirements in general during the period have increasingly represented a barrier to entry for would-be farmers.

3. Although the total quantity of inputs used in producing raw agricultural products has remained relatively stable during the post–World War II period, the total quantity of output has increased substantially. These results, taken together, imply an increase in productivity, or the ratio of output to inputs.

4. **Gross farm income** has increased erratically during the post–World War II period, while production expenses have increased steadily. The result is a highly variable level of profits, or **profitability,** from one year to the next. During the

early 1980s, net farm income after adjustments for the purchasing power of money fell to depression-era levels.

5. The financial structure of the farm sector during the post–World War II period shows that financial assets represent a considerably smaller portion of total farm assets now than was true 20 to 30 years ago. The use of loan funds to finance farm expansion during this period has increased substantially.

6. The U.S. food marketing sector is the network of processors, wholesalers, retailers, and restaurateurs that market food from farmers to consumers. Approximately 80% of the personal consumption expenditures on food went to pay for activities taking place beyond the farm gate.

7. Along the flow of products from farmers to processors and eventually on to consumers, middlemen play a vital role. Classifications of middlemen firms include merchant middlemen firms, agent middlemen firms, speculative middlemen firms, processing and manufacturing firms, and facilitative organizations.

8. In recent times, the number of **mergers** and acquisitions in food industries has increased sharply relative to historical levels. Consequently, food industries have become more concentrated. **Concentration** is particularly high in industries marketing products such as breakfast cereals, beer, candy, and soft drinks.

9. The food processing industry and the wholesale and retail trade industries are characterized by a relatively small number of firms that account for a substantial portion of total industry sales. Although aggregate concentration has increased, the number of food marketing companies has remained relatively constant.

10. Farmers and ranchers get approximately 20% of each dollar spent on food. This share varies considerably by commodity. The remaining portion goes to food processors, wholesalers, and retailers. The major categories of expenditures include labor, packaging, transportation, and advertising.

11. The transportation of food and fiber products along the marketing chain is an extremely important component. The storing and exporting of non-perishable commodities is also an important dimension of marketing agricultural commodities.

DEFINITION OF KEY TERMS

Agent middlemen firms: firms that do not own title to the products they assemble and sell. Brokers and commission firms are agent middlemen.

Asset: something of value owned by a farm or ranch. Assets are generally divided into either physical and financial assets or current (short-term) and fixed (intermediate and long-term) assets.

Balance sheet: a financial statement reporting the value of real estate (land and buildings), non–real estate (machinery, breeding livestock, and inventories), and financial (cash, checking account balance, and common stock) assets owned by farms and ranches and also outstanding debt. The difference between total farm assets and total farm debt outstanding represents the net worth of the farm.

Concentration: refers to the number and market power of firms marketing their products in a particular market. A market characterized by a small number of firms accounting for the majority of total sales is said to have a high degree of concentration.

Facilitative organizations: organizations that gather and distribute market information and establish and enforce policies and standards. Grain exchanges, stockyards, and trade associations are facilitative organizations.

Farm sector: one of the sectors of the food and fiber industry that represents an aggregation of firms that produce raw agricultural products (i.e., farms and ranches).

Food and fiber industry: consists of business entities that are involved in one way or another with the supply of food and fiber products to consumers.

Food processors and manufacturers: middlemen firms that not only transform raw farm products into final goods but also perform some of the functions identified for merchant and agent middlemen.

Food service: the dispensing of prepared meals and snacks intended for on-premise or immediate consumption.

Gross farm income: annual level of income received from farming activities before farm expenses, taxes, and withdrawals have been deducted.

Liability: refers to the amount owed by the farm or ranch to others; also called debt outstanding.

Marketing bill: the value of food expenditures contributed by firms beyond the farm gate; also called a marketing bill.

Merchant middlemen firms: firms that own title to the products they assemble and sell. Retailers and wholesalers are merchant middlemen.

Merger: the combination of two or more firms into one.

Middlemen firms: firms that perform marketing functions.

Net farm income: gross farm income minus farm expenses and taxes.

Performance: the efficiency and profitability of a farm's production activities are typical barometers of a sector's performance.

Productivity: level of output per unit of input. Crop yields per acre are a measure of productivity.

Profitability: returns on capital invested in farm assets represent a measure of profitability. Profitability may be expressed in dollar terms (e.g., net farm income) or in percentage terms (e.g., rate of return on farm capital).

Purchasing power: reflects what $1 today would have purchased in goods and services in a particular base period.

Retailer: one type of merchant middleman that purchases products in final form for resale to the ultimate consumers of goods.

Speculative middlemen firms: firms that do less in merchandising and more in seeking out and bearing price risk to profit from fluctuations in commodity prices.

Structure: the composition or makeup of a sector can be assessed by examining its physical and financial structure. Physical structure is assessed by looking at the number and size of farms, their ownership and control, and the ease of entry and exit from the sector. Financial structure is evaluated by looking at the composition of a sector's balance sheet.

Technological change: reduces the quantity of inputs required to produce a given level of output; also referred to as technical change and technical progress.

Wholesalers: operators of firms engaged in the purchase, assembly, transportation, storage, and distribution of groceries and grocery products for resale by retailers.

REFERENCES

Economic report of the president, Washington, D.C., published annually, U.S. Government Printing Office.

U.S. Department of Agriculture: *Economic indicators of the farm sector, income and balance sheet statistics* (ERS-USDA Statistical Bulletin Series), Washington, D.C., published annually, U.S. Government Printing Office.

U.S. Department of Agriculture: *Agricultural statistics,* Washington, D.C., published annually, U.S. Government Printing Office.

U.S. Department of Agriculture: *Food marketing review,* various issues, Washington, D.C., U.S. Government Printing Office.

U.S. Department of Agriculture: *Food cost review,* various issues, Washington, D.C., U.S. Government Printing Office.

EXERCISES

1. Develop an output index for Zappa's tomato processing plant (base year 1998).

Year	Quantity Processed
1996	20,000
1997	56,000
1998	60,000
1999	65,000

2. Develop a price index for hot dogs (base year is 1997).

Year	Price
1996	$.49
1997	.59
1998	.79
1999	.89

3. From the following table:
   a. Calculate the real income for Kevin Baskin.
   b. Determine which year was best for Kevin. Why?

Year	Kevin Baskin's Income	CPI (1982–84 = 1.00)
1997	$20,500	1.605
1998	21,750	1.630
1999	22,500	1.666

PART

II

UNDERSTANDING CONSUMER
BEHAVIOR

3

THEORY OF CONSUMER BEHAVIOR

We never know the worth of water, 'til the well is dry.

Proverb

TOPICS OF DISCUSSION

The biological process of photosynthesis, in which the addition of light to a plant's environment results in plant growth, can be thought of in a stimulus-response context. The stimulus is the addition of light and the response is plant growth. This process can be studied in a controlled environment using sophisticated measuring devices.

Economic behavior also can be thought of in a stimulus-response context. For example, a fall in the price of ice cream acts as a stimulus, causing consumers to purchase more ice cream. These purchases can be measured and recorded. In most respects, similarities end here. The complex process of photosynthesis can be examined and studied directly, but most economic behavior processes cannot. In fact, this example illustrates the distinction between the natural sciences (e.g., biology, chemistry, physics) and the social sciences (e.g., economics). Most economic behavior processes cannot be studied in a controlled environment.

We can examine the technical relationships of converting inputs to outputs in a production process, but we cannot observe the process of connecting the economic stimulus to an economic decision. Why does Robbin purchase more ice cream than Willis when both face the same prices and have the same income? The most prominent economic theories of consumer behavior assume that consumers are rational and seek to maximize their satisfaction while staying within their budget. In this chapter, we discuss consumer theory and how it can be used to understand the purchasing behavior of consumers.

UTILITY THEORY

Consumers typically face a broad set of choices when allocating their income among food and nonfood goods and services. Considerable attention has been given historically to the development of a theoretical framework that will help us understand the choices consumers make. In the following discussion, we will assume that consumers are rational individuals who maximize their satisfaction, or utility.

Total Utility

A consumer purchases a good or service because of the satisfaction he or she expects to receive. Early researchers of consumer behavior argued that utility was cardinally measurable.[1] They also argued that the utility derived from a given commodity is independent of the utility derived from other commodities. For example, the latter belief suggests that the consumer can determine the utility of taco consumption independently from hamburger consumption. Total utility accord-

---

[1]The term *cardinally measurable* is used in the same sense that a ruler measures distances, namely, an attempt is made to quantify the amount of satisfaction obtained from consumption. On the other hand, *ordinally measurable* implies only a ranking of distance, such as longest to shortest or vice versa.

ingly would be equal to the total utility derived from each of the individual commodities. The psychological units of satisfaction derived from consumption are generally referred to as **utils.**

A **utility function** is an algebraic expression that allows us to rank a consumption bundle by the total utility or satisfaction it provides. **Consumption bundles** refer to a particular combinations of goods being considered. The utility function describes the **total utility** derived from consuming a particular bundle. Consequently, utility or satisfaction is a function of consuming individual commodities.

To clarify the meaning and use of the utility function, consider a consumer who has the following utility function (although it is highly unlikely the consumer is aware of this mathematical representation of his or her utility function):

$$\text{total utility} = (\text{quantity of hamburgers} \times \text{quantity of pizza}) \qquad (3.1)$$

If consumption bundle *A* consists of 2.5 hamburgers and 10 slices of pizza per week, the consumer with a utility function such as Equation 3.1 would derive a total utility of 25 from the consumption of this bundle (i.e., 2.5 × 10). This bundle, two other bundles of consumer goods, and the subsequent total utility they provide are summarized in Table 3.1.

If we wanted to know whether bundle *B*, which consists of 3 hamburgers and 7 pizza slices per week, is preferred, not preferred, or indifferent to bundle *A*, we know from Equation 3.1 that the utility this consumer derives from consuming bundle *B* would be 21 (i.e., 3 × 7). Therefore, this consumer would prefer bundle *A* to bundle *B* because the utility provided by bundle *A* (25) is greater than the utility provided by bundle *B* (21). Suppose that bundle *C* consists of 2 hamburgers and 12.5 slices of pizza. The utility derived from consuming this bundle also would equal 25. Therefore, this consumer would be indifferent between bundles *A* and *C*.

The notion of a utility function may seem mysterious. In fact, it is hard to imagine a consumer thinking in terms of a specific utility function when purchasing goods and services, as suggested by Equation 3.1. Yet, the concept of satisfaction that the utility function expresses is the foundation of consumer economic analysis.

TABLE 3.1   Example of Total Utility Derived from the Consumption of Hamburgers and Pizza

Bundle	Quantity of: Hamburgers	Quantity of: Pizza	Total Utility
A	2.5	10.0	25
B	3.0	7.0	21
C	2.0	12.5	25

Marginal Utility

If utility is measurable, it is appropriate to question how total utility changes as a greater amount of a particular good is consumed. The change in total utility, associated with a specific change in the consumption of a commodity, is referred to as **marginal utility.** In economics, the term *marginal* is synonymous with the word *change.* To illustrate this, the marginal utility of hamburgers is shown in Equation 3.2, where Δ indicates the change in a value.

$$MU_{hamburgers} = \frac{\Delta \text{ utility}}{\Delta \text{ hamburgers}} \tag{3.2}$$

This measure constitutes the change in utility associated with a change in the consumption of hamburgers. This value will always be greater than zero only if we assume that the consumer's appetite never becomes totally satiated. This value will fall (rise) as hamburger consumption increases (decreases).

To illustrate the notion of marginal utility, assume that the data in Table 3.2 reflect the utility of Sue Shopper regarding hamburger consumption. The first column in this table indicates the quantity of hamburgers Sue consumes per week. The second col-

TABLE 3.2   Calculation of Marginal Utility for Sue Shopper

Quantity of Hamburgers Consumed per Week	Total Utility	Marginal Utility
1	20	
2	30	10
3	39	9
4	47	8
5	54	7
6	60	6
7	65	5
8	69	4
9	72	3
10	74	2
11	74	0
12	70	−4

umn represents her total utility associated with each specific consumption level. The third column presents the corresponding levels of marginal utility. Note that each successive increment of hamburgers increases utility by a smaller amount. When consumption of hamburgers increases from 2 to 3, utility increases by 9 utils. When consumption of hamburgers increases from 8 to 9, utility increases by only 3 utils. When marginal utility is zero, total utility is maximized. In addition, marginal utility can be negative at higher levels of hamburger consumption. Utility actually decreases by 4 utils as hamburger consumption increases from 11 to 12. Figure 3.1 shows the shape of the total and marginal utility curves associated with the data presented in Table 3.2.

A

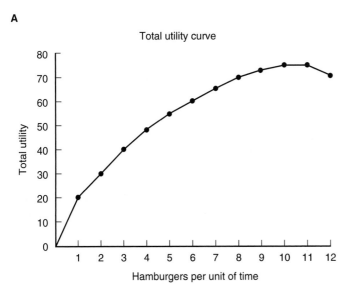

FIGURE 3.1 Total utility continues to increase as the number of hamburgers consumed increases, at least up to 11 hamburgers. At this point, total utility is maximized. Beyond 11 hamburgers, total utility decreases. (*A*). Marginal utility declines as Sue increases her consumption of hamburgers (*B*).

B

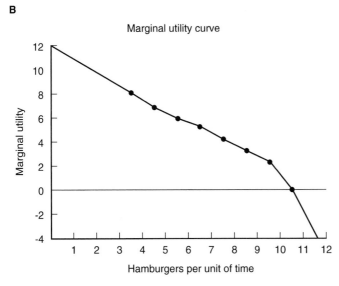

Law of Diminishing Marginal Utility

Is it clear why Sue's marginal utility declines when her consumption increases? If you consume one hamburger, then another, the second hamburger gives you less satisfaction than the first. Because there is so much truth to this notion, it has been given law-like status. The **law of diminishing marginal utility** suggests that as consumption per unit of time increases, marginal utility decreases. The fact that marginal utility eventually becomes negative in Table 3.2 suggests that a local Wendy's or McDonald's would have to pay consumers to consume more than 10 hamburgers during the week.

Does it seem logical to assume that the marginal utilities provided by different commodities are independent? Would the utility you derive from hamburger consumption depend on the amount of soft drinks, french fries, and tacos you consume? Because most people would answer in the affirmative, we must consider the consumer's consumption of all other goods and services before we can fully understand what influences consumer behavior.

INDIFFERENCE CURVES

Cardinal measurement for utility is unreasonable and unnecessary. Cardinality implies that society can add utils like it can add distances. The idea that bundle M yields a utility of 100 and bundle N provides a utility of 200 does not necessarily mean that bundle N provides twice as much satisfaction as bundle M. Instead, utility can be viewed as being ordinally measurable—that is, as a personal index of satisfaction in which the magnitude is used only to rank consumption bundles.

Modern consumption theory dismisses the notion that utility is cardinally measurable and instead measures utility in ordinal terms. All we really need to know is that bundle N is preferred to bundle M, not by how much.

Concept of Isoutility

The basic building block of modern consumption theory is the notion of an isoutility curve, which accounts for substitution in consumption for two products. The term *iso* is of Greek origin and means "equal."[2] An isoutility curve is often referred to as an **indifference curve.** A consumer is indifferent to consumption bundles that yield an equal level of satisfaction or utility.

The different combinations of goods in which indifference occurs have special significance because total utility is equal at all points along the indifference curve. The combinations of hamburgers and tacos, which represent specific levels of utility to Carl Consumer, are graphed in Figure 3.2. The curve labeled I_2 illustrates specific combinations of these two goods that yield a certain level of utility to Carl.

---

[2]For example, consider an isosceles triangle that has two equal sides.

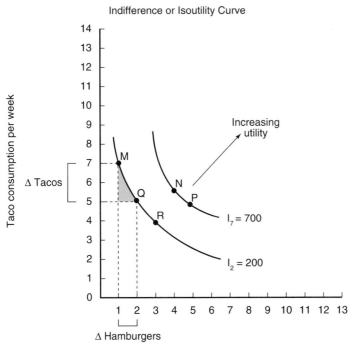

Indifference or Isoutility Curve

FIGURE 3.2 An indifference curve represents all combinations of two goods that yield an equal level of satisfaction or utility. In other words, the total utility derived from consumption is equal at all points along an indifference curve. The utility associated with consuming seven tacos and one hamburger per week (point *M*) by Carl Consumer is equal to the utility associated with consuming five tacos and two hamburgers (point *Q*). Indifference curve I_7 represents a higher level of utility than curve I_2. Why? Bundle *P* on curve I_7 corresponds to approximately the same number of tacos but more hamburgers than bundle *Q*.

Changes in the utility Carl receives from consumption would be indicated by outward (inward) shifts of an indifference curve. Indifference curve I_7 represents a *higher* level of utility than indifference curve I_2. Maximization of utility derived by Carl requires that he be on the highest possible indifference curve. Carl prefers indifference curve I_7 to I_2 because it means a higher level of utility.

Marginal Rate of Substitution

To maintain a constant level of utility, one must change consumption of one commodity to obtain additional consumption of another; that is, a consumer may substitute one commodity for another to maintain a constant level of utility. The rate at which the consumer is willing to substitute one good for another is called the **marginal rate of substitution.** The marginal rate of substitution of hamburgers for tacos, for example, represents the number of tacos Carl is willing to give up for an additional hamburger to maintain the same level of satisfaction; or, in mathematical terms, as:

$$\text{marginal rate of substitution of hamburgers for tacos} = \frac{\Delta \text{ tacos}}{\Delta \text{ hamburgers}}$$

(3.3)

The marginal rate of substitution associated with moving from point M to Q in Figure 3.2 would be approximately -2 (i.e., $-2 \div 1$). Carl is willing to give up 2 tacos for 1 additional hamburger. If he instead moved from point Q to R, we see that the marginal rate of substitution would fall to about -1 (i.e., $-1 \div 1$).

The marginal rate of substitution represents the slope for a specific segment of an indifference curve for two goods. For Carl Consumer, the cutback in taco consumption times the marginal utility of tacos is identical to the increase in hamburger consumption times the marginal utility of hamburgers. We can also equate the marginal rate of substitution of hamburgers for tacos in Equation 3.3 with the ratio of their marginal utilities, or:

$$\frac{\Delta \text{ tacos}}{\Delta \text{ hamburgers}} = \frac{\text{MU}_{\text{hamburgers}}}{\text{MU}_{\text{tacos}}} \tag{3.4}$$

The loss in utility from consuming fewer tacos is just matched by the gain in utility Carl receives from consuming more hamburgers.

Why does the marginal rate of substitution fall as we move down the indifference curve? In Figure 3.2, the marginal rate of substitution fell from -2 to -1 when Carl moved down the indifference curve. When Carl consumed 7 tacos (point M), he was willing to give up 2 tacos to eat 1 more hamburger (a movement from point M to Q). When Carl consumed 5 tacos (point Q), he was willing to give up only 1 taco to receive 1 more hamburger (a movement from point Q to R). Carl is satisfied giving up one commodity (tacos) for more of another (hamburgers).

Perhaps the most intuitive explanation we can offer at this point relies on the notion of diminishing marginal utility. As taco consumption falls, its marginal utility rises. As hamburger consumption increases, its marginal utility falls. Thus, the marginal rate of substitution *falls* as one moves *down* an indifference curve (e.g., increasing hamburger consumption and reducing taco consumption).

THE BUDGET CONSTRAINT

We often hear the phrase "I wanted to purchase it, but I just could not afford it." This phrase portrays that we are all faced with what economists call a **budget constraint;** that is, purchases by a consumer cannot exceed his or her income. If consumption decisions are made as a household, income should include all forms of family income. It should also *exclude* tax obligations to reflect the disposable income of the household.[3] We must discuss the budget constraint using a unit of time, such as the maximum expenditures per day, per week, and so on.

If all other factors remain constant, when the disposable income of a consumer increases, the percentage of income spent for food decreases. For a poor consumer, a greater percentage of income is used to purchase food. This observation

---

[3]**Disposable income** is defined as income after taxes.

is commonly referred to as **Engel's Law.** In the United States, we have a relatively high per capita national income, and we spend a relatively small percentage of our total consumption expenditures (8.4% to 10.1%) on food. In India and the Philippines, the budget share for food items is approximately 51% and 56%, respectively. Table 3.3 shows that the United States has high per capita national total personal consumption expenditures but low budget shares spent on food. The total personal consumption expenditure pattern in Table 3.3 is generally consistent with Engel's Law, but exceptions are evident. Figure 3.3 illustrates Engel's Law, using information from the household portion of the 1987–88 Nationwide Food Consumption Survey. Engel's Law states that the greater the weekly income, the lower the proportion of income spent on food. Each point in this graph corresponds to a particular household, a total of approximately 4,000.

The total expenditures made by a consumer on a number of items can be determined by multiplying the total quantity of each good or service purchased by its respective price and then totaling the value of all purchases. For example, suppose Carl Consumer had a specific amount of money to spend on nondorm food per week. If he limited this consumption to purchases of hamburgers and tacos, Carl's total expenditure would be equal to the price of hamburgers times the quantity of hamburgers he consumed *plus* the price of tacos times the quantity of tacos he consumed, or:

$$\left\{ \begin{array}{c} \text{price of} \\ \text{hamburgers} \end{array} \times \begin{array}{c} \text{quantity of} \\ \text{hamburgers} \end{array} \right\} + \left\{ \begin{array}{c} \text{price of} \\ \text{tacos} \end{array} \times \begin{array}{c} \text{quantity of} \\ \text{tacos} \end{array} \right\}$$

$$= \begin{array}{c} \text{income spent on} \\ \text{nondorm} \\ \text{food} \end{array} \tag{3.5}$$

This budget constraint limits Carl's consumption of hamburgers and tacos to no more than the total income allocated to their consumption.

When the budget constraint is graphed, it is referred to as the *budget line.* The slope of Carl's budget line is equal to the negative of the price ratio, or:

$$\begin{array}{c} \text{slope of} \\ \text{budget line} \end{array} = - \frac{\text{price of hamburgers}}{\text{price of tacos}} \tag{3.6}$$

which suggests that the budget constraint will become *steeper* (flatter) as the price of hamburgers rises (falls) relative to the price of tacos (see Figure 3.4, *D*). Similarly, the budget constraint will become steeper (flatter) as the price of tacos falls (rises) (see Figure 3.4, *C*).[4]

---

[4]The slope of the budget line can be derived by rearranging Equation 3.5 to read:

$$\begin{array}{c} \text{quantity of} \\ \text{tacos} \end{array} = \frac{\text{income}}{\text{price of tacos}} - \left[\frac{\text{price of hamburgers}}{\text{price of tacos}} \times \begin{array}{c} \text{quantity of} \\ \text{hamburgers} \end{array} \right]$$

TABLE 3.3   Percent of Total Personal Consumption Expenditures Spent on Food and Alcoholic Beverages That Were Consumed at Home, by Selected Countries, 1994

Country	Percent of Total Personal Consumption Expenditures		Personal Consumption Expenditures	
	Food	Alcoholic Beverages	Total	Food
	------------Percent------------		--------Dollars per Person--------	
United States				
ERS estimate	7.4	1.0	17,489	1,294
PCE estimate	8.4	1.7	17,489	1,469
Canada	10.3	2.4	11,581	1,193
United Kingdom	11.2	6.1	11,192	1,254
Netherlands	11.4	1.4	13,147	1,499
Hong Kong	12.3	0.7	12,602	1,550
Luxembourg (1991)	12.5	1.3	13,781	1,723
Singapore	13.8	1.6	9,268	1,279
Belgium	13.9	1.3	14,023	1,949
Sweden	14.6	2.7	12,217	1,784
Denmark	14.7	2.5	15,045	2,212
France	14.8	1.9	13,874	2,053
Australia	14.9	4.4	11,624	1,732
Austria	15.3	1.9	13,735	2,101
New Zealand	15.4	NA	8,908	1,372
Finland	15.5	3.9	10,690	1,657
Puerto Rico	16.8	2.4	6,792	1,141
Italy	17.2	1.0	10,991	1,890
Germany	17.3	NA	12,327	2,133
Japan	17.6	NA	21,830	3,842
Spain (1993)	18.2	1.4	7,753	1,411
Ireland	19.0	12.0	8,157	1,550
Iceland	19.0	2.8	13,838	2,629
Norway (1993)	19.8	3.1	12,371	2,449
Israel	20.5	0.9	9,117	1,869
Portugal (1993)	23.2	3.1	5,238	3,557
Thailand	23.3	3.8	1,360	317
Switzerland	24.4	NA	21,349	3,886
Fiji (1991)	24.4	3.6	1,352	918
Mexico	24.5	2.5	3,267	1,101
South Africa	27.5	6.5	1,846	508
Hungary	27.5	6.3	2,376	653
Cyprus	28.3	3.5	5,964	1,688
Korea, Republic of	29.1	NA	4,596	1,544
Colombia (1992)	29.6	3.7	890	263
Peru (1990)	31.0	NA	1,160	788
Greece	31.7	2.9	5,390	1,709
Malta (1993)	32.3	4.2	4,632	1,496
Ecuador (1993)	32.8	3.2	935	307
Bolivia	34.8	NA	640	240
Venezuela	38.2	NA	1,964	737
Sri Lanka	49.3	1.8	220	108
India	51.3	0.5	195	100
Philippines	55.6	NA	659	364

NA = Not available.

Source: *Food Consumption, Prices, and Expenditures,* 1999.

FIGURE 3.3 Scatter plot of weekly income and total food budget share.

TABLE 3.4 Example of Budget Constraint

Tacos ($0.50 Each)	Hamburgers ($1.25 Each)	Expenditure
10	0	$5
5	2	$5
0	4	$5

To illustrate, suppose that Carl has $5 a week to divide between the consumption of tacos and hamburgers. Tacos cost $0.50 each and hamburgers cost $1.25 each. Some of the combinations of tacos and hamburgers Carl can afford with a weekly budget of $5 appear in Table 3.4. This budget constraint is illustrated graphically in Figure 3.4, *A*. We know from Equation 3.5 that if just hamburgers are desired, taco consumption would be zero, and the quantity of hamburgers consumed would be 4 (i.e., total income [$5] ÷ price of hamburgers [$1.25]). Carl can afford a maximum of 4 hamburgers per week. If only tacos are desired (i.e., hamburger consumption is zero), Carl could afford a maximum of 10 tacos (i.e., income [$5] ÷ price of tacos [$0.50]).

All feasible consumption possibilities would, thus, appear along budget line *AB*. For example, the consumption of 2 hamburgers and 5 tacos also requires an income of $5, as indicated by point *C* in Figure 3.4, *A*.

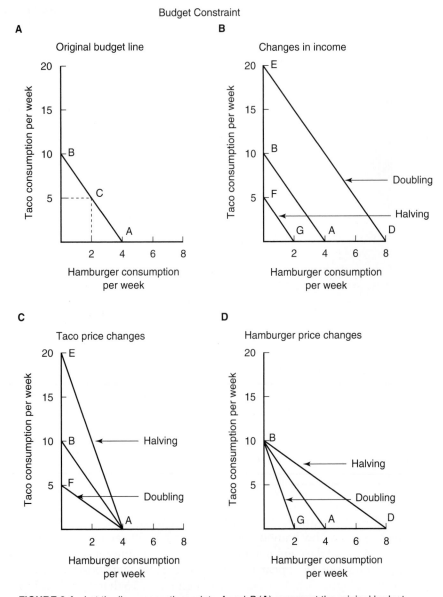

Budget Constraint

FIGURE 3.4 Let the line connecting points *A* and *B* (**A**) represent the original budget constraint or budget line. This line suggests that Carl Consumer could spend his entire weekly budget of $5 to buy 4 hamburgers costing $1.25 each, 10 tacos costing $0.50 each, or some combination of these two food items that appears along line *AB*.

What will happen to the budget line if income changes and prices remain unchanged? The answer is that the budget line will move in a parallel fashion. Suppose that the income Carl can devote to these two products doubled to $10. His maximum hamburger consumption would increase from 4 hamburgers at point *A* to 8 hamburgers at point *D* (i.e., 8 = $10 ÷ $1.25). Carl's maximum taco consumption would increase from 10 tacos at point *B* to 20 tacos at point *E* (i.e., 20 = $10 ÷ $0.50) (Figure 3.4, *B*). Thus, a line connecting 8 hamburgers on the horizontal axis with 20 tacos on the vertical axis would represent a new budget constraint (*DE*), which lies *to the right* of the original budget line. By similar logic, the budget line would take a parallel shift inward (leftward) to line *FG* if Carl reduced the amount of income he devoted to these two products by one-half. Finally, a doubling (halving) of *both* prices will also shift the budget line inward (outward) as illustrated in Figure 3.4, *B*.

Changes in the price ratio for two products will change the slope of the budget line. For example, if the price of tacos doubles, Carl's budget line will rotate to the left from line *AB* to line *AF* (Figure 3.4, *C*). This change suggests that fewer tacos can be purchased for any given level of hamburger consumption. If the price of tacos falls in half, Carl's budget line would instead rotate to the right from line *AB* to line *AE* (Figure 3.4, *C*). In both instances, the budget lines continue to have point *A* in common. At point *A*, only hamburgers are consumed; therefore, a price change in tacos would have absolutely no effect. Similarly, changes in the price of hamburgers would rotate the budget line as shown in Figure 3.4, *D*. A rightward (leftward) rotation from line *BA* to *BD* (*BG*) signifies halving (doubling) of hamburger prices.

To summarize, the slope of the budget line is given by the negative of the price ratio. This ratio indicates that the consumption of tacos associated with a one-unit increase in consumption of hamburgers is equal to the price of hamburgers divided by the price of tacos. An increase (decrease) in income will shift the budget line outward to the right (inward to the left) from the origin. This shift will be parallel in nature as long as the price ratio does not change. A change in the ratio of the two product prices, however, will alter the slope of the budget line.

SUMMARY

The major points made in the chapter may be summarized as follows:

1. The budget constraint represents the amount of income the consumer has to commit to consumption in the current period. A proportional change in all prices and income has *no effect* on the budget constraint. For this reason, economists argue that only relative price changes matter. When presented graphically, the budget constraint is frequently referred to as the budget line. The

slope of the budget line, which tells us the rate of exchange between two goods as their prices change, is given by the negative of the price ratio. A change in relative prices will change the slope of the budget line. Finally, an increase (decrease) in income will shift the budget line to the right (left).

2. We assume that consumers are rational and maximize their satisfaction, or utility. Thus, consumers are assumed to be able to rank all their choices. Furthermore, consumers are assumed to be willing to substitute commodities of equal value.

3. Early researchers of consumer behavior argued that utility could be measured. The term *utils* was used as a unit of measure. A hamburger might yield 10 utils, a soda 4 utils, and so on. Marginal utility describes the change in utility or utils as more of a good is consumed and is thought to diminish as consumption increases, but not reach zero, according to the assumption of nonsatiation.

4. Today no one really believes that utility can be measured in utils. Instead, utility is thought of in the context of a personal index of satisfaction. The magnitude of this index (or function) serves to order the consumption bundles, or combinations of goods the consumer faces.

5. All consumption points that provide the same utility form an isoutility, or indifference curve. Increases (decreases) in utility are indicated by a shift in an indifference curve to the right (left). The negative of the slope of this curve is known as the marginal rate of substitution (MRS). This rate indicates the willingness of the consumer to substitute one good for another. The declining rate as one moves down an indifference curve indicates the existence of the principle of diminishing marginal utility.

DEFINITION OF KEY TERMS

Budget constraint: defined by the income available for consumption and the prices that a consumer faces. This constraint defines the feasible set of consumption choices facing a consumer.

Consumption bundles: quantities of various goods or services that a consumer might consume.

Disposable income: personal income after the payment of tax obligations.

Engel's Law: as disposable income of a consumer increases, the percentage of income spent for food decreases if all other factors remain constant.

Indifference curve: a graph of the locus of consumption bundles that provide a consumer a given level of satisfaction.

Law of diminishing marginal utility: marginal utility declines as more of a good or service is consumed during a specified period of time.

Marginal rate of substitution: the rate of exchange of pairs of consumption goods or services to leave utility or satisfaction unchanged, or the absolute value of the slope of an indifference curve.

Marginal utility: the change to utility or satisfaction as consumption of a good is increased by one unit.

Total utility: the total satisfaction derived from consuming a given bundle of goods and services.

Utility function: a mathematical or functional representation of the satisfaction a consumer derives from a consumption bundle.

Utils: imaginary units of satisfaction derived from consumption of goods or services.

REFERENCES

Deaton A, and J Muellbauer: *Economics and consumer behavior,* Cambridge University Press, 1980.

EXERCISES

1. Based on the following table, graph the total utility curve of Robbin Denison for buffalo wings. Secondly, calculate the marginal utility between each point and plot the corresponding graph. Why is the slope of the MU curve negative?

Robbin's Utility Function

# of Wings	Total Utility
1	30
2	58
3	84
6	150
12	222
24	306

2. Robbin likes a beverage (Coca-Cola, of course) with her wings. There is only one place to buy this combination, Wings 'n' Suds. Using the graph on the following page, let the number of Coca-Colas be on the vertical axis and the number of wings on the horizontal axis. Label the graph.

   a. Robbin has her choice of getting 12 bottles of Coca-Cola and 3 wings or 3 bottles of Coca-Cola and 12 wings, free of charge. Which bundle will she choose? Why?

   b. Which would Robbin choose if she could have either 12 bottles of Coca-Cola and 6 wings or 6 bottles of Coca-Cola and 9 wings?

c. Calculate the MRS between points *A* and *E*.

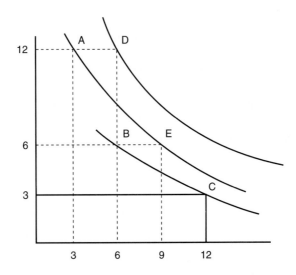

3. After careful scrutiny, Robbin budgets $12/week for Wings 'n' Suds. Consider the following:

   a. Graph and label the axes to show how much of each good Robbin is *able* to buy, if the price of a buffalo wing is $.50, while the price of Coca-Cola is $1.50.

   b. In order to attract more customers to Wings 'n' Suds, management decides to lower the price of Coca-Cola to a buck. Show what happens to Robbin's budget line compared to a.

   c. Instead of b, assume there is a sudden shortage of buffalo wings available. Now, Wings 'n' Suds has to raise the price of a wing to $3. Show how this event changes Robbin's budget line relative to a.

   d. Robbin, after winning $1,000 in the Texas lottery, decides that she can now spend $40/week at Wings 'n' Suds. Show how this event changes Robbin's budget line relative to a.

   e. What combination of wings and Coca-Colas *should* Robbin buy in a? b? c? d?

4. Given the following set of indifference curves, calculate the Marginal Rate of Substitution between the following:

   a. Points *A* and *B*.

   b. Interpret this measure.

   c. Which combination of tacos and hamburgers yields the highest level of satisfaction? Circle the correct answer(s).

    i. 7 tacos, 1 hamburger

    ii. 2 tacos, 5 hamburgers

    iii. 5 tacos, 7 hamburgers

    iv. 7 tacos, 5 hamburgers

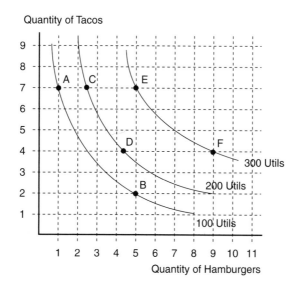

5. Given the following changes in a consumer's budget constraint, please indicate in writing to the right of each graph what caused the budget constraints to change.

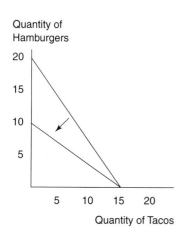

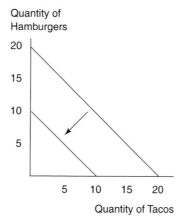

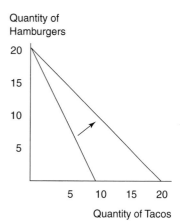

6. Assume you have interest in only two goods: cheap food and environmental quality. That is, these goods are the only ones that provide utility to you. Consider the following graph:

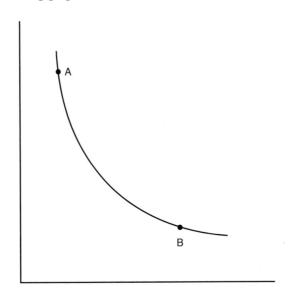

Let point *A* correspond to 80 units of cheap food and 20 units of environmental quality.

Let point *B* correspond to 50 units of cheap food and 30 units of environmental quality.

a. Label the axes. What is the technical name of the curve given above?

b. Calculate how many units of cheap food you are willing to give up to receive one more unit of environmental quality in order to maintain the same level of satisfaction.

7. Suppose Glenn Gibbs (a native of Manchester, England) has an income of $30. He derives satisfaction from the consumption of tea and biscuits. The price of tea is $3.00 per cup and the price of biscuits is $.50/unit.

a. Graphically construct the budget line for this situation. Label your axes carefully.

b. Now, suppose the price of tea increases to $5.00 per cup, and Glenn's income remains at $30. Assuming the price of biscuits remains at $.50/unit, redraw the budget line to reflect this situation.

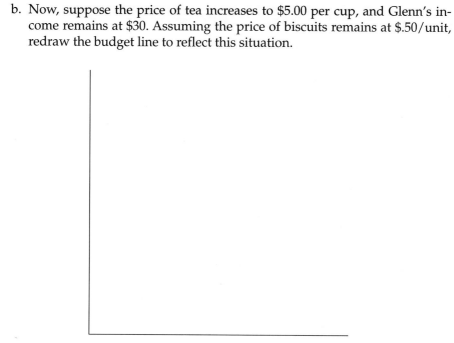

4

CONSUMER EQUILIBRIUM AND MARKET DEMAND

Economists know a lot about what makes producers tick, while they know almost nothing about the motivation of consumers.

Tibor Scitovsky
The Joyless Economy, 1977

TOPICS OF DISCUSSION

In this chapter we shall merge the concepts of the budget constraint and marginal utility presented in Chapter 3 to determine consumer equilibrium and market demand. Our goal is to understand how consumers will react to changes in prices and income when deciding about consuming specific commodities. The example of tacos and hamburgers employed in the last chapter will continue to be used when we examine the concept of consumer equilibrium.

This chapter begins with a discussion of the equilibrium demand conditions for an individual consumer, which is the point at which the consumer is satisfied when buying a specific quantity at the going price. Changes in this equilibrium when the prices of other goods change or as the consumer's income changes also are discussed. The chapter closes with a focus on the law of demand for consumers in general, including the impact that tastes and preferences have upon market demand, and with an approach to measuring changes in the economic well-being of consumers when market conditions change.

CONDITIONS FOR CONSUMER EQUILIBRIUM

If there were no budget constraint, the consumer would move toward consuming all goods and services at a point at which marginal utility of each good or service is zero.[1] With the presence of a budget constraint, consumer decisions can be thought of as the process of choosing products so that the total utility derived from consumption is maximized, subject to the amount of the budget constraint. The key assumption is that consumers try to *maximize* their satisfaction or utility.

Let us initially characterize the conditions for maximization of utility graphically. Consider the budget constraint and the set of indifference curves for Carl Consumer in Figure 4.1. If his utility is to be maximized subject to the budget constraint, it necessarily follows that we find the point on the budget line that yields the *highest* utility.

This situation occurs in Figure 4.1 at point A. Why? Point C on the I_1 curve would not exhaust the income allocated for consumption of these two goods. Carl can increase his utility to 200 by moving from point C to A on the I_2 indifference curve. Although point B on the I_4 curve represents higher utility than point A, it requires more income than Carl has to spend. Therefore, point A, which consists of five tacos and two hamburgers, represents Carl's **consumer equilibrium.** There would be economic incentive for him to move to point A if he were located anywhere else on this graph. Only at point A is the slope of the budget line equal to the slope of the indifference curve.

We can use Equations 3.3 and 3.4 to express the conditions for consumer equilibrium in mathematical terms. Setting these equations equal to each other, we see

---

[1]To go beyond this condition implies a reduction in total utility because marginal utilities are subsequently negative.

FIGURE 4.1   The rational consumer is said to be in equilibrium with regard to the consumption of two goods when the slope of the budget line (the budget constraint) is equal to the slope of the indifference curve. In the case of Carl Consumer, this equality occurs at point *A*. The indifference curve I_2 represents the highest utility curve attainable, given the nature of the budget constraint. At point *A*, where the budget line is tangent to indifference curve I_2, Carl would purchase five tacos and two hamburgers. The total cost of these items completely exhausts his $5 weekly fast food budget.

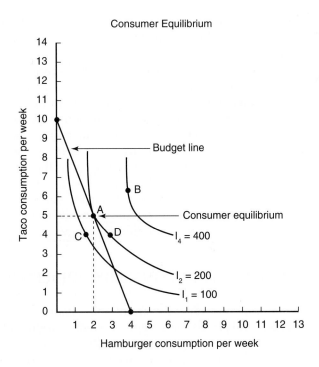

that consumer equilibrium between tacos and hamburgers for Carl Consumer is reached when the absolute values of these slopes are equal, or:

$$\frac{\text{price of hamburgers}}{\text{price of tacos}} = \frac{MU_{\text{hamburgers}}}{MU_{\text{tacos}}} \qquad (4.1)$$

The first term in Equation 4.1 represents the rate at which the market is willing to exchange tacos for hamburgers. This ratio coincides with the negative of the slope of the budget line. The second term in this equation represents the rate at which the consumer is willing to exchange tacos for hamburgers, given tastes and preferences. This ratio coincides with the slope of the marginal rate of substitution. When the equality expressed in Equation 4.1 holds, the consumer is in equilibrium. We also can use Equation 4.1 to restate the condition for consumer equilibrium as:

$$\frac{MU_{\text{hamburgers}}}{\text{price of hamburgers}} = \frac{MU_{\text{tacos}}}{\text{price of tacos}} \qquad (4.2)$$

Equation 4.2 suggests that the marginal utility derived from the last dollar spent on each good is equal. The equilibrium condition expressed in Equation 4.2 can be expanded to include all goods and services purchased by the consumer.

CHANGES IN EQUILIBRIUM

The condition previously stated for consumer equilibrium suggests that the demand for each good is influenced by consumer income and all prices. Indeed, there is a causal economic relationship that suggests that Carl Consumer's demand for hamburgers (or tacos) is a function of the price of hamburgers, the price of tacos, and his available income. In other words, changes in prices and income will lead to changes in consumer demand for goods and services.

Changes in Product Price

Economists are interested in the effects on demand of changes in income and prices, holding everything else fixed. The approach to measuring these *ceteris paribus* (i.e., all other things constant) effects is examined below. Let us first focus on the effects of changes in the price of the product.

Suppose that the price of a hamburger is varied, leaving the price of tacos and income unchanged. As denoted by point *A* in Figure 4.2, if the price of hamburgers

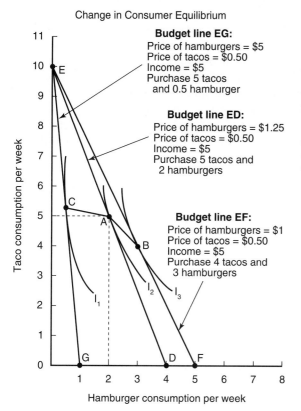

Change in Consumer Equilibrium

Budget line EG:
Price of hamburgers = $5
Price of tacos = $0.50
Income = $5
Purchase 5 tacos
and 0.5 hamburger

Budget line ED:
Price of hamburgers = $1.25
Price of tacos = $0.50
Income = $5
Purchase 5 tacos and
2 hamburgers

Budget line EF:
Price of hamburgers = $1
Price of tacos = $0.50
Income = $5
Purchase 4 tacos and
3 hamburgers

FIGURE 4.2 The consumer's equilibrium position in consumption can change as the budget line changes (see also Figure 3.1). Beginning with Carl Consumer's equilibrium at point *A*, where $5 is spent on tacos and hamburgers, we see that five tacos and two hamburgers are bought. If the price of hamburgers were to fall to $1, the budget line would rotate to the right and the new equilibrium would occur at point *B*. If the price of hamburgers rose to $5, equilibrium would occur at point *C*. The line joining points *C, A,* and *B* is called the **price-consumption curve** because it shows the amounts of hamburgers *and* tacos Carl would consume as the price of hamburgers changes.

were $1.25, the price of tacos were $0.50, and his available income were $5, Carl would consume two hamburgers and five tacos. If the price of a hamburger were to fall to $1, Carl's new equilibrium would be indicated by point *B*, where he would prefer to purchase three hamburgers and four tacos. If the price of a hamburger were varied further, other equilibria (such as point *C*) would be identified. As the price of hamburgers declines (rises), *ceteris paribus,* this consumer will purchase more (less) hamburgers.

The increase in the quantity of hamburgers Carl purchased is due to the **substitution effect** and **income effect** of the price change. The substitution effect occurs when the price of hamburgers declines relative to the price of tacos. Carl will substitute hamburgers for tacos because the price of hamburgers has declined. Yet, simultaneously, when the price of hamburgers fell, Carl's real income increased. The real income effect of a decrease in the price of hamburgers means that the consumer can buy more hamburgers or tacos (or both) even though income has not changed.

The line joining points *C, A,* and *B* in Figure 4.2 is called the **price-consumption curve.** This curve shows the quantities of hamburgers and tacos Carl will consume as the price of hamburgers changes. If the price were continuously varied, a full set of prices and quantities of tacos and hamburgers could be identified.

If we graphed this series of prices and quantities of hamburgers associated with points *C, A,* and *B,* we would get a downward-sloping demand curve, which would indicate the quantity of hamburgers demanded by Carl at alternative price levels, holding all other factors constant. The demand curve typifies the amounts of a commodity consumers are willing and able to purchase at each possible price during some specific time in a specific market, all other factors held constant. Demand may change over time, and a careful specification of demand will include the time period for which it applies and the specified market. Demand for the same commodity may be different, not only in different time periods but also in different markets.

The consumer demand curve for hamburgers in Carl Consumer's example is illustrated in Figure 4.3. This demand curve shows that if the price of a hamburger were $1.25, Carl would demand two hamburgers (see point *A*). Point *A* in both Figures 4.2 and 4.3 therefore represents the same quantity demanded. The same relationship holds for points *B* and *C* in these two figures also. Finally, as suggested by Figure 4.3, if the price of hamburgers rises above $6, the quantity of hamburgers demanded by Carl would fall to zero (see point *D*). The demand curve and the price-consumption curve are really two sides of the same coin. The demand curve represents price and quantity pairs in equilibrium. Price is on the *Y,* or vertical, axis in this graph and quantity is on the *X,* or horizontal, axis. The price-consumption curve constitutes the preliminary step to a representation of the demand curve. In this curve, equilibrium price and quantity pairs are identified using the tangencies of the indifference curves and budget lines. This curve is then transformed into a demand curve.

Box 4.1 is an account from the *Wall Street Journal,* illustrating the potential impact of price as a determinant of demand.

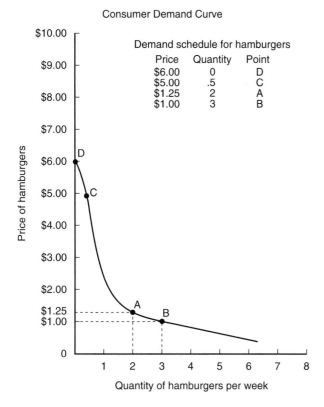

Consumer Demand Curve

Demand schedule for hamburgers

Price	Quantity	Point
$6.00	0	D
$5.00	.5	C
$1.25	2	A
$1.00	3	B

Quantity of hamburgers per week

FIGURE 4.3 If we hold the price of tacos constant at $0.50 and the budget constraint constant at $5, we can derive Carl Consumer's demand curve for hamburgers by examining the quantities of hamburgers consumed. For example, point *A* here corresponds to point *A* in Figure 4.2, where two hamburgers were purchased at a price of $1.25. If the price were to fall to $1, we see in both figures that three hamburgers would be desired (point *B*). Point *C* in both figures also indicates that Carl would consume only one hamburger every other week if the price rose to $5. Finally, at a price of $6 per hamburger, Carl would purchase no hamburgers. A line connecting these and other price-quantity combinations represents the consumer demand curve. The associated demand schedule of prices and quantities corresponding to this figure also is given.

BOX 4.1   *Price Impact on Demand*

At Fry's Food & Drug in Tempe, Arizona, shoppers were recently as thick as ants. They caused traffic jams in the parking lot. They formed checkout lines that wound all the way back to the meat department. Entire families moved through the grocery store in shopping-cart caravans.

The big attraction? A can of Coke on sale for the unheard-of price of 10 cents. A six-pack on sale at 59 cents.

Cases were disappearing from a six-foot high, nine-foot-wide cola mountain in the middle of a main grocery aisle at the rate of 2,900 a day—so quickly that Erick Thibault and five other stackers could barely keep up. Across town at Smitty's, six-packs of Pepsi, at 79 cents each, also were disappearing.

Source: Morris B: Coke and Pepsi step up bitter price war, *Wall Street Journal*, p. B1: October 10, 1988.

Changes in Other Demand Determinants

We have learned that the demand for hamburgers is a function of the price of hamburgers, the price of tacos, and the level of income. A demand curve shifts when changes in income and other prices occur. There are two major economic "shifters" of a consumer demand curve: (1) changes in the consumer's disposable income and (2) changes in the prices of other goods and services. Both changes are studied in a *ceteris paribus* context, which assumes that all other factors in the economy remain constant during the time period.

Change in Income.   When prices are held fixed and income is varied, the effects that changes in income have on consumer demand for a particular product can be assessed. This impact has led economists to classify goods into two categories: (1) normal goods and (2) inferior goods, which are defined as follows:

- **Normal goods** are those goods for which a rise (fall) in income will lead to increased (decreased) consumption. Examples of normal goods are gasoline, housing, and steak.
- **Inferior goods** are those goods for which a rise (fall) in income will lead to decreased (increased) consumption. In the past, margarine has been considered to be inferior to butter. When income rises, consumers tend to eat more butter and less margarine. Riding the bus has always been considered inferior to other modes of transportation. When income increases, consumers purchase their own car, take a taxi, or fly.

In Figure 4.4, hamburgers and tacos were both normal goods when Carl Consumer's available income increased from $5 to $6. This rise in available income increased the equilibrium consumption of hamburgers from 2.0 (point *A*) to 2.6 (point *B*). Taco consumption would change from 5.0 to 5.5. However, if income were to rise to $8, taco consumption would fall to 3.5 and hamburger consumption would rise close to 5.0 (point *C*). Carl obviously has a strong preference for hamburgers when his income increases. Although tacos were initially considered a normal good when income expanded, they became an inferior good to Carl at higher income levels.

If income varied continuously and we recorded the equilibrium choices, such as *A*, *B*, and *C* in Figure 4.4, we could plot a curve relating income to consumption of a good. Such a curve is called an **Engel curve,** named after the nineteenth-century German statistician, Ernst Engel. Recall from Chapter 3 that **Engel's law** suggests that when consumer income increases, the proportion spent on food decreases, *ceteris paribus.*

A different Engel curve exists for each commodity. In any analysis of Engel curves, income is on the vertical axis and quantity consumed is on the horizontal axis. The Engel curves for hamburgers and tacos based on this example are illustrated in Figure 4.5, *A* and *B*. Figure 4.5, *A* shows that this particular consumer will want more hamburgers per day as income increases, which suggests this good is a

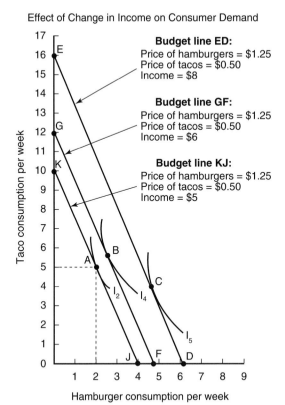

Effect of Change in Income on Consumer Demand

Budget line ED:
Price of hamburgers = $1.25
Price of tacos = $0.50
Income = $8

Budget line GF:
Price of hamburgers = $1.25
Price of tacos = $0.50
Income = $6

Budget line KJ:
Price of hamburgers = $1.25
Price of tacos = $0.50
Income = $5

Taco consumption per week

Hamburger consumption per week

FIGURE 4.4 An increase in income shifts the budget line in a parallel fashion to the right. If Carl Consumer's fast food budget increased from $5 to $6 per week, he could attain a higher indifference curve (I_4) and new equilibrium position (point *B*). If his fast food budget rose to $8, his equilibrium position would shift to point *C*. The nature of these changes in demand when income changes can be further studied, determining if both products are normal or inferior goods.

normal good. Figure 4.5, *B* shows that tacos are a normal good over the *AB* segment of the Engel curve, but inferior over the *BC* segment. Here, taco consumption per day falls when income rises.

The demand curve for hamburgers will likely shift if income changes, *ceteris paribus*. In the case of a normal good or a luxury, a rise in income leads to a rightward shift in the consumer's demand curve. In the case of an inferior good, a rise in income leads to a leftward shift in the demand curve, all other factors constant.

Changes in Other Prices. A demand for a product also may shift if the prices of other products changes. If hamburgers and tacos are substitutes, a rise in the price of tacos may cause you to eat more hamburgers and fewer tacos. Your demand curve for hamburgers would therefore shift to the right when the price of tacos rises. A decline in the price of tacos would cause the demand curve to shift to the left.

If we instead consider the relationship between soft drinks and hamburgers, we may find that a rise in the price of soft drinks leads to fewer hamburgers consumed.

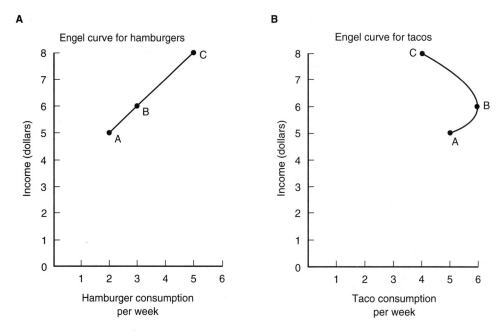

FIGURE 4.5   An Engel curve depicts the relationship between Carl Consumer's income and his consumption of hamburgers and tacos.

In this case, hamburgers and soft drinks are said to be complements. The demand curve for hamburgers would shift to the left when the price of soft drinks rises. Furthermore, a decline in the price of soft drinks would shift the demand curve for hamburgers to the right. Finally, a rise in the price of salt may cause no change in hamburger consumption. If so, hamburgers and salt are said to be independent.

THE LAW OF DEMAND

Market Demand

The concept of demand applies to a single individual or firm and to any number of individuals or firms. The economy is composed of a myriad of consumers who make expenditures on many goods. Thus, the sum of all relevant consumers comprises market demand. Consequently, it is important to distinguish demand curves for individuals from demand curves for the *market* of consumers.

To illustrate the concept of market demand, suppose that there are only two consumers in a market, Paula Purchase and Beth Buyer. The demand curves for each are represented in Figure 4.6, *A* and *B*. The market demand curve is the horizontal summation of the two individual demand curves as shown in Figure 4.6, *C*. At a price of $2.00, the quantity demanded by Paula would be one hamburger; Beth

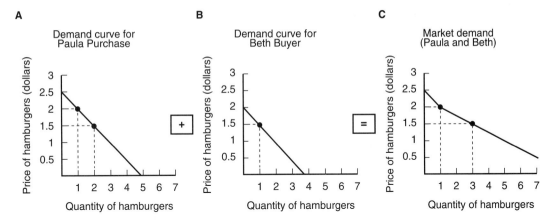

FIGURE 4.6   The market demand for hamburgers or any other consumer good is given by the summation of the quantities demanded by individual consumers at a particular price. Assume that Paula Purchase and Beth Buyer are the only two consumers in the economy. At a price of $2.50, neither consumer would want to purchase hamburgers; therefore, the market demand would be zero. If the price were $2.00, however, Paula would want to purchase one hamburger. Because Beth would still defer from hamburger consumption, the market demand curve should reflect a quantity of one. This process is then repeated for lower price levels, until the entire market demand curve is revealed (*C*).

would purchase none at this price. The combination of a price of $2.00 and a quantity of one hamburger represents one point on the market demand curve. If the price of hamburgers were $1.50, the market demand would be three hamburgers (i.e., two by Paula and one by Beth). By varying the price in this way, a continuous market demand curve can be constructed, as shown in Figure 4.6, *C*.[2]

The effects of a change in price, a change in other prices, or a change in income apply at the market or aggregate level like they do for an individual consumer. If tacos and hamburgers are substitute goods for both consumers, a rise in the price of tacos will shift each consumer's demand curve for hamburgers to the right. Thus, the market demand curve for hamburgers in Figure 4.6, *C* will shift to the right, also. Similar reasoning applies to changes in income. If the good is a normal good, increases in individual consumer incomes imply outward shifts of both the individual *and* market demand curves. We have drawn demand curves sloping downward, suggesting that consumers demand more at a lower price. This situation occurs with such regularity that it is referred to as the law of demand.[3] In other

---

[2]Some specific cases exist in which horizontally adding individual curves may not lead to the market demand curve. Suppose that the demand curve for consumer A depends on the demand by consumer B. One cannot simply add the individual demand curves *independently* to obtain the market demand curve.

[3]Is there a logical or empirical possibility of an upward-sloping demand curve? The answer to this question is yes. When this situation occurs, these goods are called Giffen goods. Although we will not develop the argument here, normal goods will never be Giffen goods. Indeed, a good must be sufficiently inferior for it to be a Giffen good. In reality, such goods are extremely rare. Even if some individuals have upward-sloping demand curves, the sum over all individuals would still probably yield a downward-sloping market demand curve.

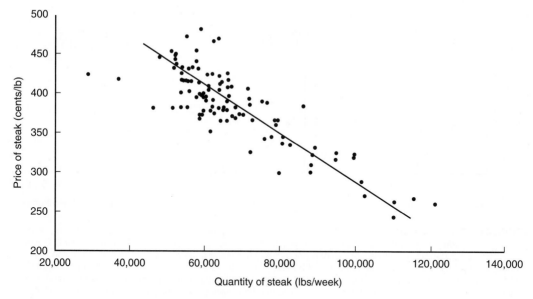

FIGURE 4.7   Scatter plot analysis of demand for steak.

words, as a general rule or law, demand curves have a negative or downward slope. An example of a market demand curve is exhibited in Figure 4.7. This relationship corresponds to the demand curve for steak in the Bryan/College Station, Texas, area. It is based on actual data using price and purchase information from area grocery stores.

Interpretation of Market Demand

Knowledge of the properties of the market demand for a particular good (e.g., food) or service (e.g., air travel) is extremely important to businesses participating in these markets. It helps them define their production schedules so they can meet demand expectations. Consumers in general also have an interest in the nature of the market demand curve, because it will ultimately have a major effect on the market price of the good or service. Policymakers also have an interest in market demand, both from a commodity policy perspective and from a macroeconomic perspective.

Before discussing the role that noneconomic factors such as tastes and preferences play in market demand, let us distinguish the difference between a **change in the quantity demanded** and a **change in demand**.

Change in Quantity Demanded.   When consumers change their expenditures because the price of the product changes, a change in the quantity demanded occurs. For example, if the price of a product falls from P_A to P_B in Figure 4.8, the

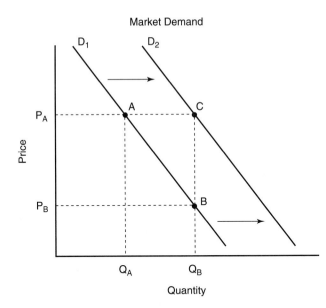

Market Demand

FIGURE 4.8 The increase in quantity from Q_A to Q_B can be the result of a change in the quantity demanded as price falls from P_A to P_B (movement from point A to B on the D_1 demand curve), or a change in demand (movement from point A on the D_1 demand curve to point C on the D_2 demand curve).

quantity demanded will increase from Q_A to Q_B. Similarly, an increase in the price of the product from P_B to P_A would cause the quantity demanded to decline from Q_B to Q_A. The movement between points A and B on the D_1 demand curve thus represents a change in the quantity demanded.

Change in Demand.   Although a change in the price of a product will result in movement along its demand curve, changes in the prices of substitutes and complements or changes in consumer income will cause the demand curve to shift. An increase in income will shift the market demand curve from D_1 to the right at D_2, assuming the good in question is a normal good. This situation suggests that quantity Q_B would now be purchased at price P_A instead of Q_A. A decrease in income would have the opposite effect. The effects on quantity here are due to a shift in the demand curve, or a change in demand.

Similarly, if the price of a competing (substitute) good increases, the demand curve for the good in question will shift to the right, *ceteris paribus.* The movement from point A to C in Figure 4.8 may be due to a rise in the price of a competing good. This situation also reflects a change in demand as opposed to a change in the quantity demanded.

TASTES AND PREFERENCES

Demand determinants include own price, prices of related goods, and incomes of consumers. Such determinants are economic factors of demand. However, noneconomic determinants also exist, namely the composition of the population, attitudes toward nutrition and health, and attitudes toward food safety, lifestyles, technological forces, and advertising. Typically, such determinants carry the label *tastes and preferences.*

Composition of the Population

The composition of the population plays a role in the demand for particular commodities. The proportion of persons in the 18- to 44-year-old and the over 65-year-old age groups is on the rise. The number of Americans aged 65 and over has doubled in the last three decades; the current total of elderly Americans is approximately 35 million, roughly 13% of the total population. The age shift of the population accounts for the decline in whole milk consumption and the increases in consumption of soft drinks, fruit drinks, and other beverages. Also, the changing racial and ethnic composition of the population is of substantial importance to the domestic food market. The fastest growing ethnic groups are Hispanics and Asians. The increase in demand for Cajun, Mexican, and Southern-style foods may be due in part to racial and ethnic elements.

Finally, the distribution of households of various sizes influences the domestic food market (Senauer, Asp, and Kinsey, 1991). Single-person households have increased dramatically. The percentage of single-person households has more than doubled from 1950 to 1980. More than one-half of all households today are composed of only one or two persons. At the same time, during the past 30 years, there has been a steady decline in the proportion of more-than-two-person households.

Attitudes toward Nutrition and Health

Medical research on the link between tobacco and cancer has led to changes in smoking habits. Concerns about calories, fitness, and health also have led consumers to change their eating habits (Figure 4.9). Medical researchers, for example, warn that consumption of too much red meat may contribute to heart disease, strokes, and cancer. Because of the emphasis on the reduction of animal fats, the demands for dairy products (e.g., eggs) and red meats may decline (the demand curve shifts to the left), while the demand for poultry and fish products may increase (the demand curve shifts to the right), *ceteris paribus.* The data in Box 4.2 support these trends between 1980 and 1990. Per capita consumption of poultry and fresh vegetables continued to increase from 1990 to 1997, while per capita consumption of red meat, eggs, and fish has leveled off.

Food Safety

Food safety issues are currently a major concern with consumers. Such issues pertain to pesticide and herbicide residues in foods, antibiotics and hormones in poultry and livestock food, irradiation, nitrates in food, additives and preservatives,

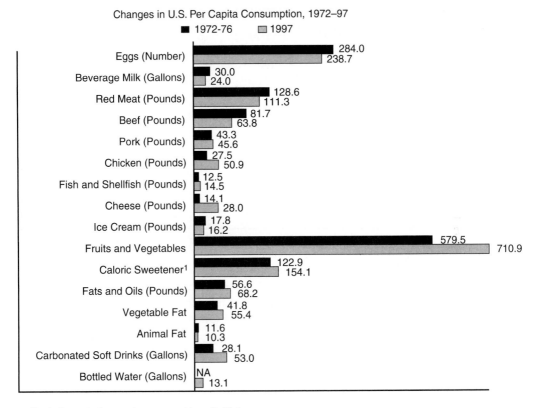

Changes in U.S. Per Capita Consumption, 1972–97

■ 1972-76   □ 1997

	1972-76	1997
Eggs (Number)	284.0	238.7
Beverage Milk (Gallons)	30.0	24.0
Red Meat (Pounds)	128.6	111.3
Beef (Pounds)	81.7	63.8
Pork (Pounds)	43.3	45.6
Chicken (Pounds)	27.5	50.9
Fish and Shellfish (Pounds)	12.5	14.5
Cheese (Pounds)	14.1	28.0
Ice Cream (Pounds)	17.8	16.2
Fruits and Vegetables	579.5	710.9
Caloric Sweetener[1]	122.9	154.1
Fats and Oils (Pounds)	56.6	68.2
Vegetable Fat	41.8	55.4
Animal Fat	11.6	10.3
Carbonated Soft Drinks (Gallons)	28.1	53.0
Bottled Water (Gallons)	NA	13.1

[1]Includes caloric sweeteners used in soft drinks.

FIGURE 4.9   The concern over cholesterol in recent years apparently has led many consumers to cut back on their consumption of red meat and to increase their consumption of fish and poultry products. (*Source: Food Consumption, Expenditures, and Prices*, 1999.)

sugar, and artificial coloring. Three examples are the case of salmonella in poultry, the chemical daminozide, sold under the name Alar and used on apples, and the outbreak of *E coli* bacteria in meat packing plants. The impact of food safety concerns typically shifts the demand curve for a commodity to the left, *ceteris paribus*.

Lifestyles

Changes in lifestyles are evident in the United States. Fashions, particularly for clothing and automobiles, serve as excellent examples of this phenomenon. The demand curve for the items that are in fashion shifts to the right; the demand curve for goods not in fashion shifts to the left, all other factors invariant. Moreover, there has been an increasing trend for both spouses to work outside the home, resulting in less time for food preparation. One of the major social trends of the last 25 years

BOX 4.2   Per Capita Consumption of Selected Products

Consumption of:	1980	1990	1997
		—Pounds per Capita—	
Red meat	126.4	112.3	111.0
Poultry	40.8	56.3	64.8
Fish	12.4	15.0	14.5
Fresh vegetables	149.3	167.2	185.6
Eggs (number)	271.1	234.3	238.7

Source: *Food Consumption, Expenditures, and Prices*, 1999.

has been the increase of women in the labor force. Concomitantly, enormous growth has occurred in the number of fast food restaurants and the increase in the demand for food away from home in general. Also, because of this social trend, there exists a rise in the demand for prepared foods.

Technological Forces

New technology in household food preparation, especially microwave ovens, and concurrent innovations in food processing continue to decrease the time needed for at-home meal preparation. Most households in the United States own a microwave oven. Industry studies show that most consumers choose foods that can be prepared in less than 20 minutes (Morris, 1985). Consequently, consumers want their food preparation to be easy and quick. During the past few decades, a myriad of convenience foods, particularly frozen items, ready-to-serve items, and mixes have been introduced into the marketplace. Convenience is presently a major attribute in food products.

Improvements and developments in processing and marketing have also contributed to the popularity of some foods. The development of single-serving, boxed fruit juices as well as an increased variety of blends has spurred the consumption of fruits. Improvements in processing techniques have permitted the production of shortening and margarine made entirely from vegetable oils. Cane and beet sugar consumption have been affected by the development of high-fructose corn syrup.

Advertising

The impact of advertising cannot be overlooked. Generic advertising campaigns are presently in full swing for fluid milk, citrus products, yogurt, cheese, butter, beef, pork, and seafood. "The Real Food for Real People" advertising campaign for beef, "The Other White Meat" campaign for pork, and the "Fabric of our Lives" campaign for cotton are testimonials. Total food-related advertising is a multi-billion-dollar industry in the United States. In 1991, for example, roughly $750 million was spent on commodity promotion in the agricultural sector (Lenz, Forker, and Hurst). Today, that figure is beyond $1 billion.

The purpose of generic advertising campaigns is to persuade consumers to buy particular products and to increase the demand for the products. Branded advertising and promotional effects attempt to make the demand curve more steep (less responsive to price) by developing brand loyalties.

CONSUMER SURPLUS

An examination of the demand curve presented in Figure 4.10 reveals that the consumer pays $6 each for five units of this product. The consumer, however, was willing to pay $10 for one unit, $9 for two units, and so on. Therefore, although the consumer actually pays $6 per unit to consume five units, he or she was willing to pay much more to purchase a smaller quantity.

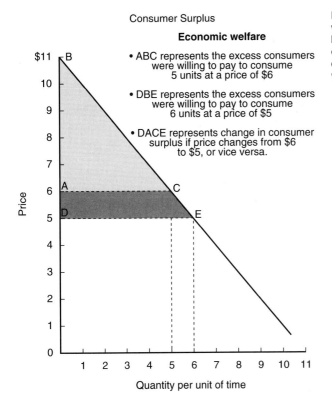

Consumer Surplus

Economic welfare

- ABC represents the excess consumers were willing to pay to consume 5 units at a price of $6

- DBE represents the excess consumers were willing to pay to consume 6 units at a price of $5

- DACE represents change in consumer surplus if price changes from $6 to $5, or vice versa.

FIGURE 4.10 The change in the economic welfare of consumers can be approximated by the concept of consumer surplus, which is equal to the difference between what the consumer was willing to pay for a product and what the consumer had to pay.

If one pursues this line of reasoning, it can be argued that the area *ABC* in Figure 4.10 is a measure of the excess the consumer was willing to pay to consume five units of this product. This difference between willingness to pay and the amount actually paid is referred to as consumer surplus. Although not without controversy, it is a "bread and butter" tool of economic policy analysis.[4]

Using this approach, consumer economic well-being would increase if the price of a product fell from $6 to $5 per unit. The amount of this gain would be equal to area *ACED*. Thus, a federal government policy that resulted in cheaper food prices would increase consumer economic well-being. A loss in consumer surplus equal to area *ACED* would occur if the market price were to rise from $5 to $6. The concept of consumer surplus allows economists to place a monetary value on changes in utility.

Often, policies change many prices. In this case, total consumer surplus changes for all markets represent the economic benefits to consumers of the policy change. The concept of consumer surplus will be discussed further in Chapters 10 and 13, when we discuss changing supply and demand conditions as well as food and fiber policy alternatives.

SUMMARY

The major points made in this chapter may be summarized as follows:

1. Factors affecting consumer demand are:
   - Price of the product,
   - Price of other products,
   - Disposable income of consumers, and
   - Tastes and preferences.
2. The consumer is at equilibrium when the marginal rate of substitution or slope of the highest attainable indifference curve is equal to the slope of the budget line, or alternatively when an additional dollar spent on each good would return the same marginal utility per dollar.
3. Each tangency point between the budget line and indifference curve when price changes identifies a new consumer equilibrium. The locus of all such tangencies is called the price-consumption curve.
4. A **demand curve** is a schedule that shows, *ceteris paribus,* how many units of a good the consumer will buy at different prices for that good during some specified time in a specified market.
5. Movement along a demand curve when the price of the good changes is referred to as a **change in the quantity demanded,** and a shift in the demand curve resulting from a change in prices of other goods, income, population and/or tastes and preferences is referred to as a **change in demand.**

---

[4]The issue is whether area *ABC* represents the consumer's actual willingness to pay for the privilege of purchasing the commodity at $6 each or only an approximation of it.

6. The market demand for a good is equal to the sum of all individual demands for the good. Individual demand curves for a product are added horizontally to obtain the market demand curve for a product. As a general law, demand curves are presumed to have a negative slope as a consequence of the law of diminishing marginal utility. The law of demand states that prices and quantities demanded are inversely related.

7. Changes in tastes and preferences refer to changes in the composition of the population, attitudes toward nutrition and health, and attitudes toward food safety, lifestyles, technological forces, and advertising.

DEFINITION OF KEY TERMS

Ceteris paribus: the assumption that all other factors that might affect demand are held constant during the time period. This term is the most often used Latin phrase by economists.

Change in demand: a shift in the demand curve generally caused by changes in the prices of complements or substitutes, income, and tastes and preferences.

Consumer equilibrium: the consumption bundle that maximizes total utility and is feasible as defined by the budget constraint. The marginal utilities per dollar spent on a good or service must be equal.

Demand curve: a schedule that shows how many units of a good the consumer will purchase at different prices for that good during some specified time in a specified market, all other factors constant.

Engel curve: the schedule that shows how many units of a good the consumer will purchase at different income levels, all other factors constant.

Engel's Law: when incomes of consumers increase, the proportion of income spent for food decreases, all other factors constant.

Income effect: decrease (increase) in the price of a product means the consumer can afford to buy more (less) of the product.

Inferior goods: goods for which consumption falls (rises) when income increases (decreases).

Normal goods: goods for which consumption rises (falls) when income increases (decreases).

Price-consumption curve: the connection or locus of all tangency points between budget lines and indifference curves. Each tangency identifies a point on the demand curve.

Substitution effect: substitution of a product for another because the price of the former has declined or increased.

REFERENCES

Becker G: A theory of the allocation of time, *Economic Journal* 74: 493–517, September 1965.

Lenz J, OD Forker, and S Hurst: U.S. commodity promotions: objectives, activities, and evaluation methods. Agricultural Economics Research Report 91-4, Department of Agricultural Economics, Cornell University, Ithaca, NY, 1991.

Morris B: How much will people pay to save a few minutes of cooking? Plenty! *Wall Street Journal:* August 5, 1985.

Senauer B, Asp E, Kinsey J: *Food trends and the changing consumer,* St. Paul, 1991, Eagen Press.

U.S. Department of Agriculture: *Food marketing review, 1989–90* (Agricultural Economics Report No. 639), Washington, D.C., 1990, U.S. Government Printing Office.

Waugh FV: *Demand and price analysis* (USDA ERS Technical Bulletin 1316), Washington, D.C., 1964, U.S. Government Printing Office.

ADVANCED TOPICS

Extension of Conceptual Framework

Recent developments and applications of consumer behavior theory have recognized that consumption goods are generally produced by the household. Labor time and effort, food, utensils, and appliances are all used to produce a meal at home. A meal consumed away from home must include travel time and expenses. Consequently, many economists have argued that resource time is extremely important in understanding much of observed economic behavior. For example, as money wages increase and more women work outside the home, will this situation alter consumption patterns due not only to increased family income but also because food preparation time is more costly? To analyze this question, it is necessary to extend the conceptual framework discussed in this chapter.

This extension of consumption theory is discussed briefly in this section to stimulate your thinking.[5] The basic building block of this theory is that time is allocated between consumption-related or household production activities and work activities (Becker, 1965). This time constraint (there are only 24 hours in a day, 168 hours in a week, and so forth), when integrated with the utility function, implies that one gives up the hourly money wage for every hour used in household production activities.

---

[5]Some examples of studies that employ this framework are: Prochaska F, Schrimper R: The opportunity cost of time and other socioeconomic effects on away-from-home food consumption, *American Journal of Agricultural Economics* 55(4): 595–603, 1973; Kinnucan H, Senauer B: The demand for home-prepared food by rural families, *American Journal of Agricultural Economics* 60(2): 338–344, 1978; and Sumner D: The off-farm labor supply of farmers, *American Journal of Agricultural Economics* 64(4): 499–509, 1982.

Suppose that one is deciding whether to consume a meal away from home or to consume a similar meal at home. At least one factor in this decision is the time factor involved in these two consumption choices. If eating at a fast-food restaurant is less time intensive than eating at home, this situation represents a relative benefit to eating away from home. Alternatively, if consuming a convenience or preprocessed product is less time intensive than a product prepared from scratch, this situation represents a relative benefit for the consumption of convenience products. In each case, the money wage multiplied by the amount of time involved in the activity measures the costs associated with time.

Of course, there are other factors that must also be weighed accordingly in decisions by consumers. However, the wages of women and the lack of available time for household production activities have been found to be crucial determinants of expenditures for food away from home and for convenience foods.[6]

EXERCISES

1. Betty and Wilma wish to buy T-shirts and shoes. Betty has a budget of $600, while Wilma has a budget of $1200. Using the indifference curve analyses below, determine prices for the T-shirts and shoes. For which commodity can you derive a demand curve? Derive this curve for Betty and Wilma. Finally, derive the market demand curve, assuming that they are the only consumers.

	Quantity of T-shirts	Pairs of Shoes	Price of T-shirts	Price of Shoes
Betty				
Wilma				

[6]See Kinsey J: Working wives and the marginal propensity to consume food away from home, *American Journal of Agricultural Economics* 65(1):10–19, 1983; Capps O, Tedford JR, Havlicek J: Household demand for convenience and nonconvenience foods, *American Journal of Agricultural Economics* 67(4):862–869, 1985; Redman BJ: The impact of women's time allocation on expenditure for meals away from home and prepared foods, *American Journal of Agricultural Economics* 62(2):234–237, 1980; and Path JL and O Capps Jr.: The demand for prepared meals by U.S. households, *American Journal of Agricultural Economics* 79, 3 (August 1997): 814–824.

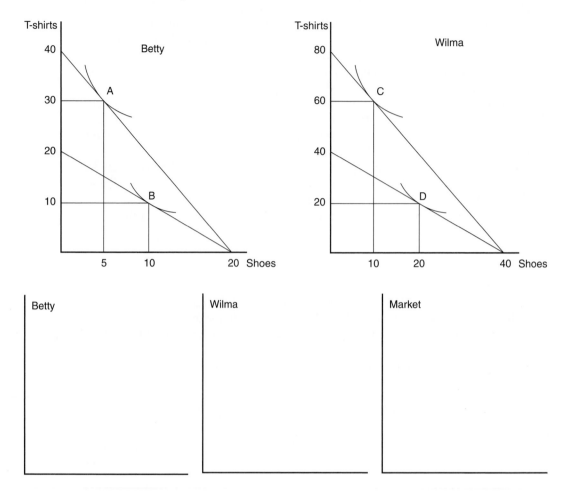

2. The marginal utility of good *A* is the same as the marginal utility of good *B*. Suppose the price of good *B* is $1. In order for a consumer to buy more of good *A*, the price of good *A* must be less than $1. True, False, or Uncertain?

3. Suppose a research study was released that provided evidence that pizza consumption reduces the chances of male pattern baldness. Show how this information may affect demand by demonstrating the direction of the demand shift, using the graph below. Also, show consumer surplus for each demand curve and the change in consumer surplus. Label the axes and respective curves.

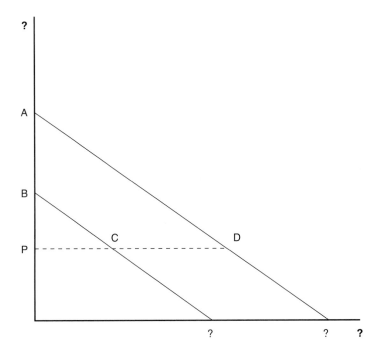

4. Derive separate Engel curves for products A and B from the following indifference curve analysis. Assume $P_A = \$1.50$ and $P_B = \$2.25$. Are these goods normal or inferior?

SEE INDIFFERENCE CURVES ON NEXT PAGE

Quantity of A	Quantity of B	Income

Engel Curve for Product A

Engel Curve for Product B

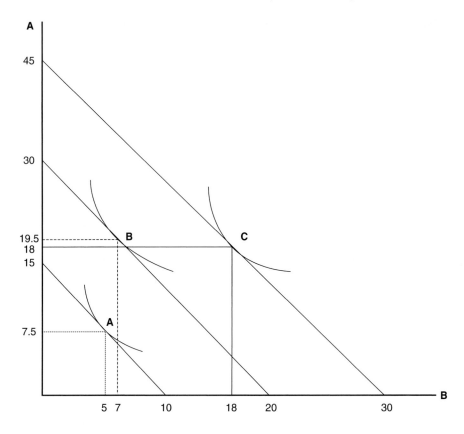

5. Following is a market demand curve for steak.

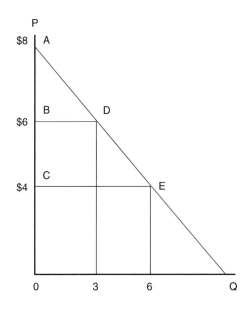

a. Clearly label the consumer surplus associated with a market price of $6 using the letters on the graph from the previous page.

b. Suppose the market price decreases to $4. Are consumers better off or worse off because of this decrease in market price? Using the letters on the previous graph, label the change in consumer surplus.

c. Calculate the dollar amount of the consumer surplus in "a."

d. Calculate the dollar amount of the change in consumer surplus in "b."

6. a.   The graph below is a demand curve for shrimp from Galveston Bay. Suppose that the Environmental Protection Agency finds that there are toxic levels of a pesticide in Galveston Bay. Show graphically what happens to the demand curve for shrimp in light of this information.

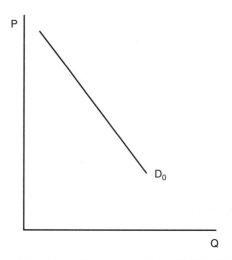

b. Is the situation described in "a" a change in quantity demanded or a change in demand? Circle your answer.

7. Consider the following diagram.

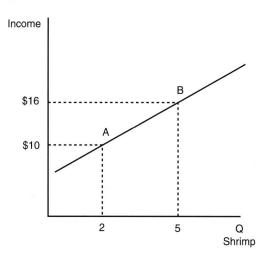

a. This diagram pertains to a(n) _____ curve.

b. Shrimp is a(n) _____ good.

5

MEASUREMENT AND INTERPRETATION OF ELASTICITIES

The elasticity of demand in a market is great or small accordingly as the amount demanded increases much or little for a given fall in price, and diminishes much or little for a given rise in price.

Alfred Marshall
(1842–1924)

TOPICS OF DISCUSSION

Own-Price Elasticity of Demand

Income Elasticity of Demand

Cross-Price Elasticity of Demand

Other General Properties

Some Real-World Examples

Applicability of Demand Elasticities

Chapters 3 and 4 discussed consumer response to a decline in a product's price by purchasing more of that product. In fact, economists are so sure of this inverse relationship between price and quantity demanded that it is referred to as the law of demand. In Chapter 4, we learned that the market demand curve for a commodity shifts to the right or the left when consumers respond to changes in prices and incomes.

What is left unsaid thus far is the degree of consumer responsiveness to change in prices and incomes. Estimates of the degree of responsiveness are expressed in what economists refer to as elasticities. The concept of demand elasticity was invented by British economist Alfred Marshall, the nineteenth-century pioneer of microeconomic theory. The purpose of this chapter is to discuss the measurement of specific widely used concepts of elasticities and provide actual estimates of these elasticities and their meaning to economic analyses.

OWN-PRICE ELASTICITY OF DEMAND

Economists compare the difference between the change in quantity demanded with the change in the price of a good in percentage terms. This comparison is called the **own-price elasticity of demand**.[1] The own-price elasticity of demand is defined as:

$$\begin{array}{c} \text{own-price} \\ \text{elasticity} \\ \text{of demand} \end{array} = \frac{\text{percentage change in quantity}}{\text{percentage change in price}} \qquad (5.1)$$

The percentage change in the quantity of hamburgers demanded, for example, is equal to the change in hamburgers divided by the average quantity of hamburgers consumed during the period. The percentage change in the price of hamburgers is equal to the change in the price of hamburgers divided by the average price of hamburgers during this period.

To illustrate the calculation of this elasticity, assume that your consumption of hamburgers drops from three hamburgers to two hamburgers when the price increases from \$1.00 to \$1.25 per hamburger. The average quantity over this range would be equal to 2.5 (i.e., [2 + 3] ÷ 2), while the average price would be \$1.125 (i.e., [\$1.25 + \$1] ÷ 2). The own-price elasticity of demand in this case would be:

$$\begin{array}{c} \text{own-price} \\ \text{elasticity} \\ \text{of demand} \end{array} = \frac{(Q_A - Q_B) \div ([Q_A + Q_B] \div 2)}{(P_A - P_B) \div ([P_A + P_B] \div 2)} = \frac{(2\text{-}3) \div 2.5}{(\$1.25 - \$1.00) \div \$1.125} = -1.8 \qquad (5.2)$$

---

[1]The elasticity of demand is an arc elasticity that applies to discrete changes in price. When the changes approach zero, a point elasticity of demand can be defined.

in which Q_A and P_A represent the quantity and price after the change, and Q_B and P_B represent quantity and price before the change. Thus, a 1% fall (rise) in the price of a hamburger will increase (reduce) quantity demanded by 1.8%. Often the minus sign is ignored (i.e., we might simply say that the own-price elasticity is 1.8). The minus sign indicates that the demand curve is indeed downward sloping.

We may simplify Equation 5.2 with some algebraic manipulation:

$$\text{own-price elasticity of demand} = \frac{\Delta Q}{\Delta P} \times \frac{\overline{P}}{\overline{Q}}, \tag{5.3}$$

where $\Delta Q = Q_A - Q_B; \Delta P = P_A - P_B; \overline{P} = \dfrac{P_A + P_B}{2}; \text{ and } \overline{Q} = \dfrac{Q_A + Q_B}{2}$

This formula, given by either Equation 5.2 or 5.3, measures average price elasticity between two points on the demand curve and is technically called the **arc elasticity.** Differential calculus permits the determination of price elasticity at a specific point on the demand curve. This measure, dealing with infinitesimal changes, is called the **point elasticity.** Because demand curves slope downward to the right, the measure of own-price elasticity is always negative. The effects of a change in the price of a good on the demand for this good are summarized in Table 5.1.

When the price elasticity of demand for a good exceeds one (in absolute value), we call the response *elastic;* that is, the percentage change in quantity demanded exceeds the percentage change in price. If the price elasticity of demand is equal to one, the curve would represent a *unitary elastic* demand. When the price elasticity of demand for a good is less than one (in absolute value), the demand is called *inelastic.* The percentage change in the quantity demanded is less than the percentage change in the product price.

If the demand curve were perfectly flat, or horizontal, it would represent a perfectly elastic demand. If the demand curve were perpendicular to the horizon-

TABLE 5.1   Own-Price Elasticity of Demand

If the Own-Price Elasticity Is:	Demand Is Said to Be:	Percentage Change in Quantity Is:
Greater than one	Elastic	Greater than percentage change in price
Equal to one	Unitary elastic	Same as percentage change in price
Less than one	Inelastic	Less than percentage change in price

tal axis, or completely vertical, it would represent a perfectly inelastic demand (see Figure 5.1, *A* and *B*).

Along the demand curve, the elasticity may be changing. Consider the case of the linear demand curve for a hypothetical product illustrated in Figure 5.2. The demand response by the consumer is elastic along the upper portion of the curve. We see this from the elasticities, calculated using Equation 5.1, that are presented in column 6 of Table 5.2. The demand response is unitary elastic at the midpoint of this curve, inelastic to the right of this point, and elastic to the left of this point.

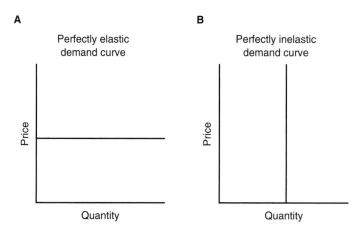

A

Perfectly elastic
demand curve

Price

Quantity

B

Perfectly inelastic
demand curve

Price

Quantity

FIGURE 5.1   *A*, A perfectly elastic demand curve is parallel to the horizontal axis. *B*, A perfectly inelastic demand curve is parallel to the vertical axis.

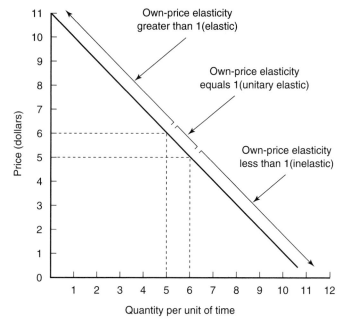

Own-price elasticity
greater than 1(elastic)

Own-price elasticity
equals 1(unitary elastic)

Own-price elasticity
less than 1(inelastic)

Price (dollars)

Quantity per unit of time

FIGURE 5.2   Graphical illustration of consumption expenditures and the own-price elasticity data given in Table 5.2.

TABLE 5.2   Consumption Expenditures and the Own-Price Elasticity

(1) Price	(2) Quantity Demanded	(3) Total Expenditure, (1) × (2)	(4) Percentage Change in Quantity % Δ (2)	(5) Percentage Change in Price % Δ (1)	(6) Own-Price Elasticity (4) ÷ (5)
$11	0	$0		−1/10.5	−21.00
			+ 1/0.5		
10	1	10		−1/9.5	−6.33
			+ 1/1.5		
9	2	18		−1/8.5	−3.40
			+ 1/2.5		
8	3	24		−1/7.5	−2.14
			+ 1/3.5		
7	4	28		−1/6.5	−1.44
			+ 1/4.5		
6	5	30		−1/5.5	−1.00
			+ 1/5.5		
5	6	30		−1/4.5	−0.69
			+ 1/6.5		
4	7	28		−1/3.5	−0.47
			+ 1/7.5		
3	8	24		−1/2.5	−0.29
			+ 1/8.5		
2	9	18		−1/1.5	−.014
			+ 1/9.5		
1	10	10			

Why does this elasticity change along a linear demand curve, although the change in quantity divided by the change in price is constant? The ratio of price to quantity is continuously changing as we move down the demand curve. In fact, the ratio of price to quantity approaches zero when price approaches zero. Therefore, in the case of a linear demand curve, we can conclude that the own-price elasticity of demand falls (rises) when the product price falls (rises). Note that the elasticity is different at each point on a linear (straight-line) demand curve, but the slope of the linear demand curve is constant.

With elastic demand, the percentage change in quantity will be greater than the percentage change in price; thus, consumer expenditures rise when prices fall. The opposite conclusion holds for a price rise; expenditures will fall when price increases. When the elasticity is one, the percentage change in quantity will be equal to the percentage change in price. There would be no change in consumer expenditures when price changes. The percentage change in quantity will be less than the percentage change in price if demand is inelastic; thus, consumer expenditures fall when price falls.

Note that in Table 5.2 total expenditures made by the consumer would be $18 if the price were equal to $9. Total expenditure would be $28 if the price fell to $7. Total expenditures would have risen by $10 (i.e., $28 − $18) if the price were to fall by $2 (i.e., $9 − $7). This relationship will always hold whenever the change in price takes place in the elastic portion of the demand curve. If the price were to fall from $4 to $2, total expenditures would fall by $10 (i.e., $28 − $18). This change in total expenditures takes place in the inelastic portion of the demand curve. The opposite conclusion holds for a price rise; a rise in price raises (lowers) expenditure if demand is inelastic (elastic).

INCOME ELASTICITY OF DEMAND

As noted earlier, it is useful to assess the effects of changes in income on changes in quantity demanded in percentage terms. This measure is called the **income elasticity of demand.** The income elasticity of demand is defined as:

$$\text{income elasticity of demand} = \frac{\text{percentage change in quantity}}{\text{percentage change in income}} \tag{5.4}$$

Consequently, the income elasticity demand is a measure of the responsiveness of the quantity of a good purchased due to changes in income, all other factors constant. In Figure 4.5, the income elasticity of demand for hamburgers over the segment AB of the Engel curve for hamburgers is equal to:

$$\text{income elasticity of demand} = \frac{(Q_A - Q_B) \div ([Q_A + Q_B] \div 2)}{(I_A - I_B) \div ([I_A + I_B] \div 2)} = \frac{(2 - 3) \div ([2 + 3] \div 2)}{(5 - 6) \div ([5 - 6] \div 2} = -2.20 \tag{5.5}$$

Therefore, a 1% increase in income leads to a 2.2% increase in the demand for hamburgers. An income elasticity greater than one implies that a 1% increase in income will cause consumption to rise more than 1%. Goods with an elasticity greater than one are called luxuries by economists. When the income elasticity is less than one but greater than zero, the good is called a **necessity,** or a **normal good.** When the income elasticity is negative, the good is referred to as an **inferior good,** which is not the same as poor quality or defective (see Table 5.3). In the example of hamburgers over the line segment AB in Figure 4.5, hamburgers are classified as a necessity. Most

TABLE 5.3 Income Elasticity Classifications

If the Income Elasticity Is:	The Good Is Classified As:
Greater than one	A luxury and a normal good
Less than one but greater than zero	A necessity and a normal good
Less than zero	An inferior good

foods are necessities, and most nonfood products, such as furniture, a physician's services, and recreation, are considered luxuries.

Again, we may simplify Equation 5.5 as follows:

$$\begin{matrix} \text{income} \\ \text{elasticity of} \\ \text{demand} \end{matrix} = \frac{\Delta Q}{\Delta I} \times \frac{\bar{I}}{\bar{Q}}, \tag{5.6}$$

where $\Delta Q = Q_A - Q_B$; $\Delta I = I_A - I_B$; $\bar{I} = \dfrac{I_A + I_B}{2}$; and $\bar{Q} = \dfrac{Q_A + Q_B}{2}$

According to Tomek and Robinson (1981), "The income elasticity for food in the aggregate, as well as for many individual products, is thought to decrease as incomes increase." Income elasticities will typically change over different income levels, and this change can be positive or negative. When incomes rise, *ceteris paribus*, demand increases for foods such as beef, poultry, shellfish, fresh fruits, and vegetables, but decreases for other foods such as sugar, processed milk, potatoes, eggs, and breakfast cereal (see Blaylock and Smallwood, 1986). In the domestic market, most foods have small, positive income elasticities. Consequently, large increases in income are necessary to generate substantial increases in consumption.

CROSS-PRICE ELASTICITY OF DEMAND

We can measure the effects of changes in the price of tacos on the demand for hamburgers by calculating the **cross-price elasticity of demand** as:

$$\begin{matrix} \text{cross-price} \\ \text{elasticity} \\ \text{of demand} \end{matrix} = \frac{\begin{matrix}\text{percentage change in quantity} \\ \text{of hamburgers}\end{matrix}}{\text{percentage change in price of tacos}} \tag{5.7}$$

$$= \frac{(Q_{HA} - Q_{HB}) \div ([Q_{HA} + Q_{HB}] \div 2)}{(P_{TA} - P_{TB}) \div ([P_{TA} + P_{TB}] \div 2)} \tag{5.8}$$

in which Q_H refers to the quantity demanded of hamburgers and P_T refers to the price of tacos. This elasticity measures the relative responsiveness of the consumption of hamburgers to the price of tacos. Once again, we may simplify Equation 5.8:

$$\begin{matrix} \text{cross-price} \\ \text{elasticity of} \\ \text{demand} \end{matrix} = \frac{\Delta Q_H}{\Delta P_T} \times \frac{\bar{P}_T}{\bar{Q}_H}, \text{ where} \tag{5.9}$$

$$\Delta Q_H = Q_{HA} - Q_{HB}; \Delta P_T = P_{TA} - P_{TB}; \bar{P}_T = \frac{P_{TA} + P_{TB}}{2}; \text{ and } \bar{Q}_H = \frac{Q_{HA} + Q_{HB}}{2}$$

TABLE 5.4   Cross-Price Elasticity Classifications

If the Cross-Price Elasticity Is:	The Goods Are Classified As:
Positive	Substitutes
Negative	Complements
Zero	Independent

We can distinguish among the three different effects that a change in the price of one good can have on the demand for another good (see Table 5.4). The effects of substitutes and complements are of interest to agricultural economists. For example, when the price for beef increases, what will happen to the demand for other products, such as poultry, fish, pork, fruit, and vegetables?

Commodities with large, positive (negative) cross-price elasticities are close **substitute** (**complementary**) commodities. Cross-price elasticities close to zero are indicative of commodities that are unrelated.

OTHER GENERAL PROPERTIES

Practical applications of the use of elasticities are described in Box 5.1. The discussion given supports the notion that the concept of elasticities is very important in business decisions. We now focus on other properties of demand curves. The larger (smaller) the number of substitutes, the more (less) elastic the demand curve. Thus, a commodity such as salt is likely to be very inelastic, and a commodity such as Hunt's catsup is likely to be very elastic. There are several substitutes for Hunt's catsup (e.g., Heinz, Del Monte, and other brands). Aggregates are generally more inelastic than their components. The demand for food is more inelastic than the demand for hamburgers. Further, the demand for Hunt's catsup is more elastic than the demand for catsup in general. The greater the number of alternative uses a commodity has, the greater its price elasticity will be.

Another general property is the budget share of the commodity. If a good or service represents a relatively large budget share or proportion of household budgets, its demand curve will be more elastic. When expenditures for a good or service are sizable, such as for automobiles, furniture, and appliances, consumers are more sensitive to changes in their prices, *ceteris paribus*. Salt expenditures comprise a relatively small percentage of total expenditures made by a consumer. Thus, salt is not likely to exhibit a high elasticity of demand. The demand for cabbage is also inelastic by virtue of its negligible budget share. A 50% increase in the price of cabbage will have very little effect on the quantity demanded, even though there are several substitutes for cabbage. There are relatively few substitutes for housing services, and its budget share is relatively large. Consequently, the elasticity of demand for housing services is expected to be large. Houthakker and Taylor (1970) estimate that the own-price elasticity of demand for housing is

BOX 5.1   Elasticities for Sale

Hugh Divine Jr. sells elasticities.

Well, actually what he sells are estimates of the demand elasticities facing individual firms. Mr. Divine's estimates are not cheap. "The price varies," he says, "but for an analysis that doesn't involve unusual complications, the price runs between $50,000 and $75,000."

Mr. Divine is executive vice president of Total, Inc., a management consulting firm in Princeton, New Jersey. The firm markets a system called PEMS, the Price Elasticity Measurement System. Total has marketed the system to clients for a year, and they have conducted 30 analyses of the price elasticity of demand facing firms in the marketplace.

A Delighted Discoverer

The concept of demand elasticity was invented by British economist Alfred Marshall, the nineteenth-century pioneer of microeconomic theory (see his biography in the January, 1987, *The Margin*). Elasticity is a measure of responsiveness. If A causes changes in B, then elasticity provides a measure of the strength of B's response to changes in A. B's elasticity with respect to A would be the percentage change in B divided by the percentage change in A that caused it.

A change in a good's price causes a change in the quantity demanded. The price elasticity of demand is the percentage change in quantity demanded divided by the percentage change in price.

While sitting on his terrace, Mr. Marshall was so pleased with his discovery he roared with delight. Thus, his discovery was noted by his neighbors and generations of economics students.

Survey Estimates

Elasticity is often estimated statistically by economists for broad categories of products, and it is a potentially useful measure for firms. The difficulty for a firm, however, is that it may not have information about its demand curve.

The PEMS approach, which was developed by Total's director of statistical research, John Morton, relies on a sophisticated interview technique.

"We define a particular product in terms of several attributes. These include various quality dimensions and the price of the product itself. We then ask the consumer how he or she would respond to changes in these attributes," Mr. Divine explains.

The analysis does not attempt to trace the firm's entire demand curve. "We take the price the firm is now charging, and look at the response of the consumer to perhaps three other prices, all relatively close together."

"You don't get a very reliable result if you try huge changes in price. We'll look at price changes on the order of about 20% at most," he says.

The system is used not only to evaluate pricing strategies for products already on the market, but also for new products as well.

"We define the new product in terms of its attributes, and can thus give our client an idea of how to price it before it hits the market."

Mr. Divine recalls an analysis for a firm that had developed a new agricultural chemical. "We gave the client an analysis that suggested the best price to charge and predicted that demand would be about 18 million gallons. The client *sold 19 million gallons.* That's pretty close for a product that had not previously existed."

Boosting Total Revenue

One important use of elasticity is to predict whether revenues will rise or fall in response to a price change. Economists say demand is *elastic* if the percentage change in quantity demanded is larger than the percentage change in price. If that is the case, a price increase will cause such a large percentage reduction in quantity demanded that total revenues of the firm will fall.

If demand is *inelastic*, the percentage change in quantity demanded will be smaller than the percentage change in price. An increase in price would still reduce quantity demanded, but by such a small percentage that total revenues would actually rise.

A firm facing inelastic demand could increase price and raise revenues at the same time. This would necessarily measure profits because the higher price would reduce quantity sold, cutting total costs.

"We found that one client, the manufacturer of a major consumer packaged good, faced inelastic demand for the product. We recommended a price increase. The client did it and increased profit by several million dollars," Mr. Divine recalls.

"In another case, though, top management thought the appeal of its brand name was so strong that it could raise price without any serious impact on sales. Our analysis showed that wasn't the case; the firm faced a high elastic demand. The client cancelled a planned price increase as a result."

The editors of *The Margin* could have profited from one of Mr. Divine's market analyses. After our first semester of publication we surveyed professors using the magazine and were assured our demand was inelastic. Not wishing to seem irrational, we boosted prices by a dollar in the second semester. The result was a sharp reduction in total revenue; demand for *The Margin* is, apparently, elastic. We have since lowered our price.

The PEMS approach is described by Mr. Divine and Mr. Morton in an article in the April, 1987 issue of *Business Marketing.*

Source: Tregarthen T: Elasticities for sale, *The Margin:* December 11, 1989.

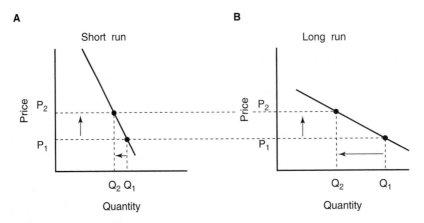

FIGURE 5.3   Consumer demand curves become more elastic (flatter) over time as consumers adjust to changing prices.

approximately equal to −1, and that the own-price elasticity of demand for cabbage is about −0.4.

Another general property of demand curves is that short-run demand is more inelastic than long-run demand. With the passage of time, consumers find that they are better able to adjust to price changes. Suppose the price of a product rises from P_1 to P_2 in Figure 5.3, A and B. In the short run, a consumer's immediate response is to reduce his or her consumption of the product only from Q_1 to Q_2 (Figure 5.3, A). As consumers make adjustments to their consumption habits over a longer period of time, however, they will cut back their consumption to Q_2 (Figure 5.3, B). During the energy crisis of the 1970s, consumers were not able to adjust their purchases of gasoline in the short run when the price of gasoline rose sharply. With the manufacturing of cars that got better gas mileage, consumers were able to adjust to higher gasoline prices by lowering gasoline consumption over the long run.

A final property of demand curves is that the price elasticity of demand for farm products is greater at the retail level than at the farm level. George Brandow, an agricultural economist at Penn State University, conducted a landmark study of selected elasticities of the demand for agricultural commodities in the United States at farm and retail levels. The differences in magnitude of the elasticities in these two markets in the food chain are primarily attributable to the relative level of prices in the two markets and to the value added to the product between these markets.

To summarize, the determinants of the elasticity of demand for a specific commodity include:

- availability of substitutes for the commodity,
- alternative uses for the commodity,

- type of market (e.g., farm level versus retail level or domestic market versus export market),
- the percentage of the budget spent on the commodity, and
- time.

Given this number of determinants, the elasticity of demand for a commodity is not a constant. Agricultural economist Fredrick Waugh stated there is no such thing as a (i.e. single) demand elasticity.

Some Real-World Examples

Economists have estimated specific own-price, cross-price, and income elasticities of demand for various products. Own-price and income elasticities for several food products at the retail level is presented in Table 5.5. In the United States, the demand for grapes is elastic (-1.3780), the demand for bananas is inelastic (-0.4002), and the demand for oranges is unitary (-0.9996).

Own-Price Elasticities. The price elasticity of demand for farm products in the United States has been very small. Estimates of this elasticity at the retail level ranged from a high of -0.34 (Brandow) to a low of -0.24 (Waugh). Therefore, elasticities of most agricultural products are in the inelastic range. Increases in the output of farm commodities because of excellent weather conditions and/or increases in productivity will depress prices rather dramatically.

Consider again the definition of the own-price elasticity expressed in Equation 5.2, or the percentage change in quantity over range *AB* divided by the percentage change in price over range *AB*. Using Brandow's estimate of the own-price elasticity for farm products of -0.34, can you use this equation to defend the statement that a 1% increase in quantity coming onto the market would depress farm prices by almost 3%? (*Hint:* Given the percent change in quantity of 1 and the elasticity of -0.34, you are left with one equation and one unknown—the percent change in price—for which to solve.)

With respect to specific commodities, George and King (1971) found that the price elasticity of demand for beef at the retail level was -0.64. Thus, a 1% fall in the price of beef at the retail level would increase the demand for beef at the retail level by 0.64%. Tweeten (1970) suggests that the short-run own-price elasticity of demand for wheat and soybeans during the 1990s was -0.475 and -0.347, and the corresponding long-run elasticities for these commodities were actually elastic (-1.220 and -1.002, respectively).

Perhaps the most comprehensive study of retail price and income elasticities is the study by Huang reported in Table 5.5. This table suggests that a 1% increase in the retail price of sweeteners would have practically no effect on demand. However, a 1% increase in the retail price of grapes would decrease demand by more

TABLE 5.5   Estimated Own-Price and Income Elasticities at the Retail Level

Commodity	Own-Price Elasticity	Income Elasticity
Beef and veal	−.6166	.4549
Pork	−.7297	.4427
Other meats	−1.3712	.0607
Chicken	−.5308	.3645
Turkey	−.6797	.3196
Eggs	−.1452	−.0283
Cheese	−.3319	.5927
Fluid milk	−.2588	−.2209
Evaporated and dry milk	−.8255	−.2664
Wheat flour	−.1092	−.1333
Rice	−.1467	−.3664
Potatoes	−.3688	.1586
Butter	−.1670	.0227
Margarine	−.2674	.1112
Other fats and oils	−.2191	.3691
Apples	−.2015	−.3514
Oranges	−.9996	.4866
Bananas	−.4002	−.0429
Grapes	−1.3780	.4407
Grapefruits	−.2191	.4588
Other fresh fruits	−.2357	−.3401
Lettuce	−.1371	.2344
Tomatoes	−.5584	.4619
Celery	−.2516	.1632
Onions	−.1964	.1603
Carrots	−.0388	−.1529
Cabbage	−.0385	−.3767
Other fresh vegetables	−.2102	.2837
Fruit juice	−.5612	1.1254
Canned tomatoes	−.3811	.7878
Canned peas	−.6926	.3295
Canned fruit cocktail	−.7323	.7354
Dried beans, peas, and nuts	−.1248	.5852
Other processed fruits and vegetables	−.2089	.6311
Sugar	−.0521	−.1789
Sweeteners	−.0045	−.0928
Coffee and tea	−.1868	.0937
Ice cream and other frozen dairy products	−.1212	.0111
Nonfood	−.9875	1.1873

Source: Huang KS: *U.S. demand for food: a complete system of price and income effects,* Washington, D.C., 1985, U.S. Department of Agriculture.

BOX 5.2   Own-Price Elasticities at the Retail and Farm Levels of the Marketing Channel

	Own-Price Elasticity of Demand	
Commodity	Retail Level	Farm Level
Turkey	−1.40	−0.92
Chicken	−1.16	−0.74
Beef	−0.95	−0.68 (cattle)
Pork	−0.75	−0.46 (hogs)
Butter	−0.85	−0.66
Cheese	−0.70	−0.53
Ice cream	−0.55	−0.11
Eggs	−0.30	−0.23
Fruit	−0.60	−0.36
Vegetables	−0.30	−0.10

Source: Brandow GE: *Interrelationships among demand for farm products and implications for control of market supply,* University Park, Penn, 1961, Agricultural Experiment Station.

---

than 1%. Own-price elasticities at the retail and farm levels of the marketing channel estimated by Brandow (1961) are exhibited in Box 5.2.

Income Elasticities.   Schultze (1971) found that the income elasticity for farm products in this country during the early 1970s was only 0.08. This elasticity was shown to vary from 0.15 in Canada to 0.75 in both West Germany and France. This relatively low income elasticity in the United States suggests that a 10% increase in income would expand the demand for farm products by less than 1%. When income increases, more is spent on nonfood products in the United States than in other developed countries, all other factors held constant. Thus, a substantial increase in consumer income would not necessarily lead to appreciable changes in the consumption of food products.

The income elasticities for major individual food items are reported in Table 5.5. This table suggests that eggs, rice, fluid milk, and other selected products are inferior goods; beef, veal, pork, chicken, and cheese are normal goods; and fruit juice is a luxury good.

Cross-Price Elasticities.   Consider the elasticities estimated by Capps, Seo, and Nichols for various spaghetti sauces reported in Table 5.6. The numbers along the diagonal in this table are own-price elasticities. The remaining elasticities are cross-price elasticities. The cross-price elasticity for Prego with respect to the price of Ragu is 0.8103. Thus, a 1% increase in the price of Ragu would have a large

TABLE 5.6   Matrix of Own-Price and Cross-Price Elasticities of Demand for Spaghetti Sauces*

Item	Prego	Ragu	Classico	Hunt's	Newman's Own	Private Label
Prego	−2.5502	.8103	.0523	.3918	.1542	.1386
Ragu	.5100	−2.0610	.1773	.1381	.0750	.0448
Classico	.2747	.9938	−2.6361	.1432	.2496	.4194
Hunt's	1.0293	.5349	.0752	−2.7541	−.0605	−.0316
Newman's Own	1.0829	.9066	.5487	−.0861	−3.4785	.3562
Private Label	.6874	.4368	.6430	−.0111	.2469	−2.8038

*Values along the diagonal = own-price elasticities; other values = cross-price elasticities

Source: Capps, Jr., O, SC Seo, and JP Nichols, On the estimation of advertising effects for branded products: an application to spaghetti sauces, *Journal of Agricultural and Applied Economics* 29, 2 (December 1997): 291–302.

effect on the quantity of Prego demanded (i.e., Ragu spaghetti sauce is a very close substitute for Prego). According to Capps and associates, most spaghetti sauces are substitutes for each other. As well, the own-price elasticities are in the elastic range.

Applicability of Demand Elasticities

Estimates of own-price, cross-price, and income elasticity of demand have a variety of applications. They can be found in policy debates, wage contract negotiations, and trade negotiations at the macroeconomic level.

Applicability to Policymakers.   One of the means the U.S. Secretary of Agriculture has historically had to support farm prices and incomes of farmers is to change the percentage of land that farmers must set aside or idle if they are to receive federal farm program benefits.[2] The secretary, for example, could increase the amount of wheat land idled (i.e., increase set-aside requirements) if surplus production was expected to increase stocks and depress wheat prices and income. This policy action would lower current production and eventually lead to higher wheat prices and incomes. Importantly, if the demand curve was highly inelastic (i.e., the demand curve is very steep), a relatively small amount of land would need to be idled to achieve a specific price level. The less inelastic the demand curve, the more land the secretary would have to idle to achieve a specific price objective. Policymakers should not idle land if demand is elastic, because this action would cause revenue to fall.

---

[2]The historical features of federal government farm programs and how they have historically affected the levels of production, farm commodity prices, farm incomes, and other aspects of the nation's food and fiber industry will be discussed in depth in Chapter 13. A major change enacted in 1996 decoupled planting decisions from government subsidies.

TABLE 5.7   Own-Price Elasticity and Impacts of Supply Change on Farm Revenues

If the Own-Price Elasticity Is:	Increase in Supply Will	Decrease in Supply Will
Elastic	Increase revenue	Decrease revenue
Unitary elastic	No change in revenue	No change in revenue
Inelastic	Decrease revenue	Increase revenue

Applicability to Farmers.   The Secretary of Agriculture's actions have historically had a direct impact on farmers. If the own-price elasticity of demand for wheat is less than one in absolute value (i.e., inelastic), actions taken to limit the quantity coming into the market will have the desired effect of raising wheat prices by a greater percentage than the cutback in quantity, thus raising the revenue of wheat farmers. If the own-price elasticity of demand for wheat is greater than one in absolute value (i.e., elastic), and the federal government takes actions to limit the quantity coming onto the market to support farm prices and incomes, the opposite will happen. Here we get the undesired outcome of a drop in revenue accruing to farmers (see Table 5.7).

Applicability to Consumers.   Another obvious application of the own-price elasticity is predicting what a change in price will mean for consumer expenditures. In Table 5.1, we saw that consumer expenditures fell when the own-price elasticity of demand declined. In the wake of inelastic demands, increases in supply will, *ceteris paribus*, lower the cost of food and fiber products to consumers.

According to Table 5.5, apples at the retail level have a highly inelastic own-price elasticity of −0.2015. Therefore, a plentiful crop of apples should mean much cheaper apple and apple product prices for consumers. A hard freeze in apple-producing areas would mean substantially higher prices for consumers. Specifically, a 10% increase (decrease) in the quantity of apples will lead to a nearly 50% decrease (increase) in the price of apples.

Applicability to Input Manufacturers.   Estimates of demand elasticities also can guide farm input manufacturer and supplier decisions by indicating the potential degree to which their market might change because of the derived nature of the demand for farm inputs. These manufacturers and suppliers depend upon a healthy farm input demand to promote the growth of their businesses. Given an inelastic demand for farm products, policies that idle land to support prices at a specific level also mean that input purchases will decline by a smaller amount than would occur if the farm level own-price elasticity of demand were more elastic.

A good example of the derived relationship between farm production and the level of farm input use is the effect the federal government's payment-in-kind (PIK) program in 1983 had upon input demand. This program made income support payments to farmers denominated in bushels of wheat, corn, and other surplus commodities rather than in dollars. This policy action dramatically reduced production and the sales of manufactured inputs to farmers in 1983.

Applicability to Food Processors and Trade Firms.   We can also draw conclusions about the impacts changing market conditions have on food processing firms and wholesale and retail trade firms based on published own-price, cross-price, and income elasticities. Box 5.2 shows that vegetables have an inelastic own-price elasticity of demand at the retail level. Thus, an increase in vegetable production will decrease the retail price of vegetables and, *ceteris paribus*, will increase the quantity of vegetables purchased. But the revenue received by retail food businesses will fall because the percentage drop in retail prices will exceed the percentage increase in vegetable consumption. The price elasticity of demand for farm products is greater at the retail level than at the farm level.

Continuing with the example of the inelastic own-price elasticity of demand for vegetables at the retail level, the demand for vegetables at the wholesale level will be even more inelastic. Thus, changes in vegetable production not only affect the revenue received by retailers, but also wholesalers and food processors. Most food products have an income elasticity substantially less than one, and some are negative (an inferior good) (Table 5.5). Thus, a rapid growth in consumer income nationwide will not necessarily translate into a market expansion in the demand for food products. Note the income elasticity for nonfood goods and services at the bottom of this table is greater than one.

SUMMARY

The major points made in the chapter may be summarized as follows:

1. The own-price elasticity of demand measures the percentage change in the quantity demanded for a good given a 1% change in price. If this elasticity is greater than one, demand is said to be elastic (i.e., the percentage change in quantity exceeds the percentage change in price). If this elasticity is less than one, demand is said to be inelastic (i.e., quantity changes by a smaller percentage than price). If this elasticity is equal to one, demand is said to be unitary elastic (i.e., quantity changes by the same percentage as price).

2. The income elasticity of demand measures the percentage change in the quantity demanded for a good given a 1% change in income. When the income elasticity of demand is between zero and one, the good is classified as a normal good; when this elasticity exceeds one, the good is classified as a luxury or superior good. When the income elasticity of demand is negative, the good is classified as an inferior good.

3. If demand is inelastic, a rise (reduction) in price will lead to increased (decreased) consumer expenditures. If demand is elastic, a rise (reduction) in price will lead to a reduction (increase) in consumer expenditures. Finally, if demand is unitary elastic, expenditures are unchanged as price changes.

4. A cross-price elasticity measures the change in the demand for one good in light of a 1% change in the price of another good. If this elasticity is positive (negative), the two goods are said to be substitutes (complements). If this elasticity is equal to zero, the two goods are independent in demand.

5. Determinants of the elasticity of demand of a commodity include availability of substitutes for the commodity, alternative uses for the commodity, type of market (e.g., farm level versus retail level or domestic market versus export market), the percentage of the budget spent on the commodity, and time.

DEFINITION OF KEY TERMS

Arc elasticity: a measure of elasticity between two distinct points on a demand curve.

Complements: goods *A* and *B* are complements if the cross-price elasticity of demand is negative.

Cross-price elasticity: a measure of the response of consumption of a good or service to changes in the price of another good or service. It is defined as the percentage change in the quantity of good *A* demanded divided by the percentage change in the price of good *B*.

Income elasticity: a measure of the relative response of demand to income changes. It is defined as the percentage change in the quantity demanded divided by the percentage change in income.

Inferior goods: goods for which consumption falls (rises) when income increases (decreases).

Luxury: normal goods whose income elasticity exceeds one.

Necessity: normal goods whose income elasticity is less than one.

Normal goods: goods for which consumption rises (falls) when income increases (decreases).

Own-price elasticity: a measure of the relative response of consumption of a good or service to changes in price. It is defined as the percentage change in the quantity demanded divided by the percentage change in price.

Point elasticity: a measure of elasticity at a given point on a demand curve.

Substitutes: goods *A* and *B* are substitutes if the cross-price elasticity of demand is positive.

REFERENCES

Blaylock JR, Smallwood DM: U.S. demand for food: household expenditures, demographics, and projections, *ERS Technical Bulletin* 1713: February 1986.

Brandow GE: Interrelationships among demand for farm products and implications for control of market supply, University Park, Penn, 1961, Agricultural Experiment Station.

Capps, Jr., O, S Seo, and JP Nichols, "On the Estimation of Advertising Effects for Branded Products: An Application to Spaghetti Sauces," *Journal of Agricultural and Applied Economics* 29, 2 (December 1997): 291–302.

Dunham D: Food cost review, 1987, *USDA, ERS, Agricultural Economics Review* No. 596: September 1988.

George S, King GA: *Consumer demand for food commodities in the United States with projections for 1980*, Giannini Foundation Monograph 26, Davis, Calif, 1971, California Agricultural Experiment Station.

Houthakker HH, Taylor L: *Consumer demand in the United States: analyses and projections*, Cambridge, Mass, 1970, Harvard University Press.

Huang KS: U.S. Demand for food: a complete system of price and income effects, *USDA Technical Bulletin* 1714: December 1985.

Schultze CL: *The distribution of farm subsidies: who gets the benefits?* Washington, D.C., 1971, The Brooking Institution.

Tomek WG, Robinson KL: *Agricultural product prices*, ed 2, Ithaca, N.Y., 1981, Cornell University Press.

Tregarthen T: Elasticities for sale, *The Margin:* 11, December 1989.

Tweeten LG: *Foundations of farm policy*, Lincoln, 1970, University of Nebraska Press.

Waugh FV: Demand and price analysis, *ERS Technical Bulletin* 1316: November 1964.

EXERCISES

1. If McDonald's launches a successful ad campaign for Big Macs, what will happen to Big Mac demand? Draw a graph to show this effect. What other determinants for demand are there?

2. Mabel Cranford only eats syrup with pancakes.

   a. What is the technical name for this relationship between syrup and pancakes?

   b. Suppose the price of pancakes goes up.

      i. Represent the effect of this price increase on Mabel's pancake demand curve.

      ii. Represent the effect of this price increase on Mabel's syrup demand curve.

   c. What can we say about the cross-price elasticity between syrup and pancakes for Mabel Cranford?

3. Based on the graph below, estimate the own-price elasticity between points *A* and *B* and the own-price elasticity between points *B* and *C*. Are they elastic or inelastic? Why are the elasticities different? To increase revenue, at least in the short run, would you recommend a price increase or a price decrease?

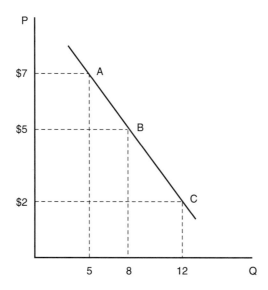

4. The Dixie Chicken currently sells 1,500 burger platters per month for $3.50 and the own-price elasticity for this platter has been estimated to be −1.3. If the Chicken raises prices by 70 cents, how many platters will be sold?

5. Calculate the income elasticity from the following graph between points *A* and *B* and between points *B* and *C*. Define as specifically as possible the type of good represented by each income elasticity.

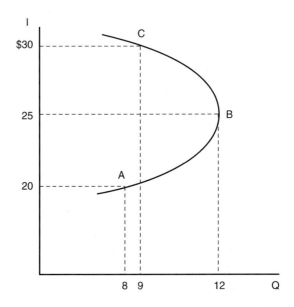

6. The cross-price elasticity for hamburger demand with respect to the price of hamburger buns is −.6. If the price of hamburger buns rises by 5% *ceteris paribus*, what change will occur for hamburger consumption? What is the relationship of these goods? Why?

7. Assume that a retailer sells 1,000 six-packs of Pepsi per day at a price of $3/six-pack. You, as an economic analyst, estimate that the cross-price elasticity between Pepsi and Coca-Cola is .7. If the retailer raises the price of Coca-Cola by 5%, how would sales of Pepsi be affected, *ceteris paribus?* Why?

P A R T

III

Business Behavior and Market Equilibrium

6

ASSESSING BUSINESS PERFORMANCE

To find men capable of managing business efficiently and secure to them the positions of responsible control is perhaps the most important single problem of economic organization on the efficiency scale.

Frank H. Knight
(1885–1972)

TOPICS OF DISCUSSION

The U.S. food and fiber industry is composed of many types and sizes of businesses producing different goods and services that, by definition, are directly related to the supply of food and fiber products to consumers. As indicated in Chapter 2, some of these businesses provide goods and services that represent inputs to farming and ranching operations. John Deere, the world's leading producer of farm tractors, combines, planters, and other equipment, is a good example of this type of business. A local machinery dealership that provides repair and maintenance services to farmers and ranchers is another example.

Most farms and ranches in the United States are involved in the production of a narrow range of commodities and buy almost all of the variable inputs they use from farm input supply businesses. A small 60-cow dairy farm in northern Wisconsin that buys its feed from a local feed dealership is a perfect example of this type of business.

Many businesses are processors or manufacturers of food and fiber products. An example of such a company is the Sara Lee Corporation, which manufactures products such as Sara Lee pastries, Jimmy Dean sausage products, and Hillshire meat products. Archer Daniels Midland, a major biochemical producer that "adds value" to farm commodities such as corn by producing products such as ethanol, is another example of a processor business. Blended gasoline, which contains ten parts of either ethanol or methanol, has been phased into the nation's gasoline supplies, as mandated by the 1990 Clean Air Act. Smaller "value added" businesses, such as the local grain elevator and the local cotton gin, are also a part of the food and fiber industry.

Some large food processing corporations, such as Del Monte Foods, are also involved in growing their own crops or livestock as an input to their processing operations. Economists refer to this as vertical integration, which we will discuss in Chapter 9.

Many businesses in the U.S. food and fiber industry are also multidimensional in the products they produce and often multinational in their processing and marketing operations. Nestle International, for example, is headquartered in Geneva, Switzerland, but sells less than 2% of its annual production in Switzerland. Some of its products, such as Nestle's tollhouse morsels (i.e., chocolate chips), are marketed almost exclusively in the United States. Nestle increased its presence in the U.S. food and fiber industry in the mid-1980s when it purchased Carnation Foods.

Some businesses are involved in marketing and transporting these food and fiber products in wholesale markets, such as the McLane Company, and retail markets, such as Kroger Foods (see Table 2.3 for other examples).

Each of these businesses has several things in common. They each want to make money from their business operations and must, therefore, decide how best to organize their operations to accomplish this goal. Some of these decisions must be made during the current period, while others may be made over time. Regardless of whether the business is an owner operation (i.e., a sole proprietorship), a partnership, or a vastly held corporation, it must be familiar with its costs of production and other factors influencing its current economic performance.

The purpose of this chapter is to discuss the general properties of annual business reports, some approaches to analyzing these reports, and several alternative perspectives on the costs and profit of businesses.

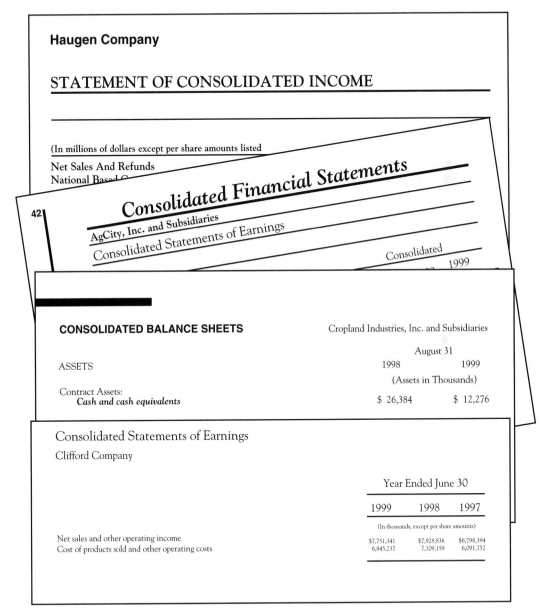

FIGURE 6.1   Major publicly held corporations active in the nation's food and fiber industry publish an annual report and a Form 10-K, which is required by the Securities and Exchange Commission. The annual report presents the business's income statement and balance sheet, which document the business's most recent economic performance and its financial position.

ANNUAL BUSINESS REPORTS

What does Nestle International have in common with a small northern Wisconsin dairy farmer? They are both in business to make a profit. To make a profit, revenue must exceed expenses. The profitability of a business and its financial strength can be assessed by examining the business's annual reports (Figure 6.1). One company's annual report had a mission statement "to be a well-managed, profitable, consumer-driven, innovative agricultural company."[1] The report also states "The system seeks to maximize members' profits as owners and customers."

One of the principal sources of information on the performance of a business is its annual financial statements. Large corporations generally prepare an annual report for their stockholders. Like large corporations, farmers must prepare financial statements. Most annual reports consist of at least three financial statements: (1) an **income statement** or statement of operations, (2) a **balance sheet,** and (3) a **cash flow statement.**[2] The general structure of the first two financial statements is presented in this section.[3] In discussing these statements, we will provide an abbreviated version of the actual financial statement for an existing company—the 1999 balance sheet for Farmland Industries (presented in its 1999 annual report) and the 1999 income statement for Deere & Company (presented in its 1999 annual report).

Business Balance Sheets

A balance sheet presents a picture of a business's financial position at a particular point in time, usually at the end of the business year. The term *balance* in the name of this financial statement refers to the fact that the business's total assets must, by definition, balance or *equal* its debts plus equity.

Table 6.1 presents the general structure of a balance sheet. Assume the business has current assets (cash on hand, bank deposits, etc.) of $100 and other assets totaling $500. Further assume that the business has current liabilities (current loan payment due, fuel bill due, etc.) of $75 and other liabilities totaling $200. Therefore, the business's total assets would be $600 and its total liabilities would be $275. The general form of the balancing equation underlying the structure of a balance sheet requires that:

$$\text{equity} = \text{total assets} - \text{total liabilities} \tag{6.1}$$

which, for the simple balance sheet depicted in Table 6.1, suggests that:

$$\text{equity} = \$600 - \$275 = \$325 \tag{6.2}$$

---

[1]*Deere & Company 1996 annual report,* Moline, Ill, 1996, Deere and Company.
[2]Corporations must also prepare a 10-K form, which contains much of the same information as is found in their annual report, for the Securities and Exchange Commission.
[3]For a more comprehensive discussion of business financial statements and their analysis, see Fraser L: *Understanding financial statements,* Reston, Va., 1988, Reston Publishing Company. You may also write to major food and fiber manufacturers for a copy of their annual report and 10-K report.

TABLE 6.1 Structure of a Balance Sheet

Current assets	$100.00	Current liabilities	$ 75.00
Other assets	500.00	Other liabilities	200.00
		TOTAL LIABILITIES	275.00
		Equity	325.00
TOTAL ASSETS	$600.00	TOTAL LIABILITIES PLUS EQUITY	$600.00

This value is entered in the right-hand column of Table 6.1, which makes both columns total $600. Trends in the value of assets, liabilities, and equity in the farm sector were presented in Figure 2.8, *A*.

Finally, two ratios commonly used to assess or analyze a balance sheet are the current ratio and the solvency ratio. The current ratio assesses the **liquidity** of the business at the date the balance sheet was prepared, or whether the business's current assets exceed its current liabilities. If the business can convert all its current assets to cash and pay off its current liabilities, the business is said to be *liquid* (i.e., current assets > current liabilities). In the case of the hypothetical business depicted in Table 6.1, we can say that its current ratio of 1.33 (1.33 = 100 ÷ 75) suggests the business is in a liquid financial position. The solvency ratio, on the other hand, illustrates whether the business can convert all assets to cash, retire all of its liabilities, and still have some cash remaining. The solvency ratio for the balance sheet presented in Table 6.1 is 2.18 (i.e., 2.18 = 600 ÷ 275), which indicates a strong solvent position. Both ratios should exceed 1.0 at minimum.

Table 6.2 presents the 1997 consolidated balance sheet for Farmland Industries, Inc., a large multidimensional company headquartered in Kansas City, Missouri. Farmland Industries is perhaps best known for its double circle cooperative supply services to farmers, AMPRIDE service stations in the Midwest, and Farmland Bacon and other food products.

The term *consolidated* in the context of financial statements means that the assets, liabilities, and equity of all the business's subsidiaries are captured in the statement. The timing of the balance sheet is August 31, the end of its accounting year.

Several entries in the balance sheet presented in Table 6.2 need some explanation. **Cash equivalents** refer to other financial assets such as checking account balances and certificates of deposit, which can be converted to cash quickly. **Accounts receivable** refer to the funds the business expects to receive in the short run from other businesses it has billed for payment. **Inventories** are the business's current stocks of unsold production and its stock of unused variable production inputs. **Depreciation** refers to the reduction in the value of assets due to their consumption or wear-out during business operations over time. **Notes payable** refers to payments that the business must make to its lenders in the short run. **Accounts payable** represents unpaid expenses that must be paid to others in the near future (i.e., the opposite of accounts receivable). **Capital shares and equities** refer to the shares of common stock, preferred stock, and earned surplus (net income) from operations.

TABLE 6.2   Farmland Industries Consolidated Balance Sheet

	August 31, 1999	August 31, 1998	Percent Change
	—Million Dollars—		
Current assets:			
Cash and cash equivalents	.5	7.3	−93.2
Accounts receivable	794.2	596.4	33.2
Inventories	840.5	725.9	15.8
Other current assets	203.3	206.9	−1.7
Total current assets	1,838.0	1,536.7	19.6
Other assets:			
Plant and equipment	1,744.3	1,680.4	3.8
Less depreciation	911.0	853.2	6.8
Net plant and equipment	833.2	827.1	0.7
Investments and long-run receivable	329.7	298.4	10.4
Other assets	256.6	212.4	20.8
Total assets	3,257.6	2,874.6	13.3
Current liabilities:			
Current loans and notes payable	591.0	447.5	32.1
Accounts payable	463.3	330.0	40.4
Other current liabilities	333.4	323.6	3.0
Total current liabilities	1,387.6	1,101.2	26.0
Other liabilities:			
Total long-term debt	808.4	728.1	11.0
Deferred income taxes	63.1	65.2	−3.2
Other liabilities	40.2	31.9	26.0
Capital shares and equities	917.3	912.7	0.5
Total liabilities and equity	3,257.6	2,874.6	13.3

Source: *Farmland Industries 1999 annual report,* Kansas City, Mo, October 1999.

Table 6.2 suggests that Farmland Industries was in a strong financial position entering 2000. Its current assets were 32.4% greater than its current liabilities, which meant Farmland Industries could convert its current assets to cash (i.e., sell them), retire (i.e., pay off) its current liabilities, and have $450.4 million (i.e., $1,838.0 − $1,387.6) remaining.

Table 6.2 also shows that Farmland Industries' total assets grew by 13% in value during 1999, which included a 3.8% growth in its plant and equipment (trucks, machines, buildings, etc.).

Business Income Statements

The income statement contains two major components, the revenue component and the expense component. The revenue component accounts for the business's sources of revenue during the accounting period, including cash received from the

TABLE 6.3   Structure of an Income Statement

Cash from sales	$100.00
Other revenue	25.00
TOTAL REVENUE	*125.00*
Variable expenses	50.00
Fixed expenses	25.00
TOTAL EXPENSES	*75.00*
Income before tax	50.00
Provision for tax	10.00
NET INCOME	*$40.00*

sale of goods or services by the business. The expense component captures the business's expenses during the period, including those that vary with the quantity of goods sold, or variable expenses. Table 6.3 presents the general structure of a typical income statement. This hypothetical income statement suggests the business had $125 in total revenue and $75 in total expenses, leaving a before-tax income of $50.

Net income, an important performance statistic, is defined as:

$$\text{net income} = \text{total revenue} - \text{total expenses} - \text{taxes} \qquad (6.3)$$

The data presented in Table 6.3 suggests that the net income of this business would be $40, or $50 in before-tax income minus a provision for taxes of $10. Trends in revenue, expenses, and net income in the farm sector were illustrated in Figure 2.7.

Two performance statistics typically used to assess the income statement are the gross ratio and the **rate of return on equity.** The gross ratio, which is used to assess the economic efficiency of a business, is given by the ratio of total expenses to total revenue. It indicates the expenses per dollar of revenue. Table 6.3 indicates that the gross ratio for this business would be 0.60 (i.e., $0.60 = 75 \div 125$). Businesses would prefer to minimize the value of this ratio. The rate of return on equity capital uses information on equity from the balance sheet in Table 6.1 and on net income from the income statement in Table 6.3. The rate of return on equity for this hypothetical firm would be 12.3% (i.e., $0.123 = 40.0 \div 325$). A rate of return on equity of this magnitude would be attractive to investors. Higher rates of return are preferred more than lower rates of return. Negative rates of return will occur in periods when a business experiences a net loss (total expenses > total revenue).

Table 6.4 is the 1999 income statement for Deere & Company, or John Deere as it is often referred to. This business, headquartered in Moline, Illinois, is the world's largest manufacturer of farm tractors, combines, and other farm equipment. It also manufactures construction equipment, lawn mowers, and related equipment. In addition to its manufacturing operations, John Deere operates a credit and an insurance subsidiary.

TABLE 6.4   Deere & Company Income Statement

	Year Ending 1999	October 31, 1998	Percent Change
	—Million Dollars—		
Net sales and revenues:			
Net sales of farm and industrial equipment	$ 9,701	$11,926	−18.7
Finance and interest income	1,104	1,007	9.6
Other income	945	889	6.3
Total net sales and revenue	11,750	13,822	−15.0
Costs and expenses:			
Cost of goods sold	8,178	9,234	−11.4
Research and development	458	444	3.2
Selling and general expenses	1,362	1,309	4.0
Interest expense	557	519	7.3
Other expenses	831	755	10.1
Total costs and expenses	11,386	12,262	−7.1
Income before taxes	365	1,560	−76.6
Provision for income taxes	135	554	−75.6
Income of consolidated group	230	1,006	−77.1
NET INCOME	239	1,021	−76.6
NET INCOME PER SHARE	1.03	4.20	−75.4

Source: *Deere & Company 1999 annual report,* Moline, Ill, 1999, Deere & Company.

Several entries in the income statement presented in Table 6.4 may also need further explanation. **Net sales** represents the business's total sales less returns and allowances and sales discounts. **Cost of goods sold** represents the cost of manufacturing the goods (machinery and equipment in this case) the company actually sold during the year. **Net income** is the final entry in an income statement and reflects what is remaining of total revenue after all expenses have been deducted, (i.e., profit).

The 1999 consolidated income statement for Deere & Company presented in Table 6.4 shows that John Deere's net sales and revenue fell 15% in 1999, while its costs and expenses fell by 7.1%. Net income was $239 million in 1999, a decrease of 76.6% from year earlier levels. Net earnings per share to John Deere's stockholders fell by 75.4% to $1.03 per share.

Use in Business Planning

The income statement and balance sheet, along with other information on the business's economic performance and financial position, assist the business's management in deciding how to best organize its portfolio of assets and liabilities and how to utilize its assets to make a profit. These and other financial statements should represent an input to evaluating the relative performance of the business with respect to

- its liquidity at the end of the year,
- its solvency at the end of the year,
- the economic efficiency of its current operations, and
- the profitability of its current operations.

Assessing Liquidity.   The liquidity of a business refers to its ability to convert its assets to cash quickly and without loss in the value of the business, and the use of this cash to meet all current claims on the business. The current ratio, or the ratio of current assets to current liabilities in the business's balance sheet, is a widely used measure of liquidity. Table 6.2 suggests that Farmland Industries was in a strong liquidity position as of August 31, 1999. A current ratio of 1.33 (i.e., $1,838 million ÷ $1,387.6 million) means this business's current assets exceed current liabilities by 20% and, therefore, that Farmland Industries is in a *liquid financial position.*

Assessing Solvency.   The **solvency** of a business can be assessed with information from a business's balance sheet. The solvency ratio, which is equal to total assets divided by total liabilities, indicates the ability of a business to repay its total liabilities if the business were dissolved. Again, a ratio of one suggests the business is solvent. A ratio of less than one would mean the business is insolvent. Table 6.2 indicates that Farmland Industries is in a solvent financial position; its solvency ratio is 1.39 ($3,257.6 million divided by $2,340.3 million). As mentioned in Chapter 2, many farmers and ranchers became insolvent during the mid-1980s and are no longer in business. Similarly, some equipment manufacturers, such as those that manufactured Oliver tractors and Minneapolis-Moline tractors, are also no longer in business.

Assessing Economic Efficiency.   The **economic efficiency** of business operations depends on the degree to which the business uses its resources to achieve a desired result with little or no wasted effort.[4] Gross ratio is one economic efficiency ratio; it is equal to the ratio of total costs and expenses to net sales and revenue. A review of Table 6.4 suggests that Deere & Company had a gross ratio of 96.9% in 1999 ($11,386 million divided by $11,750 million). This is down 11% from the gross ratio Deere & Company achieved just one year earlier, and is substantially higher than the gross ratio of 60% suggested in Table 6.3 for the hypothetical firm.

---

[4]Physical efficiency ratios cannot be assessed from the business's financial statements. Some examples are pigs per litter, yield per acre, and feed fed per pound of gain in livestock production.

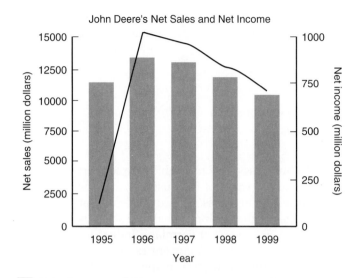

John Deere's Net Sales and Net Income

FIGURE 6.2   The left axis shows the levels of net sales achieved by Deere & Company over the 1995 to 1999 period. The right axis reflects the level of net income. Deere's profitability as measured by net income fell sharply in 1999 as a result of further declines in crop prices resulting from a weak export demand.

▨ Net sales   —— Net income

Assessing Profitability.   The **profitability** of business operations can be assessed in a variety of ways. One is to look at the trend in the level of net income for the past several years. Figure 6.2 presents the trend in net sales and revenue, and net income for Deere & Company from 1995 to 1999. Such historical comparisons help us to understand the business's relative performance in the current period. For example, Figure 6.2 shows that John Deere experienced a sharp decline in profits in 1999 stemming in part from the third straight year of declining crop prices. After recording net losses in 1991 and 1993 because of low sales, the company earned a profit in 1994 through 1999 as commodity prices rose in the mid-1990s and as federal government payments to farmers propped up net incomes later in the decade.

One can also evaluate the profitability of the business by examining the rate of return on existing equity capital. This rate of return should compare favorably with the alternative rate of return investors could earn by investing their capital elsewhere.

Figure 6.3 suggests that the rate of return on equity capital invested in the business fell to 5.8% in 1999. This graph clearly illustrates the concept of derived demand and the importance of a healthy agriculture to the stability of farm input manufacturers like Deere & Company.

Comparative Analysis.   In addition to evaluating a business's current performance relative to its past performance (a historical analysis), economists also compare performance measures, such as the rate of return on equity capital or the current ratio and leverage ratio discussed for Farmland Industries, with the performance ratios achieved by similar firms over the same time period. This is referred to as a comparative analysis. For example, while John Deere's return on eq-

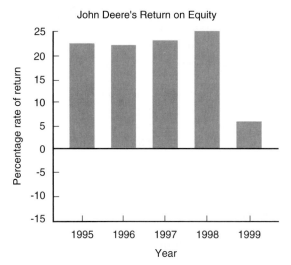

FIGURE 6.3   The left axis shows the percentage rates of return on John Deere's existing equity capital. Although this figure tells much the same story as Figure 6.2, it goes one step further by expressing the level of net income or returns as a *percent* of the amount of equity capital invested in Deere & Company. The decline in 1999 shows the net effect of a weakening export market.

uity capital appears relatively stable until 1999, the company suffered negative returns to equity in 1991 and 1993, and earlier in 1986 and 1987.

Although comparative analysis may mean very little to some investors looking for a more stable rate of return, it helps verify the existence of common trends that affect all firms in an industry. We know from Chapter 2, for example, that many of the nation's farmers and ranchers were under considerable financial stress during the mid-1980s and chose to postpone further investments in new farm equipment. Without substantial government payments to farmers in 1997, 1998, and 1999, equipment sales would have fallen off much sooner.

COST OF DOING BUSINESS

A business must know its annual cost of production. This is true for Deere & Company, as it tries to economize on the cost items reported earlier in Table 6.4. It is equally important for a Texas cotton farmer, who is attempting to minimize the business's planting and harvesting costs, to know his or her annual cost of production.

Two cost concepts that help to explain how different people assess a business's economic performance differently are the differences between **explicit and implicit costs** and the difference between costs in the short run versus the long run. Other cost concepts will be discussed in Chapter 7.

Explicit versus Implicit Costs

Explicit business costs are those costs that ordinarily appear on a business's income statement. For example, the $8,178 million spent in 1999 by Deere & Company to manufacture the goods it sold represents an explicit cost. Explicit costs include the cost of fuel and electricity and wages paid to hired labor. Other explicit costs appearing on Deere & Company's 1999 income statement are research and development costs, selling costs, and general expenses. In fact, *all* the costs and expenses appearing in this income statement totaling $11,386 million are explicit costs. For farmers and ranchers, explicit costs are the costs of production, such as feed expenses or fertilizer expenses. Explicit costs occur when a business makes a payment to others in exchange for goods and services it receives.

Implicit business costs are the opportunity costs a business charges itself for using its own existing resources. For example, the income foregone by farmers not exercising their next best alternative represents an implicit cost to them. A cost is implicit when there is no payment of money.

Costs in Short Run versus Long Run

Costs of production have a time dimension, which influences the nature of the decisions that can be made. The short run is the length of time that is short enough for some costs and resources to be fixed. For example, it takes time for a business to expand its manufacturing capacity (e.g., to build and operationalize a new dairy parlor or modernize a canning factory). The long run is anything longer than the short run. Thus, if we define the short run to be one year, then the long run is anything beyond one year. In the long run, all resources and claims on the business are variable.

All costs can be modified in the long run, but some costs are fixed in the short run.[5] Those costs incurred by a business that do not vary with the level of production are referred to as fixed costs in the short run, which are costs the business must meet regardless of the level at which it chooses to operate.

Three examples of fixed costs are property taxes, insurance premiums, and the interest payments on a mortgage. Although these costs may change over time, they do not vary directly with the level of the business's current operations. The current property tax assessment owed by a business will be the same whether it produces at its capacity or produces nothing at all.

Costs that do not meet the definition of fixed costs in the short run are classified as variable costs. Three examples of variable costs are wages paid to hired labor, fuel expenses, and fertilizer expenses. Wages paid to hired labor are considered a variable cost because these costs vary with the level of production. A factory, for example, will need more workers if its management decides to add a night shift. Fuel expenses and—if applicable—herbicide, pesticide, and fertilizer expenses are also considered variable expenses. Suppose a corn farmer decides to make an additional pass over his field to reduce pest infestation. More fuel will be

---

[5]These inputs may be fixed because of extremely large and time-consuming adjustment costs. Land is often treated as a fixed input in the short run. It also takes time to construct a new packing plant or milking parlor. The short run is typically defined to be one production cycle. We will assume the short run to be equal to one year. In the long run, however, all inputs are variable.

needed to power the tractor, and more pesticides and fertilizer will be used. A business's total costs are therefore given by:

$$\text{total cost} = \text{variable costs} + \text{fixed costs} \tag{6.4}$$

which is also equal to the business's total explicit costs.

Table 6.4 identified several categories of costs and expenses for Deere & Company in 1999. An economist will not only want to group these entries into specific categories for further analysis, but will also want to see what these costs might have been if the business had either expanded or contracted its operations. Both of these topics are addressed in Chapter 7.

MEASURING BUSINESS REVENUE

We will discuss the determination of the level of revenue for businesses operating under different market structures (i.e., perfect competition versus several forms of imperfect competition) in the nation's food and fiber system in Chapters 8 and 9. It is important here, however, that we define what we mean by the term *revenue* as an input to the discussion of net income and profit. To do so, we will introduce the hypothetical business TOP-AG, Inc., a small manufacturer of farm equipment parts. This firm will be examined extensively in the next several chapters.

A business's total revenue associated with its current level of operations can be defined as:

$$\text{total revenue} = \text{product price} \times \text{output} \tag{6.5}$$

or the price of the product multiplied by the quantity of output produced by the business.[6] Assume that TOP-AG produced 12 units of output an hour and that the price of its product was $45.00 per unit. Equation 6.5 would suggest that this business's total revenue would be:

$$\begin{aligned} \text{total revenue} &= \$45.00 \times 12 \text{ units of output} \\ &= \$540.00 \end{aligned} \tag{6.6}$$

or $540 per hour. Given a fixed market price for its product, TOP-AG's total revenue would fall if it produced fewer than 12 units per hour. What would happen to TOP-AG's total revenue if it attempted to expand its production beyond 12 units per hour? Total revenue may be higher; however, the answer to this question is not as clear-cut as you might think.

---

[6]If the business is involved either in the merchandising of products manufactured by others or in providing services to others, total revenue would be equal to the price of its services times the amount of merchandise or services provided to others.

ASSESSING PROFIT

Table 6.4 presented the 1999 income statement for Deere & Company. The bottom line in this statement was net income. Because many businesses like Deere & Company explicitly state a desire to maximize profits, we must understand the two alternative perspectives taken in its measurement.

The Accountant versus the Economist

The general notion of profit suggests that we are interested in knowing if total revenue exceeds total costs. Two alternative concepts of profit are accounting profits and economic profits.

Accounting Profits.   The level of accounting profits for a particular level of production is found by subtracting the explicit costs of production paid by the business from total revenue. This is the approach an accountant would use to define net income as it appeared at the bottom of Deere & Company's income statement in Table 6.4. It also represents the accountant's definition of profits, or:

$$\text{accounting profit} = \text{total revenue} - \text{total cost} \tag{6.7}$$

If we assume that TOP-AG's total fixed costs per hour are $75 and that its total variable costs per hour are $310, we can conclude that this business's profit from an accounting standpoint would be:

$$\begin{aligned} \text{accounting profit} &= \$540 - (\$310 + \$75) \\ &= \$155 \end{aligned} \tag{6.8}$$

or $155 per hour of operations. TOP-AG's returns above variable costs in the short run would be $230 (i.e., $155 + $75 in explicit fixed costs).

Economic Profits.   Economists view profits somewhat differently. Economic profits for a particular level of production are found by subtracting both explicit and implicit costs from the business's total revenue, in which implicit costs reflect the opportunity costs (i.e., income or wealth foregone) associated with employing the business's resources (land, labor, capital, and management) in their current use rather than using these resources in their next best alternative.

The level of profit in the eye of the economist therefore is equal to accounting profits minus implicit costs, or:

$$\text{economic profit} = \text{total revenue} - \text{total cost} - \text{implicit costs} \tag{6.9}$$

in which total costs also represent the explicit costs of production, which appear in a business's income statement.

If we assume that TOP-AG's total implicit costs are $25 per hour, the economist would assert that this business's profit is equal to:

$$\text{economic profit} = \$540 - (\$310 + \$75) - \$25$$
$$= \$130 \qquad\qquad (6.10)$$

or $130.00 per hour. This figure suggests that TOP-AG can cover both its explicit and implicit costs of production and have $130 per hour remaining to augment the business's existing equity capital.

The term *economic profit* will recur many times throughout the remainder of this book. It represents tangible proof to a business's management and owners that the business is not only covering its costs as an accountant would define them, but also is exceeding the return investors could have achieved by investing their money elsewhere.

Why the Fuss over Profit?

Many people feel business profits are higher than they need to be. Consumer prices would be lower if it were not for businesses attempting to maximize their profits. What do you think the profit margin is for the major businesses participating in the nation's food and fiber industry? You might have been surprised by what the consolidated income statement of Deere & Company suggests as shown in Figure 6.3. This major corporation earned a -10.2% on its existing equity in 1986 and -5.0% in 1987. This was offset by the 14.8% rate of return by 1990 and illustrates the variability of returns in ag-related sectors.

Like any business, farms, input manufacturers, food processors, fiber manufacturers, and others involved with the transportation and trade of food and fiber products at the wholesale and retail levels are in the business to make a profit. The same can be said of the nation's farmers and ranchers.

Throughout this textbook, we will assume businesses are motivated by the goal of maximizing profits. The economic objective helps us to understand the economic decisions businesses make in the short run and the long run. We are not suggesting that these businesses ignore other meaningful objectives, such as personal, social, or environmental objectives. However, businesses' main concerns will always be with economic profits.

WHAT LIES AHEAD?

Accountants are responsible for assembling an information system that accurately measures the costs of doing business and for preparing the business's balance sheet, income statement, and cash flow statement, among other reports. Unfortunately, some farmers and ranchers serve as their own accountants. Where possible, however, an outside audit (i.e., external to the business's accountants) of these statements is both beneficial, and sometimes a precondition, for obtaining outside funding.

The role of the economist, which in many cases in agriculture is the individual farmer or rancher armed with the economic lessons learned in previous years, is to assess all available information to provide answers to the following questions:

- What products should be produced, and how much should be produced?
- What level of specific inputs will be needed to produce these products?
- How should the acquisition of these inputs be financed?
- How large should the business become?

In addition to the information contained in these financial statements, the economist will also require other information such as the prices of the products sold by the business, the prices of inputs purchased by the business, and the trends in interest rates on loans and rates of return on alternative forms of investment necessary to address these questions.

SUMMARY

The purpose of this chapter was to introduce the importance of information for small and big businesses. Most businesses maintain a set of financial statements that can be used to assess the economic performance of the business and the strength of its financial position. Information on the cost of operations can be categorized for further analysis, including partitioning costs into fixed versus variable costs, and the calculation of profit associated with the business's current operations.

The major points made in this chapter may be summarized as follows:

1. The various revenue and cost concepts introduced in this chapter can be summarized as follows:

   total variable cost = costs that vary with the level of production (e.g., fuel expenses)

   total fixed cost = costs that do not vary with the level of production (e.g., property taxes)

   total cost = total variable cost + total fixed cost

   total revenue = product price × total output

   accounting profit = total revenue − total costs

   economic profit = accounting profit − implicit costs

2. A balance sheet for a farm, elevator, processing plant, or other business in the nation's food and fiber industry is structured the same. Total assets equal, or balance, total liabilities (or debt) and equity (or net worth).

3. An income statement is also the same for different types and sizes of businesses. This financial statement reports the total revenue of the business, its total costs of production, and its net income.

4. A consolidated financial statement captures the performance of all subsidiaries or components of a business. Many corporations present their consolidated financial statements in an annual report to stockholders.

5. A business's financial statements can be evaluated for what they say about the business's liquidity, solvency, economic efficiency, and profitability. These perspectives help both management and its owners and creditors determine the business's relative current economic performance and financial strength.

DEFINITION OF KEY TERMS

Accounts payable: unpaid expenses that must be paid to others in the short run.

Accounts receivable: payments the business expects to receive in the short run from others it has billed for payment.

Balance sheet: a financial statement reporting the value of real estate (land and buildings), non-real estate (machinery, breeding livestock, and inventories), and financial (cash, checking account balance, and common stock) assets owned by farms and ranches and also outstanding debt. The difference between total farm assets and total farm debt outstanding represents the net worth of the farm.

Capital shares and equities: shares of common stock, preferred stock, and earned surplus (net income) from operations.

Cash equivalents: noncash financial assets that can be converted to cash quickly to help pay off expenses that are due.

Cash flow statement: a financial statement that reports the sources and uses of cash by the firm during the period.

Cost of goods sold: cost of manufacturing the goods a business actually sold during the year.

Depreciation: an implicit cost of doing business; it reflects the wear and tear or consumption of plant and equipment during the year.

Economic efficiency: the ability of a business to use its resources to achieve a desired result with little or no wasted effort.

Explicit costs: those expenses such as wages paid to hired labor where a cash payment to others is required.

Implicit costs: those expenses such as depreciation that do not involve the payment of money.

Income statement: a financial statement that reports the revenue and expenses incurred by the firm during the period, and the net income (loss).

Inventories: value of unsold production and/or inputs on hand as of the balance sheet date.

Liquidity: the ability of a business to convert its assets to cash quickly to meet claims on the business without disrupting its ongoing production operations.

Net income: the bottom line in the income statement, represents the accountant's version of after-tax profits.

Net sales: total sales less returned unsold merchandise and allowances discounts.

Notes payable: payments the business must make to its lenders in the short run.

Profitability: the ability of a business to generate revenue above total costs of production.

Rate of return on equity: ratio of net income appearing on the income statement to the level of equity in the business appearing on the balance sheet.

Solvency: the ability of a business to convert all its assets to cash and retire (pay off) all of its liabilities and have some cash remaining.

REFERENCES

Deere & Company 1999 annual report, Moline, Ill, 1999, Deere & Company.

Farmland Industries 1999 annual report, Kansas City, Mo, 1999, Farmland Industries, Inc.

EXERCISES

1. Briefly describe the basic differences between explicit costs and implicit costs. Which concept is more important to economists than to accountants? Why?

2. What are the two major components of total cost? What is the essential difference between these two components?

3. Identify and describe the major performance measures you would examine if you were asked to evaluate a firm's financial statements.

4. Given the following values, calculate the firm's liquidity, solvency, and profitability:

Current assets	$100
Other assets	500
Current liabilities	75
Total liabilities	275
Variable expenses	50
Provision for taxes	10
Total expenses	75
Total revenue	125

7

INTRODUCTION TO PRODUCTION AND RESOURCE USE

Maximum economic production does not lead necessarily to maximum economic satisfaction.

Sir Josiah Stamp
(1880–1941)

TOPICS OF DISCUSSION

Businesses employ a wide variety of resources during the process of producing a particular good or service. A wheat farm, for example, uses a different combination of resources than does a cotton gin, or a meat packing firm. Some businesses use labor on a more seasonal basis than others do. And some businesses require more land when producing their product than others do.

Despite the major differences in resources used by certain types of businesses involved in the nation's food and fiber industry, there are a number of features associated with the use of resources by these businesses that can be generalized. First, inputs can be grouped into several specific categories that facilitate their description and analysis. All these firms also share specific production and cost relationships that provide the foundation for the economic decisions made by a business, including the level of output to produce and the level of variable inputs to employ as it utilizes its current productive capacity.

In this chapter, we will assume the existence of **perfect competition.** The farm sector comes closer than any other sector of the economy to satisfying the conditions of perfect competition. We will also focus initially on understanding the effect that varying the use of a single input has upon the production of a *single product.* Both of these assumptions will be reviewed in later chapters when we examine multiple input choices, multiple product choices, and the nature of decisions under **imperfect competition,** which exists in several sectors of the food and fiber industry.

The focus of this chapter is on introducing some important physical and economic relationships, which will be broadened in later chapters. A thorough understanding of the concepts presented in this chapter is essential before proceeding with the remaining chapters in this book.

CONDITIONS FOR PERFECT COMPETITION

What does it take to have a perfectly competitive economic situation? Before we discuss how economic decisions are made, let us look at the economic environment in which they are made. An input or product market structure can be classified as perfectly competitive if the following conditions hold:

- The product sold by businesses in a sector is homogeneous. In other words, the product sold by one business is a perfect substitute for the product sold by the other businesses. This enables buyers in the market to choose from a number of sellers.

- Any business can enter or leave the sector without encountering serious barriers for entry. Resources must be free to move into the sector without encountering barriers to entry (e.g., patents, licensing). The same condition holds for resources leaving the sector.

- There must be a large number of sellers of the product. No single seller has a disproportionate influence on price; each is a price taker.

- Perfect information must exist for all participants regarding prices, quantities, qualities, sources of supply, and so on.

When all four conditions hold, we can say a market's structure is perfectly competitive. Businesses that satisfy these conditions are also, by definition, perfectly competitive. A perfectly competitive business is a price taker, or it accepts the price of the product it receives in the product market or pays in the input as given. A corn farmer is a good example of a perfect competitor. There are thousands of corn producers that produce a homogeneous product (e.g., no. 2 yellow corn), each with equal access to corn market information, and each with no ability to control the price of corn they receive, or the price they pay for fuel.

CLASSIFICATION OF INPUTS

It is helpful to have some broad classifications in mind when discussing production relationships. These classifications not only promote efficient communication but also help to conduct economic analyses. Although not uniformly accepted, classification of inputs into land, labor, capital, and management has proven useful.

Land

Land includes not only the land forms associated with the earth's crust but also resources such as minerals, forests, groundwater, and other resources given by nature. Such resources are classified as either *renewable* (e.g., forests), or *nonrenewable* resources (e.g., minerals). An example of a key land input in farming activities is productive topsoil, which has many of the attributes of a nonrenewable resource identified in Chapter 2.

Labor

Labor includes all labor services used in production with the exception of managerial activities. In crop production, labor activities include seed bed preparation, planting, irrigation, chemical applications, and harvesting. Labor activities in a canning plant include receiving and grading of fruit and vegetables arriving from the field, blanching, inspection, canning, and warehousing.

Capital

The term *capital* takes on different meanings in different contexts. When using the term *capital*, a banker is referring to stockholders' equity appearing on a bank's balance sheet. In a discussion of input use in the context of production, however, capital refers to manufactured goods such as fuel, chemicals, tractors, trucks, and buildings that provide productive services to their users.[1]

---

[1]Jargon often comes under attack by those outside a discipline of study. For example, Edwin Newman in *Strictly Speaking*, New York, 1980, Warner Books, has satirized the excessive use of such language. Economics seems to be no better or worse than other disciplines. Indeed, in most introductory classes, learning the jargon is an important portion of the course.

A key aspect of capital goods is that they do not provide consumer satisfaction directly but rather aid in the production of other goods and services. Nondurable capital inputs such as fuel and chemicals are entirely used up during the current production period. Durable capital inputs such as machinery and buildings, on the other hand, are utilized over a period of years.

Management

The final input category is management. Its functions are varied and are easier to conceptualize than to measure. Like the leader of an orchestra, farmers and agribusinesses must make decisions as to how, when, and what to produce when organizing their inputs, when and how to market the business's output, how large to grow, and how to finance business expansion.

In this chapter, we abstract from most of management's differences and instead highlight some concepts common to all inputs, with particular emphasis on technical relationships. Input and product prices will be meshed with these technical relationships in this chapter when we discuss those input-output combinations that achieve a specific economic goal, such as profit maximization.

IMPORTANT PRODUCTION RELATIONSHIPS

Several key relationships between the level of output and the level of input use must be understood before we consider the prices of these inputs and outputs. These relationships include the concept of a production function that reflects this input-output relationship and the concepts of marginal and average product.

The Production Function

A production function characterizes the physical relationship between the use of inputs and the level of output. Suppose you are a salesperson for a fertilizer company, and a farmer asks you to recommend the amount of fertilizer to apply per acre to maximize profit. Before you can recommend the quantity of fertilizer the farmer should apply, you must have some knowledge of the physical relationship between yields and the level of fertilizer use. If the application of more fertilizer has no effect on crop yields, the answer is simple: a profit-maximizing farmer obviously should not apply any additional fertilizer.

In the general case, where there are n number of identifiable inputs, a production function may be expressed as:

$$\text{output} = f(\text{quantity of input 1, quantity of input 2,} \ldots, \text{quantity of input } n) \qquad (7.1)$$

which, in words, simply states that the level of output is a function of (i.e., depends on) how much of input 1, input 2, . . ., and input *n* you use. For example, in an early 1880s agricultural setting, crop output was a function of the services provided by labor, land, seed, a workhorse, a few basic implements, and management.

A production function is a rule associating an output to given levels of the inputs used. Output is measured in physical units such as bushels of wheat, gallons of milk, cases of canned peas, etc.[2] If one input in a production is varied, and all the other inputs are held fixed at their existing level, we can rewrite Equation 7.1 as:

$$\text{output} = f(\text{labor} \mid \text{capital, land, and management}) \qquad (7.2)$$

in which the bar separating the first input from all other inputs indicates that only the first input is being varied and the other inputs are held fixed at existing levels.[3] This enables us to examine the relationship between labor and output as opposed to the other inputs employed in the firm's production process.

To understand the concept of a production function, let's return to TOP-AG, the hypothetical business introduced in Chapter 6. Column 2 in Table 7.1 reports the potential output levels TOP-AG can achieve per day with specific levels of labor use. One input (labor) is varied here, and all other inputs (capital, land, and management) are held fixed at their current levels. Column 1 in Table 7.1 indicates the level of daily labor use associated with the levels of output in column 2. For example, if the daily use of labor is 10 hours a day, column 2 in Table 7.1 indicates the level of output would be one unit.

Total Physical Product Curve

If we were to connect this and other combinations of output and labor use reported in columns 1 and 2 in Table 7.1, we would obtain the input-output relationship known as the **total physical product curve** presented in Figure 7.1. By reading along the *X*, or horizontal, axis to a particular input level, reading up to the total physical product curve, and then reading over to the *Y*, or vertical, axis, you can determine the level of output associated with this input use.

The total physical product curve typically will initially increase at an increasing rate, then increase at a decreasing rate, and finally decrease over a full range of potential input use levels. These and other properties of the total physical product curve can be better understood by calculating two additional product curves.

---

[2]Another way of characterizing a production function is to think of it as a cooking recipe. For example, it takes a specific combination of inputs to bake a cake. Baking a cake requires flour, eggs, water, and other ingredients. It also requires labor to blend the ingredients and capital (in the form of an oven and energy) to bake the ingredients.

[3]The terms *output* and *total physical product* can and will be used interchangeably throughout this and subsequent chapters.

TABLE 7.1   Production Relationship for TOP-AG, Inc.

Point on Figure 7.1	(1) Daily Labor Use	(2) Daily Output Level	(3) Marginal Physical Product, Δ(2) ÷ Δ(1)	(4) Average Physical Product, (2) ÷ (1)
A	10.0	1.0		.10
B	16.0	3.0	.33	.19
C	20.0	4.8	.45	.24
D	22.0	6.5	.85	.30
E	26.0	8.1	.40	.31
F	32.0	9.6	.25	.30
G	40.0	10.8	.15	.27
H	50.0	11.6	.08	.23
I	62.0	12.0	.03	.19
J	76.0	11.7	−.02	.15

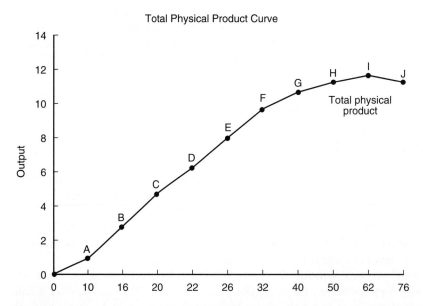

FIGURE 7.1   The marginal physical product curve for TOP-AG crosses the horizontal axis (i.e., negative) when the total physical product curve has reached its peak and begins to decline.

Marginal Physical Product Curve

If a farmer adds another pound of fertilizer per acre, will corn yields increase? If so, by how much? Or would more fertilizer "burn out" the crop and cause yields to decline? Does the addition of another employee at a grain elevator expand its output? These questions give rise to the important concept of **marginal physical product.**[4]

The marginal physical product for an input is the change in the level of output associated with a change in the use of a particular input, where all other inputs used in the production process remain fixed at their existing levels. Stated in equation form, the marginal physical product is equal to:

$$\text{marginal physical product} = \frac{\Delta \text{ output}}{\Delta \text{ input}} \qquad (7.3)$$

in which the "Δ" sign stands for "change in."

To illustrate the calculation and interpretation of the marginal physical product, consider the data for TOP-AG in Table 7.1. There may be some confusion regarding which level of output to associate with a marginal physical product as you read this table. Why are the values in column 3 on a different line than the other columns in this table? While the other columns represent levels of activity, column 3 reflects changes in levels of activity.

For example, when the daily labor use at TOP-AG is increased from 10 hours to 16 hours, the marginal physical product would be 0.33 units of output (i.e., 0.33 = [3.0 − 1.0] ÷ [16.0 − 10.0]). Thus, if labor use is increased by one hour, the business can complete assembling one-third of another spare part. This value is listed between the rows associated with 10 and 16 hours of labor and one and three units of output. If 20 hours of labor were used, the marginal physical product would be 0.45 units of output (i.e., 0.45 = [4.8 − 3.0] ÷ [20.0 − 16.0]) and so on. The marginal physical product curves for TOP-AG are plotted in Figure 7.2.

An important relationship exists between the total and marginal physical products. The slope of the total physical product curve (with respect to the use of labor) is approximately equal to the marginal physical product. If the change in labor use is very small, the marginal physical product is exactly equal to the slope of the total physical product curve. In other words, the marginal physical product curve measures the rate of change in output in response to a change in the use of labor.

The marginal physical product curve takes certain twists and turns as we move along the total physical product curve. For example, the marginal physical product curve cuts the average physical product from the top at approximately 30 hours of labor, where the total physical product curve in Figure 7.1 began to increase at a decreasing rate.[5] A particularly important twist is that, when the total physical product curve is *decreasing*, the marginal physical product curve will be *negative*. This can be

[4]Some refer to marginal physical product as simply "marginal product." We will follow the time-honored tradition of using the word *physical* in our discussion, which makes it clear that the units of measurement are in physical units rather than dollars.

[5]The nature of the linear segments comprising the total physical product (TPP), marginal physical product (MPP), and average physical product (APP) curves precludes the MPP curve from intersecting the APP curve at its exact maximum and the MPP curve from being precisely zero at the peak of the TPP curve in Figure 7.1. A smoothing of these relationships or small linear segments would more closely approximate these conditions.

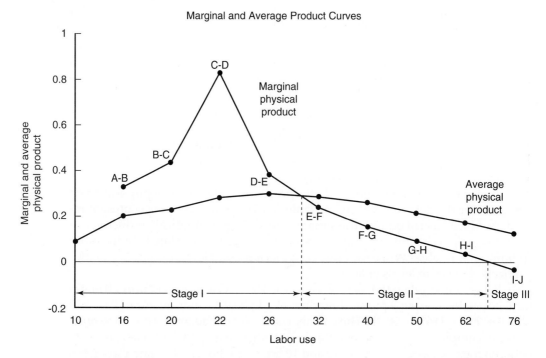

FIGURE 7.2   The marginal physical product curve illustrates the change in output associated with a change in labor input use by TOP-AG. The marginal physical product curve for labor falls below zero at approximately 70 hours of labor use.

seen in Table 7.1 at the point at which TOP-AG increases its use of labor from 62 hours a day to 76 hours. Column 3 shows that the marginal physical product becomes negative at about 70 hours. The point at which the marginal physical product becomes negative corresponds to the point where the total physical product curve begins to decline.[6] Remember the marginal physical product approximates the slope of the total physical product curve. This characteristic will help us identify the rational range of production over the total physical product curve. This issue will be addressed shortly when we discuss the stages of production.

Average Physical Product Curve

A final input-output relationship is the **average physical product.** The average physical product is related to the level of output relative to the level of input use instead of their incremental change. Stated in equation form:

[6]You can find the point at which the total physical product curve changes from increasing at an increasing rate to increasing at a decreasing rate by drawing a ray emanating from the origin and seeing where this ray is tangent to the total physical product curve. At this point of tangency, TOP-AG's marginal physical product is equal to its average physical product.

$$\text{average physical product} = \frac{\text{output}}{\text{input}} \tag{7.4}$$

In the context of our example, the average physical product represents the output per hour of *labor* with all other input levels held constant.

Column 4 in Table 7.1 presents the value of the average physical product for labor use for TOP-AG. The average physical product is shown to rise as output rises, but then falls as output expands at a decreasing rate. The average physical product curve for TOP-AG in Figure 7.2 is intercepted from above by the marginal physical product curve as output begins to increase at a decreasing rate.

A review of Table 7.1 and Figure 7.2 suggests the following conclusions:

- If the marginal physical product curve is above the average physical product curve, the average physical product curve must be rising.
- If the marginal physical product curve is below the average physical product curve, the average physical product curve must be falling.
- The marginal physical product curve therefore cuts the average physical product curve from above at that point where the average physical product curve reaches its maximum.

We now have the three input-output relationships we need to evaluate ranges of rational and irrational regions of production.

Stages of Production

To understand the production relationships illustrated in Table 7.1, you may divide them into stages of production. Stage I is the point at which the marginal physical product lies above the average physical product curve. Stage II is the point at which input use begins at the end of stage I and continues until the value of the marginal physical product becomes equal to zero. Stage III is the point at which input use lies to the right of stage II, or at which the marginal physical product is negative.

Returning to the example in Figure 7.2, the stages of production associated with the use of labor by TOP-AG are outlined in Table 7.2. The irrational nature of stage III under normal economic conditions can be easily explained. If input is increased beyond 62 units of labor, output will fall. It is irrational to increase the use of an input if it only leads to less output. Would you recommend that TOP-AG increase its daily use of labor to 76 hours daily knowing that output would fall?

Electing to stop in stage I is also irrational, although it may be more difficult to see why. A good grade on your last examination (marginal physical product) that raises your semester average (average physical product) is an example of an outcome occurring during stage I. If you were permitted to take a make-up test for the last examination and substitute this grade for your original test score, you

TABLE 7.2   Stages of Production

Stage	Usage of Labor	Operate?
I	Between 0 and 30	Yes
II	Between 30 and 70	Yes
III	Greater than 70	No

would likely take this make-up test only if you felt you could increase your semester average grade.

In the context of Table 7.1, look at the values in columns 3 and 4. A marginal physical product of 0.85 units per hour increases the average physical product from 0.24 units to 0.30 units. And a marginal physical product of 0.40 units increases the average physical product from 0.30 units to 0.31 units. But a lower marginal physical product of 0.25 units brings the average physical product down to 0.30 units. The first two observations are in stage I, and the third observation would occur in stage II.

Why should you stop producing if the output of your business is increasing at an increasing rate? As long as the average physical product is rising, you should expand input use. We will leave the presentation of the economic rationale for support of this argument for later in this chapter. At this point, we recognize that stage II appears to hold primary interest for a firm that wishes to maximize its profits.

Figure 7.2 shows that the marginal physical product of labor use is falling throughout stage II. This phenomenon is so frequently observed that it is called a law—**the law of diminishing marginal returns.** This law states that:

> as successive units of a variable input are added to a production process with the other inputs held constant, the marginal physical product eventually decreases.

Therefore, in the region of greatest economic interest, we would expect to observe diminishing marginal physical product for variable inputs.

A real world input-output relationship between lime and alfalfa estimated by Free and Hall is illustrated in Figure 7.3. The total physical product curve illustrated here contains only stage II and stage III of production. This curve shows that, when lime is increased beyond three tons per acre, the total physical product curve actually begins to decline—evidence that stage III is present. No rational farmer would want to increase the use of lime beyond this point, which implies a "burning out" of the crop when too much lime is applied.

A farmer obviously would not knowingly apply lime to the point where stage III occurs. If a farmer asked you to identify the application rate in stage II that makes the most economic sense, what would you advise this farmer? Do you have all the information you need to provide the farmer an answer? The answer is no. Although knowledge of the production relationships discussed thus far are extremely important, we must also know the costs of production.

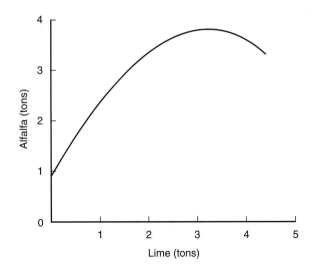

FIGURE 7.3   Input-output relationship between lime and alfalfa. (**Source:** Hall H, Free W: On evaluating crop response to lime in the Tennessee Valley Region, *Southern Journal of Agricultural Economics:* 75–81, December 1979.)

ASSESSING SHORT-RUN BUSINESS COSTS

The total cost of production is the costs associated with the use of *all* inputs to production. In Chapter 6, we learned that a business's total costs in the short run can be divided into **fixed costs** and **variable costs.** We learned that fixed costs are those costs that *do not* vary with the level of input use and that variable costs are those costs that *do* vary with the level of input use. It is important to understand the measurement of these and other concepts of cost and their relationship to the level of production activity when making short-run economic decisions.

Two additional cost concepts related to the level of production that are extremely important to economic decision making are marginal costs and average costs. Each of these cost concepts, and how they are related to the total, average, and marginal product curves, is addressed in this section.

Total Costs and the TPP Curve

Figure 7.1 shows that the total physical product (TPP) curve typically first increases at an increasing rate, then increases at a decreasing rate, and finally decreases. In light of this curve, what do the **total variable cost** curve and **total fixed cost** curve look like?

A complete cost schedule for our hypothetical business TOP-AG is presented in Table 7.3. We have assumed that TOP-AG pays $5 per hour for labor and has fixed costs of $100 per hour, as we did earlier in Equation 6.5. This includes $75 of explicit costs and $25 of implicit costs. No matter what happens to output in column 1, fixed costs in column 2 remain the same. The current property tax assessment owed by a business will be the same whether it produces at its capacity or produces nothing at all.

TABLE 7.3   Short-Run Cost Schedule for TOP-AG, Inc., and Selected Cost Concepts

Point on Figure 7.1	(1) Total Output	(2) Total Fixed Cost	(3) Average Fixed Cost (2) ÷ (1)	(4) Total Variable Cost	(5) Average Variable Cost (4) ÷ (1)	(6) Total Cost (2) + (4)	(7) Marginal Cost Δ(6) ÷ Δ(1)	(8) Average Total Cost (3) + (5)
A	1.0	100.00	100.00	50.00	50.00	150.00		150.00
B	3.0	100.00	33.33	80.00	26.67	180.00	15.00	60.00
C	4.8	100.00	20.83	100.00	20.83	200.00	11.11	41.67
D	6.5	100.00	15.38	110.00	16.92	210.00	5.88	32.31
E	8.1	100.00	12.35	130.00	16.05	230.00	12.50	28.40
F	9.6	100.00	10.42	160.00	16.67	260.00	20.00	27.08
G	10.8	100.00	9.26	200.00	18.52	300.00	33.33	27.78
H	11.6	100.00	8.62	250.00	21.55	350.00	62.50	30.17
I	12.0	100.00	8.33	310.00	25.83	410.00	150.00	34.17
J	11.7	100.00	8.55	380.00	32.48	480.00	n/a	41.03

Total variable costs per hour for our hypothetical business TOP-AG are shown in column 4 of Table 7.3. Looking at both columns 1 and 4, we see that when the level of output rises, the level of total variable costs rises. A factory will need more labor and will incur more labor costs if management decides to expand production.

Figure 7.4, *A* graphically illustrates the nature of the total cost, total variable cost, and total fixed cost series reported in Table 7.3. The total fixed cost curve is parallel to the horizontal axis, thus illustrating its fixed nature when output rises. The total variable cost curve, on the other hand, rises when the level of output rises. Finally, the total cost curve, which reflects both fixed and variable costs, rises when output rises. The constant gap between the total cost curve and total variable cost curve when output rises is equal to $100, or the level of fixed costs for our hypothetical business.

If you compare the total variable cost curve in Figure 7.4 with the TPP curve in Figure 7.1, you will see that they are mirror images of each another. Where the TPP curve is concave when viewed from below, the total variable cost curve is convex, and vice versa.[7] Both curves have an output axis in common, but output is on the vertical, or *Y*, axis in the TPP curve in Figure 7.1 and on the horizontal, or *X*, axis in Figure 7.4.

These total cost measures serve as the basis for average variable, average fixed, and average total costs, and for marginal costs.

---

[7]This relationship will hold as long as the per unit price of the input (wage rate paid for labor in this instance) does not change as the use of this input is expanded. A demonstration of this relationship for the two-input case is presented in the Advanced Topic section at the end of Chapter 8.

Cost Relationships

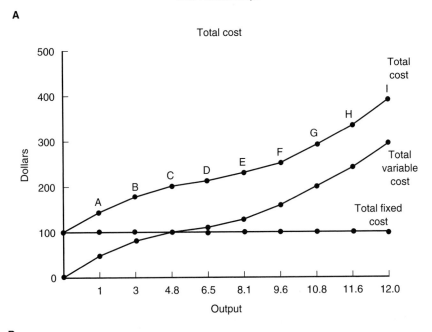

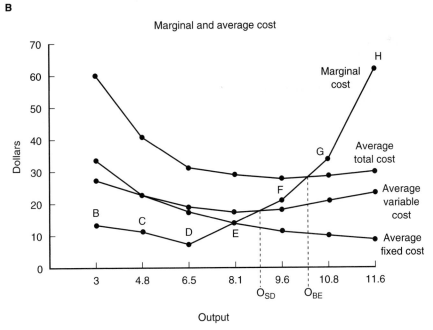

FIGURE 7.4 The cost relationships calculated for TOP–AG in Table 7.3 are plotted above. Output levels O_{BE} AND O_{SD} illustrate that the marginal cost curve intersects the average variable and average total cost curves at their minimum.

Average Costs and the APP Curve

The concept of **average cost** involves measuring costs per unit of output, or the level of cost associated with the level of output. The concepts of total costs, fixed costs, and variable costs discussed may be expressed in terms of average costs as:

$$\begin{array}{c}\text{average}\\\text{total}\\\text{costs}\end{array} = \frac{\text{total costs}}{\text{output}} \tag{7.5}$$

$$\begin{array}{c}\text{average}\\\text{fixed}\\\text{costs}\end{array} = \frac{\text{total fixed costs}}{\text{output}} \tag{7.6}$$

$$\begin{array}{c}\text{average}\\\text{variable}\\\text{costs}\end{array} = \frac{\text{total variable costs}}{\text{output}} \tag{7.7}$$

Average fixed costs for TOP-AG are calculated in column 3 of Table 7.3. **Average variable costs** for this business are calculated in column 5. Average total costs, which are equal to average fixed costs plus average variable costs, are calculated in column 8 in Table 7.3. Figure 7.4, *B* illustrates the general nature of these three short-run average-cost curves. Average fixed costs decline over the entire range of production because they do not vary with output. Average variable costs, on the other hand, normally decrease up to a certain output level and then increase when output expands further.

If you compare the average variable cost curve in Figure 7.4 with the APP curve in Figure 7.2, you will see that these two curves are mirror images of each other. Although the APP curve for TOP-AG is convex, its average variable cost curve is concave. The maximum of the APP curve of 0.31 units in column 4 of Table 7.1 is attained at 26 hours of daily labor use and 8.1 units of output. The minimum of the average variable cost curve of $16.05 in column 5 of Table 7.3 was also attained at 8.1 units of output. The reason for this inverse relationship is that, when output per unit of labor rises, average variable costs must necessarily decline. (Review Equation 7.7 if this is unclear.)

Marginal Costs and the MPP Curve

Marginal cost is perhaps the most important concept. Marginal cost is the change in the business's total costs per unit of change in output. Marginal cost is measured as:

$$\text{marginal cost} = \frac{\Delta \text{ total cost}}{\Delta \text{ output}} \tag{7.8}$$

in which Δ represents the change in a particular item (e.g., costs, output). Marginal cost also represents the slope of the total cost and total variable cost curves.

The level of marginal costs for TOP-AG is calculated in column 7 of Table 7.3. The marginal cost curve for this business is plotted in Figure 7.4, *B*. Like average variable cost, marginal cost first falls and then rises. It cuts the average total cost and average variable cost curves at their minimums (see outputs O_{BE} and O_{SD} in Figure 7.4, *B*).[8]

Once again, the cost curve is a mirror image of a product curve. If you compare the marginal cost curve in Figure 7.4, *B* with the MPP curve in Figure 7.2, you will see that they are the inverse of one another. The maximum of the MPP curve in column 3 of Table 7.1 occurs at 0.85 units of output as labor use is expanded from 20 to 22 hours per day and total output jumps from 4.8 to 6.5 units of output. The minimum of the marginal cost curve in column 7 of Table 7.3 is $5.88, and also occurs as TOP-AG's total output is expanded from 4.8 to 6.5 units.

ECONOMICS OF SHORT-RUN DECISIONS

Now that we have gained an understanding of the physical aspects and cost aspects of production, the next logical issue is to determine the level of output and input use that will maximize the business's current economic profit. Before we can do this, however, we must discuss two additional revenue concepts, marginal revenue and average revenue.

Marginal and Average Revenue

In the last section, we discussed the calculation of marginal and average costs of production. They have their counterpart on the revenue side. The change in total revenue is called **marginal revenue** and it represents the change in revenue from producing more output, or:

$$\text{marginal revenue} = \Delta \text{ total revenue} \div \Delta \text{ output} \qquad (7.9)$$

which, under perfect competition, will also be equal to the per unit sales price of the product (price per bushel, per ton, per pound, etc.) for the business's product.[9]

If TOP-AG increased its production from 11.6 units per hour to 12 units per hour, its total revenue would increase from $522 an hour (i.e., 11.6 units of output multiplied by a product price of $45 per unit) to the $540 that we calculated back in Chapter 6. (See Equation 6.3.)

---

[8]The "BE" and "SD" subscripts refer to the *break-even* and *shutdown* levels of output. These levels of output will take on special significance later in this chapter when we examine how changing market price levels affect the level of output desired by profit-maximizing businesses.

[9]Under imperfect market structures, marginal revenue will differ from market price, which we will discuss in Chapter 11.

If TOP-AG expands its output from 11.6 units to 12 units per hour, its marginal revenue would be equal to:

$$\text{marginal revenue} = (\$540 - \$522) \div (12.0 - 11.6)$$
$$= \$18 \div 0.40 \text{ units of output}$$
$$= \$45 \qquad (7.10)$$

which is identical to the $45 market price assumed for TOP-AG's product.

The concept of **average revenue** reflects the revenue per unit of output the business receives for its product, or

$$\text{average revenue} = \text{total revenue} \div \text{output} \qquad (7.11)$$

which simply represents another way of looking at the price of the product. If TOP-AG produced 12 units of output an hour and received $45 for each unit it produced, its total revenue would be $540 per hour. TOP-AG's average revenue under these circumstances would be:

$$\text{average revenue} = \$540 \div 12 \text{ units of output}$$
$$= \$45 \qquad (7.12)$$

which is identical to the market price this business receives when it sells its product in the market place and to the marginal revenue calculated in Equation 7.10. The marginal revenue and average revenue curves under conditions of perfect competition assumed in this chapter will be perfectly flat, which we will illustrate shortly. This reflects the notion that a business is a price taker; nothing it does will change the price received for its output. In the example presented above, the intercept of these two curves on the vertical, or Y, axis would be $45.

Level of Output: MC = MR

Expansion of the business's variable input use in the current period is profitable at the margin, or as long as the marginal revenue exceeds the marginal cost. A business should not increase the use of an input if marginal cost exceeds marginal revenue, or the change in cost of purchasing additional inputs is greater than the revenue the business would receive from their use. Furthermore, as long as a higher profit is preferred over a smaller profit, a business should not stop expanding production if marginal revenue exceeds marginal cost.

This logic leads to the following economic strategy, under conditions of perfect competition in the short run, in which you produce at the point at which:

$$\text{marginal revenue} = \text{marginal cost} \qquad (7.13)$$

or to the point at which the marginal revenue from the sale of another unit of output equals the marginal cost of producing this unit.

Let us expand our discussion of TOP-AG developed in Tables 7.1 and 7.3 to determine its profit-maximizing level of output. Rows F and G of Table 7.4 suggest that when TOP-AG expands its production from 9.6 units of output to 10.8 units of output, the business will achieve an economic profit of $186 per hour. TOP-AG's total revenue would be $486 per hour, while its total costs would be $300 per hour. The entry in column 5 of row G of $186 is the largest entry in this column, which suggests that profit is maximized at 10.8 units of output. Is it?

It is very important that you understand the economic rationale underlying the profit-maximizing level of output. If TOP-AG expanded its output to 11.6 units of output, the business's total revenue would fall from $186 to $172 an hour (see row H in Table 7.4). Obviously, TOP-AG's management would not wish to expand its operations in the current period if its goal is to maximize profits.

As TOP-AG expanded its output from 9.6 units of output to 10.8 units, its marginal cost was $33.33 as shown in column 6, and its marginal revenue was $45 as shown in column 7. The net benefits from this expansion are positive. But if TOP-AG further expands its operations from 10.8 units per hour to 11.6 units, the marginal cost of doing so would be $62.50 as compared with a marginal revenue of only $45. The level of output that maximizes profit, or the point at which the marginal revenue associated with the expansion just equals the marginal cost associated with the expansion, occurs when marginal cost equals $45. This will occur somewhere between rows G and H, or between 10.8 and 11.6 units of output as shown in Figure 7.5.

Reviewing Figure 7.5, we see that the marginal revenue curve is perfectly flat, reflecting the fact that this business is a price taker. The business thinks that the level of its production is small enough not to have a perceptible impact on the market price. Because we have assumed the presence of perfect competition in this chapter, the marginal revenue curve also reflects the average revenue. The intersection of this marginal revenue curve and the marginal cost curve indicates

TABLE 7.4   Determination of TOP-Ag's Profit-Maximizing Level of Output

Point on Figure 7.1	(1) Total Output	(2) Market Product Price	(3) Total Revenue (1) × (2)	(4) Total Costs	(5) Economic Profit (3) − (4)	(6) Marginal Cost Δ(4) ÷ Δ(1)	(7) Marginal Revenue Δ(3) ÷ Δ(1)
A	1.0	$45.00	$45.00	$150.00	−$105.00		
B	3.0	45.00	135.00	180.00	−45.00	$15.00	$45.00
C	4.8	45.00	216.00	200.00	16.00	11.11	45.00
D	6.5	45.00	292.50	210.00	82.50	5.88	45.00
E	8.1	45.00	364.50	230.00	134.50	12.50	45.00
F	9.6	45.00	432.00	260.00	172.00	20.00	45.00
G	10.8	45.00	486.00	300.00	186.00	33.33	45.00
H	11.6	45.00	522.00	350.00	172.00	62.50	45.00
I	12.0	45.00	540.00	410.00	130.00	150.00	45.00
J	11.7	45.00	526.50	480.00	46.50	n/a	n/a

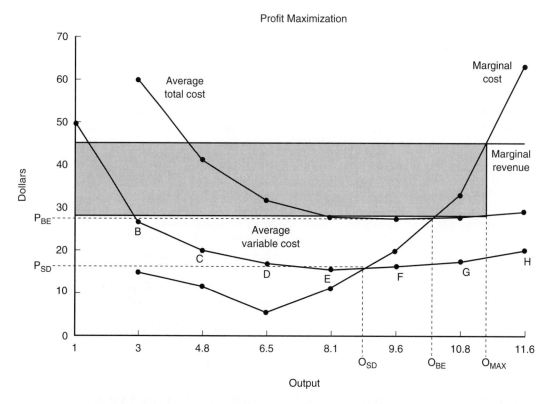

FIGURE 7.5   We can determine graphically what Table 7.4 could only hint at given the levels of production studied in that table. Given the MC = MR criterion expressed in Equation 7.13 for profit maximization, we see that profits would be maximized at O_{MAX}, or slightly more than 11 units of output per hour.

that output O_{MAX}, or slightly more than 11 units of output, would maximize TOP-AG's profit.

Two additional output levels deserve special mention here—the break-even level of production and the shutdown level of production. If the marginal revenue curve falls in a parallel fashion to the point where it is just tangent with the minimum point on the average total cost curve, the business's average total costs will be exactly equal to its average revenue. The business would be able to meet both its fixed and variable costs of production with the revenue it received during the current period.

If the marginal revenue curve were to fall further (again in a parallel fashion) to the point at which it is just tangent to the minimum point on TOP-AG's average variable cost curve, TOP-AG would be just able to cover its variable costs but none of its fixed costs. Further declines in marginal revenue would cause the business to cease operations in the current period. TOP-AG could no longer pay its fuel bill, meet its hired labor payroll, or pay other expenses that vary with the level of production. Either the fuel supplier will stop making deliveries, hired workers will leave, or other factors critical to production will be curtailed. Thus, output level O_{SD} represents the shutdown level of production, and the minimum point on TOP-AG's effective marginal cost curve, or its supply curve.

Finally, the level of economic profits in Figure 7.5 is equal to the shaded rectangle formed by the difference between TOP-AG's average revenue and average total cost per unit at output O_{MAX}, or \$45 minus approximately \$28, multiplied by the quantity of output at O_{MAX}, or approximately 11.1 units. The level of economic profit here would be approximately \$189 (i.e., [\$45 − \$28] × 11.1).

Level of Resource Use: MVP = MIC

Now that we have determined the level of output at which a business should operate if its management wishes to maximize its profit, we must next determine what this means for the level of resource use. In the single variable input case, such as in our labor example for TOP-AG in Table 7.1, this is a relatively simple process. If you have determined the profit-maximizing level of output, you simply go to column 2 and read over to column 1 and observe the level of input use associated with this output.

An alternative approach to determining profit-maximizing input demands involves comparing the marginal benefit for a given level of input use with the **marginal input cost.** Because revenue is equal to the product price times output, it is clear that the marginal benefit from input use is equal to the change in total revenue per unit change in the input. This marginal benefit is called the **marginal value product,** and for labor is equal to:

$$\begin{array}{l} \text{marginal value} \\ \text{product for} \\ \text{labor} \end{array} = \text{MVP}_{labor} = \text{MPP}_{labor} \times \text{product price} \qquad (7.14)$$

The optimum, or profit-maximizing, level of input use occurs when the marginal value product equals the marginal input cost. For the case of labor, the optimal level of labor use is given by:

$$\text{MVP}_{labor} = \text{wage rate} \qquad (7.15)$$

If additional labor were employed beyond this point, the marginal cost (i.e., the wage rate) would exceed the marginal benefits (i.e., the marginal value product for labor).

Let us illustrate the determination of the profit-maximizing level of input use by applying Equation 7.15 to the information presented in Table 7.5. This table illustrates that the price of TOP-AG's product is \$45 per unit and the cost of labor is equal to \$5 per hour. Multiplying the marginal physical product in column 2 by \$45 gives us the marginal value product reported in column 3, or the marginal benefit from adding another hour of labor.

TABLE 7.5   Determination of TOP-AG's Profit-Maximizing Level of Labor Use

Point on Figure 7.6	(1) Use of Labor	(2) Marginal Physical Product*	(3) Marginal Value Product (2) × $45	(4) Wage Rate	(5) Marginal Net Benefit (3) − (4)	(6) Cumulative Net Benefit
A	10					
B	16	.33	$14.85	5.00	$9.85	9.85
C	20	.45	20.25	5.00	15.25	25.10
D	22	.85	38.25	5.00	33.25	58.35
E	26	.40	18.00	5.00	13.00	71.35
F	32	.25	11.25	5.00	6.25	77.60
G	40	.15	6.75	5.00	1.75	79.35
H	50	.08	3.60	5.00	−1.40	77.95
I	62	.03	1.35	5.00	−3.65	74.30
J	76	−.02	−.90	—	—	—

*Column 3 in Table 7.1.

Column 4 in Table 7.5 indicates that TOP-AG is a price taker in the labor market because its increased use of labor had no effect on hourly wage rates. The contribution to profit (marginal net benefit) is reported in column 5. This value is found by subtracting the marginal cost in column 4 from the marginal benefit reported in column 3.

The information in Table 7.5 illustrates that TOP-AG should use slightly more than 40 hours of labor. The wage rate in column 4 would equal the marginal value product of using additional labor in column 3. If labor were expanded to 50 hours, the cumulative net benefit in column 6 would be declining.

The marginal value product and marginal input cost relationships calculated in Table 7.5 are plotted in Figure 7.6. Because the marginal value product curve is nothing more than the marginal physical product curve multiplied by a fixed product price, the marginal value product curve looks very much like a marginal physical product curve. As Table 7.5 suggests, the labor use that maximizes profits is about 41 hours, the point at which the marginal value product curve intersects the marginal input cost curve.

The analysis presented in Table 7.5 and Figure 7.6 can be extended to other inputs. It represents a general way of characterizing the profit-maximizing level of input use. Profit maximization requires that the marginal value product (marginal benefit) of each variable input equals its marginal input cost simultaneously.

A Real-World Application

In Figure 7.3, we examined the impact of lime on alfalfa yields. Let us now examine the following question: given this partial production relationship, what quan-

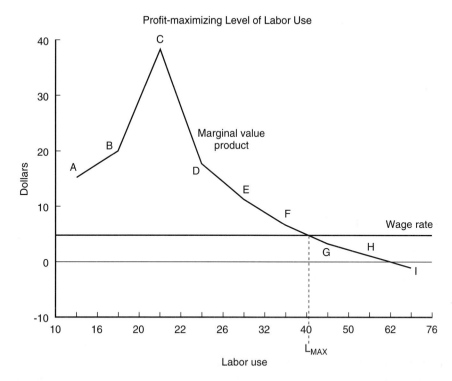

FIGURE 7.6   Equation 7.15 indicates that the profit-maximizing level of input use in the short run occurs at the point at which the marginal value product of labor equals the marginal input cost. In the case of TOP-AG, we see that this occurs at approximately 41 hours of labor per day if the wage rate is $5 an hour, or at quantity L_{MAX}.

tity of lime should the farmer apply if he or she wants to maximize profit? The answer to this question will hinge on the impact of lime applied this year on future soil acidity. Let us consider the simplest case possible: the farmer leases the land for only one year; therefore, only *this year's* benefit is relevant. Table 7.6 presents the yields and marginal physical products calculated for selected quantities of lime that underlie the curve plotted in Figure 7.3. In Table 7.6, the marginal physical product of lime falls from 0.605 to 0.083 when lime use is increased from 2 tons per acre to 3 tons per acre. If the farmer wished to maximize yield, more than 3 tons per acre would be applied (one would continue to apply lime until its marginal physical product was zero). However, lime is not free. Therefore, the farmer will want to weigh the costs and benefits of lime use.

Assume that the price received for alfalfa (net of other expenses) is $30 per ton and that the price of lime is $10 per ton. We see in Table 7.6 that the profit-maximizing level of lime use is given by the point where the marginal physical product of lime times the price of alfalfa is equal to $10. The marginal value product of lime is found by multiplying column 3 in Table 7.6 by $30 (the price

TABLE 7.6   Determination of Profit-Maximizing Level of Lime Use

Lime Use	Yield	Marginal Physical Product of Lime per Acre	Marginal Value Product	Price of Lime per Ton	Marginal Net Benefit
2.0	3.3300				
2.1	3.3905	0.605	18.15	10.00	8.15
2.2	3.4452	0.547	16.41	10.00	6.41
2.3	3.4941	0.489	14.67	10.00	4.67
2.4	3.5372	0.431	12.93	10.00	2.93
2.5	3.5745	0.373	11.19	10.00	1.19
2.6	3.6060	0.315	9.45	10.00	−.55
2.7	3.6317	0.257	7.71	10.00	−2.29
2.8	3.6516	0.199	5.97	10.00	−4.03
2.9	3.6657	0.141	4.23	10.00	−5.77
3.0	3.6740	0.083	2.49	10.00	−7.51

of alfalfa).[10] If lime use were to increase from 2.0 to 2.1 tons per acre, the marginal value product would be $18.15, which exceeds the marginal input cost of $10 for a marginal net benefit of $8.15. Hence, profit can be increased by $8.15 by increasing lime use. Following this line of reasoning, we find the profit-maximizing lime use to be approximately 2.58 tons per acre. Adding lime beyond this point would lead to a situation in which the marginal value product of additional lime use is less than the marginal input cost, and losses would actually occur.

WHAT LIES AHEAD?

The focus of this chapter was on the economic decisions faced by a business in the short-run or current period. We looked at the effects that varying the use of one input (hired labor) would have on output and the business's costs of production. Following up on the explicit statements made by many firms eluded to in Chapter 6, that a major objective of the business is to maximize profit, this chapter determined the profit-maximizing level of output in the short run and the profit-maximizing use of a variable input.

Chapter 8 will broaden the focus of the business's decisions by examining the determination of the least-cost combination of variable inputs in the short run and the optimal expansion path for labor and capital over the long run. Chapter 9 will address the profit-maximizing combination of products to produce.

---

[10]Generally, the costs of other inputs are deducted from the price of the product so that the marginal benefit is expressed in net terms.

SUMMARY

The purpose of this chapter was to illustrate the various physical relationships that exist between inputs and outputs with which agricultural economists must be familiar. The major points of this chapter may be summarized as follows:

1. Farm inputs can be classified into land, labor, capital, and management.
2. A production function captures the causal physical relationship between input use and the level of output.
3. The total physical product reflects the level of output of a given level of input use. Marginal physical product represents the change in the level of output associated with a change in the use of a particular input. Finally, average physical product reflects the level of output per unit of input use. In each case, all other inputs are held fixed. The value of the marginal physical product represents the slope of the total physical product curve. No rational farmer would want to produce beyond the point at which the marginal physical product equals zero, because further input use would cause the level of output to fall.
4. There are three stages of production:

   - Stage I is the point at which the marginal physical product curve for a particular input is rising but still lies above the average physical product curve.
   - Stage II is the point at which the marginal physical product equals the average physical product and continues until the marginal physical product for the input in question reaches zero.
   - Stage III is the point at which stage II left off, or where the total physical product curve begins to decline and the marginal physical product curve becomes negative.

5. The law of diminishing marginal returns states that as the use of an input increases, its marginal physical product will eventually fall.
6. Marginal cost is the change in total cost with respect to a change in output. Average cost is total cost divided by total output. Fixed costs are those costs that do not vary with output.
7. The profit-maximizing level of output occurs in the short run at the output level at which MC = MR. The competitive business takes the market price (MR) as given by the marketplace and makes its production decisions by equating MC = MR. (See Equation 7.13.)
8. The profit-maximizing level of input use occurs in the short run at the input level at which MVP = MIC. The competitive business takes the per unit price of the variable input (MIC) as given by the marketplace and makes its purchasing decisions by equating MVP = MIC. (See Equation 7.15.)
9. The business will break even (TR = TC) in the short run at the output level where the price the business receives for its product falls to the point at which AR =

ATC, or where average profit per unit of output is zero. (See O_{BE} in Figures 7.4 and 7.5.) The business may continue to operate in the short run if AR < ATC, because it can minimize its losses (i.e., cover at least some of its fixed costs).

10. The business will cease operations, or shut down, in the short run if the price the business receives for its product falls to the point at which AR < AVC. When this occurs, the business will no longer be able to cover its variable costs of production (e.g., pay its fuel bill) and will be unable to acquire additional inputs. (See O_{SD} in Figures 7.4 and 7.5.)

DEFINITION OF KEY TERMS

Physical Production Relationships

Average physical product (APP), or simply **average product:** the level of output or total product produced by a business per unit of input used. Average physical product is calculated as: APP_{labor} = output ÷ labor; $APP_{capital}$ = output ÷ capital; etc. Examples include yield per acre, gain per pound of feed fed, etc.

Marginal physical product (MPP), or simply **marginal product:** the change in output or total product the business would achieve in the current period by expanding the use of an input by another unit. Marginal physical product is calculated as MPP_{labor} = Δ output ÷ Δ labor; $MPP_{capital}$ = Δ output ÷ Δ capital; etc. Examples include an increase in alfalfa production from the application of additional lime, and the additional number of cases of fruit canned in the current period from the addition of another canning line.

Total physical product (TPP), or simply **total product:** the total output of goods or services produced by the firm during the current period. The total product of a wheat farmer is the yield per acre multiplied by the number of acres harvested. Examples include total wheat produced by a wheat producer, total pounds of milk produced by a dairy farmer, and total number of cases canned by a canning factory.

Cost and Revenue Concepts

Average fixed costs (AFC): the fixed costs incurred by the business in the current period per unit of output. Average fixed costs are calculated as AFC = TFC ÷ output, or AFC = ATC − AVC. The average fixed cost curve, or AFC associated with specific levels of output, declines as output is expanded. (See Figure 7.4, *B*.)

Average revenue (AR): the level of revenue earned per unit of output from the production of the business is expanded. Average revenue is calculated as AR = revenue ÷ output. Average revenue is also equal to the market price under the conditions of perfect competition. This suggests that the revenue the business receives per unit is identical no matter how much the business produces.

Average total costs (ATC) or **average costs:** the total costs incurred by the business in the current period per unit of output. Average total costs are calculated as ATC = TC ÷ output, or ATC = AFC + AVC. The average total cost curve, or ATC, associated with specific levels of output plays an important role in determining total profit. Figure 7.5 showed that the difference between average revenue (which is the same as marginal revenue) or AR and ATC at O_{MAX} represents the average profit or profit per unit at this level of output. This difference (AR − ATC) multiplied by the level of output O_{MAX} represents the level of total profit. The minimum point on the ATC curve, where the MC curve intersects this curve from below, represents the break-even level of output. (See O_{BE} in Figures 7.4 and 7.5.)

Average variable costs (AVC): the variable costs incurred by the business in the current period *per unit of output.* Average variable costs are calculated as AVC = TVC ÷ output, or AVC = ATC − AFC. The average variable cost curve, or AVC, associated with specific levels of output also plays an important role in assessing the economic performance of a business. Figure 7.4, *B* showed that the minimum point on the AVC, where the MC curve intersects this curve from below, occurs at O_{SD}. This level of cost per unit of output corresponds to the lowest the business can afford to see the market price (and AR) fall and continue to operate in the short run.

Fixed costs (FC): specific form of current production costs that do *not* vary with the level of output or input use. This is a short-run cost concept; all costs are considered variable in the long run. Fixed costs are calculated by outside entities. Examples include the individual value of the business's current property tax bill, the insurance premium due this year, or the interest portion of the business's current mortgage payment.

Imperfect competition: market structure when one or more of the characteristics of perfect competition are not present.

Law of diminishing marginal returns: as successive units of a variable input are added to a production process with the other inputs held constant, the marginal physical product eventually decreases.

Marginal cost (MC): the change in total cost of production as the output or total product of the business is expanded. Marginal cost is calculated as MC = Δ cost ÷ Δ output. Marginal cost represents the total cost of producing another unit of output. Marginal cost to a wheat producer is the change in total costs of producing another acre of wheat. This is an important statistic because the profit-maximizing level of output for a business under conditions of perfect competition occurs at the point where the marginal cost of production is identical to the price of the product, or marginal revenue. The portion of the business's marginal cost curve lying above AVC will be shown in later chapters to represent the firm's supply curve.

Marginal input cost (MIC): the change in the cost of a resource used in production as more of this resource is employed. Marginal input cost is set by the market for the resource. The marginal input cost for labor, for example, is equal to the wage rate the business faces in the hired labor market. The marginal input cost for fertilizer is the price of fertilizer in the marketplace.

Marginal revenue (MR): the change in the revenue earned from the production of the business is expanded. Marginal revenue is calculated as MR = Δ revenue ÷ Δ output. Marginal revenue under conditions of perfect competition is identical to the price the business takes in the marketplace. This means that the additional revenue received from the marketplace will be unaltered by changes in the quantity the business produces; it is a price taker. For example, suppose the price of corn is $2.50 per bushel. This price will remain unchanged by the production decision of an individual producer.

Marginal value product (MVP), or **value of marginal product:** the change in the revenue earned by the business as it employs an additional unit of a resource, holding other resource use constant. Marginal value product is calculated as MVP = MPP × market price of product. The marginal value product of labor to a wheat producer, for example, is equal to the MPP_{labor} or change in wheat output resulting from the employment of an additional farm laborer, multiplied by the price of wheat. Stated another way, this represents the change in revenue the wheat producer would receive by hiring another laborer.

Perfect competition: market structure characterized by large number of producers selling a homogeneous product, each with perfect information, and no barriers to entry or exit.

Total costs (TC): sum of all individual categories of production costs during the current period. Total costs are calculated as TC = TVC + TFC. See examples for total fixed costs and total variable costs. Total costs are an important statistic used to calculate accounting profit, as shown in Equation 6.4.

Total fixed costs (TFC): sum of all current production costs that do *not* vary with the level of output or input use. Total fixed costs are calculated by adding up all individual fixed costs. Total fixed costs can also be measured residually by subtracting total variable production costs from total costs. Examples include the total value of all fixed costs, or the business's current property tax bill, plus its current insurance premium due, plus the current mortgage interest payment due, etc.

Total revenue (TR): sum of all money received by the business from the sale of the products it markets during the current period. Total revenue is calculated as TR = $(P_{corn} \times Q_{corn}) + (P_{wheat} \times Q_{wheat}) + \ldots$, or the sum of the cash receipts from marketings of the business's individual products. Examples include cash receipts from wheat marketed by a wheat farmer, cash receipts from flour marketed by a miller, cash receipts from bread marketed by a baker, etc. Total revenue is another important statistic used to calculate accounting profit, as shown in Equation 6.4.

Total variable costs (TVC): sum of all individual categories of production costs that do vary with the level of output or input use. Total variable costs are calculated by adding up all individual variable costs. Total variable costs can also be measured residually by subtracting total fixed costs from total costs. Examples include the total value of all variable costs, or the current fuel bill, plus the fertilizer bill, plus wages paid to hired labor, plus the rental payment on farmland, etc.

Variable costs (VC): level of specific current production costs that *do* vary with the level of output or input use. It is a short-run cost concept; all costs are considered variable in the long run. Variable costs are calculated by taking the price of the input multiplied by the quantity of the good or service used (i.e., hourly wage rate multiplied by the hours of hired labor employed by the business). Examples include current fuel bill, fertilizer bill, wages paid to hired labor, rental payment on farmland, and repair costs for machinery and motor vehicles.

EXERCISES

1. Please insert the appropriate labels in the blanks in the graph below. Examine the graph carefully to note all labels. Then clearly indicate below the graph the particular significance of point *A*, point *B*, and point *C*.

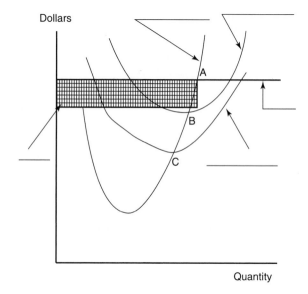

2. The following graph presents selected cost functions for a typical firm. The dashed lines are inserted for easy reference to give you the corresponding values on the dollar and quantity axes. Please answer the following questions based on this graph.

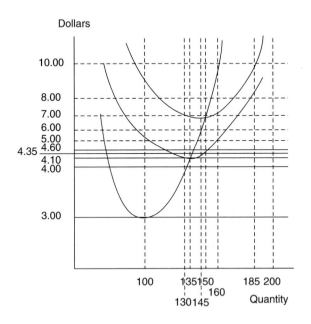

a. At a price of $7, what would be the level of fixed costs that would be covered by the firm? Show all prices, quantities, and the method of calculation (i.e., what is the dollar amount of the fixed costs that are covered?).

b. What is the level of the breakeven price?
   The shutdown price?
   The break-even quantity?
   The shutdown quantity?

c. At a price of $6 and assuming that the firm is a profit maximizer, what would be the level of fixed costs *that would be covered* by the firm? What is the *total* level of fixed costs for the profit maximizing quantity? Show all prices, quantities, and the method of calculation (i.e., what is the dollar amount of the fixed costs that are covered?).

d. What would the level of production and profits for this firm be if the price of the product were $4 per unit?

3. The partial table is for a firm operating in a perfectly competitive market. Please complete the table and answer the following questions based on the answers that you provide in the table.

Input Usage	Output	MPP	APP	Output Price	TR	MR	AR	Input Price	FC	TVC	TC	MC	ATC	AVC	AFC	Profit
	160	XXX	8.0		1200	XXX						XXXX				
30		25								900						
40			4.5													−150

In the preceding example, would the profit maximizing level of output be less than 20 units, between 20 and 30 units, between 30 and 40 units, or greater than 40 units? Why? (You might find it helpful to include a graph *with* your explanation.)

4. At what point do firms maximize profits?

5. Define the shutdown point of a perfectly competitive firm.

6. Define the break-even point of a perfectly competitive firm.

7. Define the supply curve of a perfectly competitive firm.

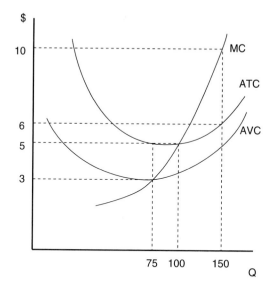

Use the graph on page 169 to answer questions 8 through 10.

8. Find the shutdown point. What is the quantity produced, average total cost, average variable cost, total cost, total variable cost, and profit (loss) at this point?

9. Find the break-even point. What is the quantity produced, average total cost, average variable cost, total cost, total variable cost, and profit (loss) at this point?

10. If MR = 10, then what is the quantity produced, average total cost, average variable cost, total cost, total variable cost, and profit (loss) at this point?

11. Define in words and write the formula for TFC, TC, TVC, MC, AVC, ATC, and AFC. There may be more than one formula for each one.

12. Fill in the missing cells. Assume the firm operates in a perfectly competitive environment in both the input and output markets. Calculate the profit (loss) when the firm receives $0.40 for the product.

L	Q	P(L)	TFC	TVC	TC	MC	ATC	AVC	AFC
2	40	5	110						
	65					.4			
	80							.375	
	90				150				

13. List four conditions for perfect competition.

   a.
   b.
   c.
   d.

14. The following information pertains to a production schedule for sorghum from a west Texas farm.

Land (Acres)	Fertilizer (Pounds)	Sorghum Yield (Tons)	MPP	APP	Stage of Production
4 4	40	68 75	—	 1.25	—

   a. Which input is the variable input?
   b. Which input is the fixed input?
   c. Fill in the blanks in the table.

15. Complete the following table:

INPUT	OUTPUT	TFC	TVC	TC	MC	AFC	AVC	ATC
2	20 40	 —	100	125	— 10			

8

ECONOMICS OF INPUT SUBSTITUTION

The essential requisites of production are three—labor, capital, and natural agents; the term capital *including all external and physical requisites which are products of* labor, *the* term natural agents *all those which are not.*

John Stuart Mill
(1806–1873)

TOPICS OF DISCUSSION

The example of labor use by TOP-AG in Chapter 7 focused on varying use of only one input. This allowed us to introduce a number of important production concepts, their relationship to the cost of production, and the profit-maximizing level of output and input use in the short run. Let us now expand this discussion to include two variable inputs and input substitution. This requires shifting the bar appearing after the first input (labor) in Equation 7.2 so that it appears after the second input (capital).

In virtually every setting, a business can alter the combination of capital and labor used in production. For example, weeds can be pulled or hoed (a labor-intensive practice) as they were at the turn of the century, or they can be killed with herbicides (a capital-intensive practice).[1] The choice between capital-intensive and labor-intensive operations becomes an issue in the long run and is influenced by such things as the relative cost of capital and labor, and changes in technology.

As illustrated in Chapter 2, farming operations have become much more capital intensive during the post–World War II period. This trend not only has implications for farm input manufacturers and farm laborers but also has environmental consequences, which will be discussed in Chapter 12.

The purpose of this chapter is to explain the economics of input substitution in the short and long run. In the short run, we determine the least-cost combination of labor and variable capital inputs, given the business's existing fixed resources and technology. Because all inputs are variable in the long run, the business will also have an interest in the optimal expansion path of labor and all capital over time.

CONCEPT AND MEASUREMENT OF ISOQUANTS

If we attempted to graph a total physical product curve for two inputs, it would take three dimensions: two dimensions for the two inputs and one dimension for output. However, three-dimensional figures are difficult to draw and understand; therefore, in this chapter two-dimensional figures will be used. This can be done by focusing on the combination of two inputs that, when used together, result in a specific level of output.

A curve that reflects the combinations of two inputs that result in a particular level of output is called an **isoquant** curve. The term *iso* here has the same meaning (i.e., equal) as it did in Chapter 3 when we were discussing iso-utility, indifference curves for two goods faced by consumers. An isoquant consists of a locus of points that correspond to an equal or identical level of output. Along any isoquant, an infinite number of combinations of labor and capital that result in the same level of output are depicted. As the quantity of labor (capital) increases, less capital (labor) is necessary to produce a given level of output.

---

[1]Remember the term *capital* can include both variable inputs such as fuel, fertilizer, and rented land or machinery, and fixed inputs such as owned machinery, buildings, and land.

To illustrate this point, think of quantities of capital as being divisible units of fuel and machinery (e.g., hours of tractor use). When the tractor and its complementary equipment and fuel are increased in quantity, fewer hours of labor are required to produce a given level of wheat production, for example. Similarly, with less capital available, more hours of labor are required to produce the same amount of wheat. We can conclude from this discussion that capital and labor are **technical substitutes.**

Rate of Technical Substitution

To determine the rate of substitution between two inputs, which represents (the negative of) the slope of an isoquant, we must measure the **marginal rate of technical substitution.** This concept is illustrated in Figure 8.1. As we move from range *A* to range *B* on the isoquant corresponding to ten units of output, we see that less capital and more labor are required.

Consider the three separate one-unit changes in labor illustrated: ranges *A, B,* and *C* each represent different reductions in the use of capital for three separate one-hour increases in labor use on the isoquant associated with 10 units of output. Figure 8.1 implies that the marginal rate of technical substitution of capital for labor falls from

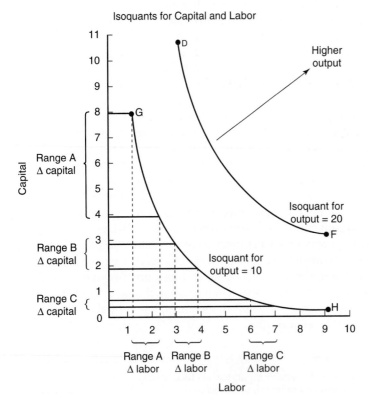

FIGURE 8.1 The slope of an isoquant for a particular level of output typically changes over the full range of the curve.

approximately 4 over range A to 1 over range B, and to 0.25 over range C. The rate of substituting capital for labor can be expressed mathematically as:

$$\frac{\Delta \text{ capital}}{\Delta \text{ labor}} = \frac{\text{MPP}_{\text{labor}}}{\text{MPP}_{\text{capital}}} \tag{8.1}$$

in which $\text{MPP}_{\text{capital}}$ and $\text{MPP}_{\text{labor}}$ represent the marginal physical products for capital and labor, and Δ represents the change in a variable.

The expression in Equation 8.1 indicates that changes in labor must be compensated by changes in capital, if the level of output is to remain unchanged.[2] For example, if output is to remain unchanged and the marginal rate of technical substitution of capital for labor is equal to three, capital use must be reduced by three hours if labor is increased by one hour.[3]

These observations illustrate that when labor is substituted for capital along an isoquant (output remaining unchanged), the marginal rate of technical substitution of capital for labor falls. A declining marginal rate of technical substitution is a consequence of the law of diminishing returns (discussed in Chapter 7). When labor increases, its marginal physical product falls. Reductions in capital imply an increase in its marginal physical product.

How do the isoquants in Figure 8.1 relate to the stages of production discussed in Chapter 7? Focusing on the isoquant for 10 units of output, the marginal physical product of capital is negative above point G and to the right of point H. You will recall that the marginal physical product was negative in stage III. Because stage III is not of economic interest, the economic region of production is bounded by points G and H for the isoquant corresponding to an output of 10 units, and by points D and F for the isoquant associated with an output of 20 units. Thus, only certain regions of input-output relationships are of interest to businesses seeking to maximize their profit.

Increases in output are reflected in Figure 8.1 by isoquants that lie farther away from the origin. In this figure, the isoquant for an output of 20 units lies farther from the origin than the isoquant associated with an output of 10 units.

Finally, isoquants at the extreme can be either perfect substitutes or perfect complements. Each case is illustrated in Figure 8.2.

A set of isoquants for perfect substitutes is a straight line, which implies a constant marginal rate of technical substitution. This differs from the imperfect substitution implied by the isoquants in Figure 8.1, which have a decreasing marginal rate of technical substitution as one moves down the isoquants. A set of isoquants for perfect complements forms 90-degree angles, indicating that both capital and labor are required to produce a specific level of output. That is, it takes a certain proportion of labor and capital to produce a product.

---

[2]If output is to remain unchanged (i.e., remain on the same isoquant), the loss in output from the decrease in labor must equal the gain in output from the increase in capital $-\Delta$ labor $\times \text{MPP}_{\text{labor}}$ $= + \Delta$ capital $\times \text{MPP}_{\text{capital}}$ Equation 8.1 simply represents a rearrangement of this statement.

[3]Because the marginal rate of technical substitution is negative in all rational areas of production (i.e., stage II), most economists do not bother to include the minus sign.

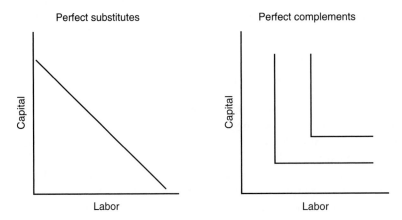

FIGURE 8.2   A graphical illustration of extreme perfect substitutes and complements.

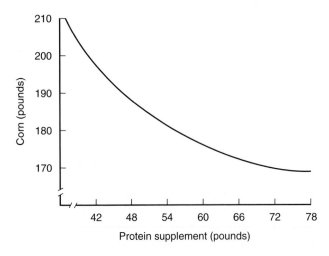

FIGURE 8.3   Relationship between the weight of swine, corn consumption, and protein supplement consumption.

A Case Example

Swine production involves making choices between corn and protein supplement in feed rations. Output in this case is the amount of weight gain for swine, and the inputs are the corn and protein supplement. Consider the isoquant corresponding to an initial weight of 60 pounds and a closing weight of 200 pounds (i.e., a weight gain of 140 pounds) as illustrated in Figure 8.3.

The marginal rate of technical substitution of corn for the protein supplement is positive, indicating that the level of output would be unchanged if corn were decreased and protein supplement increased by an appropriate amount. The ending weight of the swine (200 pounds) can be achieved with 48 pounds of protein supplement and 193 pounds of corn. This same weight gain can also be achieved using 75 pounds of protein supplement and 170 pounds of corn.

We can calculate the marginal rate of technical substitution for different combinations of corn and protein. The substitution rate is about 3.4 pounds when 48 pounds of protein supplement and 193 pounds of corn are used. This substitution rate would fall to 1.7 if the protein supplement were increased to 75 pounds and corn were decreased to 170 pounds. This illustrates the principle of diminishing marginal rate of technical substitution of corn for the protein supplement.[4] If a particular level of output is desired (e.g., 10 units of output in Figure 8.1), it becomes an economic issue as to which of the many combinations of labor and capital to use.

THE ISO-COST LINE

Assume that a business uses two inputs (labor and capital) to produce a particular product. The total cost of production in this case would be equal to the wage rate times the hours of labor used plus the cost of capital times the amount of capital used. The concept of wage rates paid to labor is familiar, but the cost of capital will require further explanation.

We have learned that capital includes both variable and fixed inputs. We cited fuel as an example of a variable input and land as an example of a fixed input. The cost of capital therefore equals the price of fuel and other variable inputs purchased times the amount purchased and the **rental rate of capital** for using fixed inputs such as tractors and other machinery, buildings, and land. In the short run, the business can rent an additional tractor or land. The annual rental payment for leased fixed inputs is a variable cost of production. Owned capital has its costs, too, as we initially mentioned in Chapter 6 when we discussed the concept of implicit costs. The owner of a building, for example, has the option of leasing or selling the building to someone else and using those monies in their next best alternative. The revenue forgone from not selling or leasing the building is a cost. Economists call this an implicit or opportunity cost. Thus, our cost of capital is a composite variable that reflects in the short run both the cost of variable inputs, such as labor, and the cost of renting capital. The prices for these two inputs, or the wage rate for labor and the rental rate for capital, are treated as fixed in the short run; they will not vary with the level of input use by a single firm.

Suppose Frank Farmer has $1,000 available daily to finance a business's production costs. The wage rate for labor is $10 per hour, and the rental rate for capital is $100 per day. The business's daily budget constraint therefore is:

$$(\$10 \times \text{use of labor}) + (\$100 \times \text{use of capital}) = \$1,000 \qquad (8.2)$$

---

[4]Although the marginal rate of technical substitution is a useful measure of substitution when one moves along an isoquant, it is similar to the marginal physical product because it is not free of the units of measurement problem that hinders comparisons of unlike physical quantities. A measure that is unit free is the elasticity of substitution because it is expressed in percentage terms. For two inputs on a particular isoquant and at a particular input ratio (e.g., capital to labor), this elasticity is given by the percentage change in the input ratio (i.e., the percentage change in labor divided by the percentage change in capital) divided by the percentage change in the marginal rate of substitution of capital for labor. Of special interest is the elasticity of substitution between capital and labor in agriculture. For the United States, estimates indicate considerable substitution possibilities, with an elasticity of substitution near one in some studies.

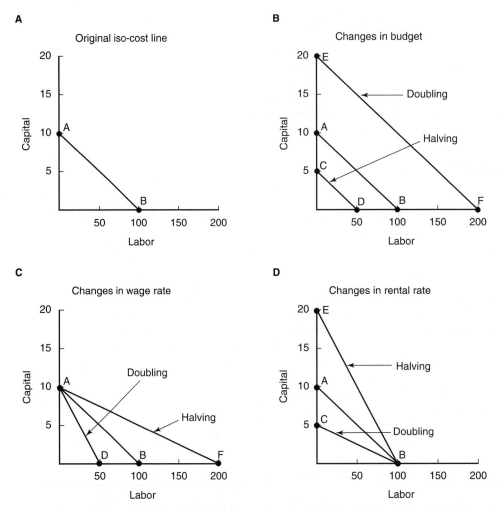

FIGURE 8.4   The iso-cost line plays a key role in determining the least-cost combination of input use. A, The slope of curve *AB* is given by the ratio of the wage rate for labor to the rental rate for capital. B, A doubling of the production budget changes both intercepts but not the slope of the iso-cost line. C, A doubling of the wage rate (holding the rental rate constant) would make the iso-cost line steeper as shown by line *AD*. D, A doubling of the rental rate (holding the wage rate constant) would make the iso-cost line flatter as shown by line *CB*. Declines in these input prices would have the opposite effect.

Frank's choice of how much capital and labor to employ must be no more than $1,000. The combination of labor and capital Frank can afford for a given level of total cost is illustrated by line *AB* in Figure 8.4. This relationship is referred to as an **iso-cost line.**

The slope of the iso-cost line is equal to the negative ratio of the wage rate to the rental rate of capital,[5] or:

$$\frac{\text{slope of}}{\text{iso-cost line}} = -\frac{\text{wage rate}}{\text{rental rate}} \qquad (8.3)$$

If these two input prices change by a constant proportion, the total cost will change, but the slope of the iso-cost line will remain constant.

To illustrate the nature of the iso-cost line, suppose the budget the firm allocated to these two inputs was doubled. Total costs may double, but the iso-cost line *EF* would still have the same slope as line *AB* (Figure 8.4, *B*). Only changes in the relative price of inputs (or input price ratio) will alter the slope of the iso-cost line.

For a given total cost, a rise (fall) in the price of capital relative to that of labor will cause the iso-cost line to become flatter (steeper) (Figure 8.4, *D*). If labor's wage rises (falls), the iso-cost line would become steeper (flatter) (Figure 8.4, *C*).

Suppose that the wage rate was $20 an hour instead of $10. The new iso-cost line *AD* would be steeper than line *AB* (Figure 8.4, *C*). The new iso-cost line would still intersect the capital axis at point *A*, because a maximum of 10 units of capital can be purchased, if the producer's total budget is limited to $1,000. If the capital price rose to $200 per unit, the iso-cost line *CB* would be flatter than the original iso-cost line *AB* (Figure 8.4, *D*).

LEAST-COST USE OF INPUTS FOR A GIVEN OUTPUT

There are essentially two input decisions a business faces in the short run that pertain to input use. One concerns the least-cost combination of inputs to produce a given level of output. Here, the level of production is not constrained by the business's budget. The other is the least-cost combination of inputs and output constrained by a given budget. This section focuses on the first of these two concerns.

Short-Run Least-Cost Input Use

The first of these two perspectives on the least-cost use of inputs requires that we find the lowest possible cost of producing a given level of output with a business's existing plant and equipment. Technology and input prices are assumed to be known and constant. We know from Figure 8.1 that the alternative combinations of capital and labor produce a given level of output that forms an isoquant, and that the relative prices of inputs help shape the iso-cost line in Figure 8.4.

[5]Equation 8.3 can be rearranged algebraically to give the iso-cost line and its slope as follows:

$$\text{hours of capital} = \frac{\$1,000}{\text{rental rate}} - \frac{\text{wage rate}}{\text{rental rate}} \times \text{hours of labor}$$

We need to find the least-cost combination of inputs that will allow the business to produce a given level of output in the current period. Any additional capital is rented (variable cost) through a short-term leasing arrangement rather than owned (fixed cost), or represents nonlabor variable inputs (e.g., fuel and chemicals). Graphically, the least-cost combination of inputs is found by shifting the iso-cost line in a parallel fashion until it is tangent to (i.e., just touches) the desired isoquant. This point of tangency represents the least-cost capital/labor combination of producing a given level of output and the total cost of production.

Figure 8.5 can be used to determine the least-cost combination of labor and capital to produce 100 units of output using the business's current productive capacity. Assume that iso-cost line *AB* reflects the existing input prices for labor and capital and current total costs of production. The least-cost combination of labor and capital to produce 100 units of output is found graphically by shifting line *AB* out in a parallel fashion to the point where it is just tangent to the desired isoquant.

Figure 8.5 shows that line *A*B** is tangent to the isoquant associated with 100 units of output at point *G*. The new total cost at point *G* in Figure 8.5 can be determined by multiplying the quantity of labor (L_1) times the wage rate and adding that to the product of the quantity of capital (C_1) times the rental rate for capital.

A fundamental interpretation to the conditions underlying the least-cost combination of input use is illustrated in Figure 8.5. The slope of the isoquant is equal to the slope of the iso-cost line at point *G*. At this point, the marginal rate of technical substitution of capital (fertilizer, fuel, feed, rental payments, etc.) for labor, or the negative of the slope of the isoquant, is equal to the input price ratio, or the negative of the slope of the iso-cost line. Thus, the least-cost combination of inputs requires that the market

Least-Cost Input Choice for Given Output

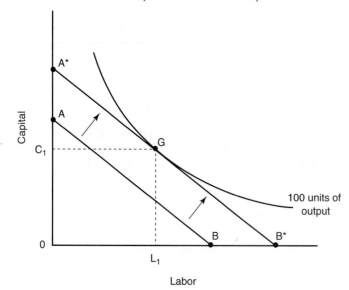

FIGURE 8.5 The least-cost choice of input use is given by the point where the iso-cost curve is tangent to the isoquant for the desired level of output. If the iso-cost line is line *AB*, the least-cost combination would occur at point *G* (where 100 units of output are produced).

rate of exchange of capital for labor (i.e., the ratio of input prices) equal their rate of exchange in production (i.e., their marginal rate of technical substitution).

We can express the foregoing conditions for the least-cost combination of labor and capital in mathematical terms as:

$$\frac{\text{MPP}_{\text{labor}}}{\text{MPP}_{\text{capital}}} = \frac{\text{wage rate}}{\text{rental rate}} \qquad (8.4)$$

We can rearrange Equation 8.4 as:

$$\frac{\text{MPP}_{\text{labor}}}{\text{wage rate}} = \frac{\text{MPP}_{\text{capital}}}{\text{rental rate}} \qquad (8.5)$$

Equation 8.5 suggests that the marginal physical product per dollar spent on labor must equal the marginal physical product per dollar spent on capital. This is analogous to the condition for consumer equilibrium described in Equation 4.2, and it represents a recurring theme in economics. In the present context, a firm should allocate its expenditures on inputs so the marginal benefits per dollar are spent on competing equally.[6]

The discussion presented above can be summarized as follows: input use depends on input prices, desired output, and technology. Such cost-minimizing input use is often referred to as conditional demand because it is conditioned by the desired level of output.

Effects of Input Price Changes

Now let us see what would happen to these input demands if we allow the price of an input to change. Because total production costs equal the sum of expenditures on each input, total cost will also change. A fundamental principle of economic behavior is that a firm will use less of an input as its per unit cost rises (Figure 8.6).

Figure 8.6 shows that as the relative price of labor (wage rate divided by price of capital) falls, the iso-cost line becomes flatter, as illustrated by the shift of iso-cost line *AB* to line *AB**. We know from the previous discussion that our next step must be to find the point of tangency with the desired isoquant. Moving line *AB** inward in a parallel fashion to the point where it (let's use a new label; line *DE*) is tangent to the isoquant, we see that the least-cost combination of inputs for 100 units of output shifts from point *G* to point *H.*

---

[6]Another way to think of this equilibrium is that marginal benefit equals marginal cost. Suppose that the marginal value product of labor usage (marginal physical product times the price of output) is $5 and the corresponding marginal benefit is $7 for capital. The opportunity cost of expending a dollar on increased labor usage is the $7 gain if this expenditure were instead used to purchase another unit of capital services. Therefore, the marginal benefit ($5) is less than marginal cost ($7), and labor usage should be reduced. If output is to remain constant when labor is reduced, capital must be expanded until marginal benefit equals marginal cost.

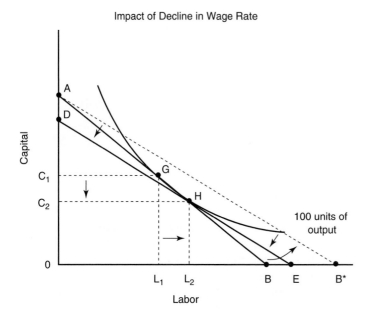

Impact of Decline in Wage Rate

FIGURE 8.6   The shift in the iso-cost line from *AB* to *AB** caused by a lower wage rate suggests that labor should be increased to L_2 and capital use should be reduced to C_2. Line *DE* represents a parallel shift of line *AB** to a point of tangency with the desired isoquant.

Therefore, when the price of labor falls (rises) relative to the price of capital, labor is substituted for capital, causing the capital/labor ratio to fall (rise). Because of diminishing marginal products, equilibrium is attained by reducing capital (from C_1 to C_2) use and using more labor (L_2 instead of L_1).

LEAST-COST INPUT USE FOR A GIVEN BUDGET

The previous section illustrated how to determine the least-cost combination of inputs in the current period to produce a given level of output. A somewhat different twist to this analysis is to determine the least-cost combination of inputs and output in the current period for a given production budget. We will continue to use the concept of the iso-cost line and the isoquant for specific levels of output.

Assume that a firm has a specific amount of money to spend on current production activities and wants to know the least-cost combination of capital (currently owned plus rented) and labor to employ. Figure 8.7 contains four isoquants that present all the information we need to answer this firm's question. The isoquant for 50 units of output shows all the combinations of labor and capital that are needed to produce this level of output. Similar isoquants are shown for 75, 100, and 125 units of output.

Line *MN* in Figure 8.7 represents a total cost of production that completely exhausts the amount of money the firm desires to spend. The point of tangency between this iso-cost line and the highest possible isoquant will indicate the least-cost combination of inputs associated with the firm's current budget constraint. This

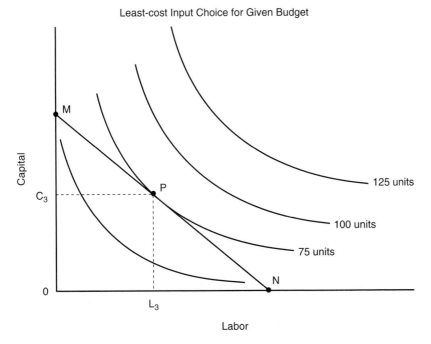

Least-cost Input Choice for Given Budget

FIGURE 8.7 The least-cost choice of input use for a given budget is found by plotting the iso-cost line associated with this budget and observing the point of tangency with the highest possible isoquant.

occurs at point P in Figure 8.7, which suggests that the firm would utilize C_3 units of capital and L_3 units of labor. The economic conditions set forth in Equation 8.4 are satisfied by this combination of inputs (remember, the left-hand side of this equation represents the slope of the isoquant, and the right-hand side represents the slope of the iso-cost line).

Point P also suggests that the firm would produce 75 units of output. The firm simply could not afford to operate on a higher isoquant in the current period. The only way the firm could move out to the isoquant associated with 100 units of output is if it were able to attract additional funds or if both input prices declined to the point where the iso-cost line became tangent with this higher isoquant.

LONG-RUN EXPANSION OF INPUT USE

In the previous section, we learned that some costs are variable in the short run, and other costs are fixed. In the long run, however, a business has the time to expand the size of its operations, and all costs become variable. The purpose of this section is to discuss the long-run average cost curve and the factors that influence its shape.

Long-Run Average Costs

Figure 8.8 depicts three short-run average cost (SAC) curves. The presence of fixed inputs in the short run ensures that these short-run average cost curves are U-shaped. Each short-run average cost curve reflects the full average cost of the business for three separate sizes. Size A is the smallest, with costs represented by SAC_A. This curve might correspond to TOP-AG, using a specific amount of capital to produce 12 units of output in the example discussed previously.

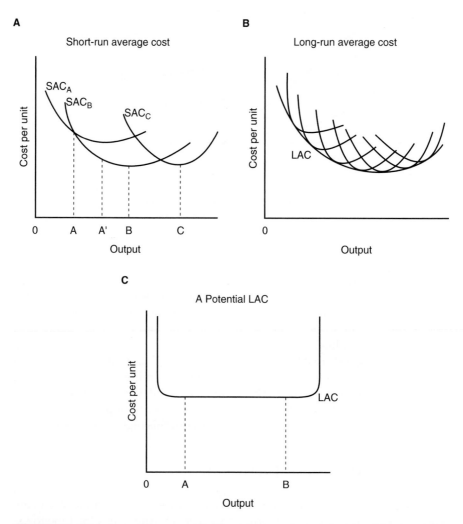

FIGURE 8.8   The long-run average cost (*LAC*) curve plays a key role in determining the minimum costs of operation in the long run. Often referred to as the *planning curve*, the *LAC* curve represents an envelope of a series of short-run average cost curves (*A* and *B*).

Size *B* is larger and can operate at much lower costs. Curve SAC_B is much lower, except at its extreme left end. This curve might reflect the capital needed by TOP-AG to minimize its cost of producing more than 12 units of output. Size *C* is still larger, but the curve SAC_C represents a higher cost structure than size *B*.

A business desiring to minimize its production costs will wish to operate at size *A* in Figure 8.8 if an output equal to *OA* or less is desired. The average cost of production will be substantially lower than size *B* at outputs less than *OA*. If output *OB* or *OC* is desired, however, the business will prefer size *B* or size *C*, respectively.

If the business desires to produce an output somewhat larger than *OA*, size *B* would be better than size *A* because its costs are lower. Output *OA'* has special significance for a business of size *A*, because it represents the minimum cost of operation. Obviously, it would be better, when producing more than *OA*, to operate a business of size *B* at less than capacity than it would be to continue operating a business of size *A*, because the costs of production would be lower.

The Long-Run Planning Curve

When deciding how big the business should be in the long run, management must consider a relevant range of minimum cost. Management may be aware of this range, either from its own experience or from economic feasibility studies conducted for other businesses of a similar nature. The short-run average cost curves associated with different sizes over this range enable the business to determine its long-run average cost (*LAC*) curve.

Often referred to as the long-run planning curve, the long-run average cost curve illustrates to the business how varying its size will affect the business's economic efficiency. It also indicates the minimum per unit cost at which any output can be produced after adjusting the business's size.

What causes the long-run average cost curve to decline, become relatively flat, and then increase? The answer lies in what economists call returns to size. If an increase in output is exactly proportional to an increase in inputs, constant returns to size are said to exist. This means that a doubling or a tripling of inputs used by the firm will cause a doubling or tripling of its output.[7] If the increase in output is more (less) than proportional to the increase in input use, we say the returns to size are increasing (decreasing). Decreasing (increasing) returns to size will exist if the firm's long-run average costs are increasing (decreasing) when the firm is expanded.

---

[7]A word of caution: the phrase "the economies of mass production" carries several meanings, some of which are irrelevant here and therefore are potential sources of confusion. For example, the greater efficiency frequently observed for larger production units (in contrast to smaller ones) is often caused because larger units are newer and use better production techniques than the older and smaller units. However important this may be, improvements in technology are not part of the concept of returns to size, which assumes a given technology.

Increasing Returns to Size.   Some of the physical causes of increasing returns to size are purely dimensional in nature. If the diameter of a pipe is doubled, the flow through it is more than doubled. The carrying capacity of a truck also increases faster than its weight. After some point, such increases in dimensional efficiency stop. As the size of the pipe is increased, it has to be made out of thicker and stronger materials. The size of the truck will also be limited by the width of streets, the height of overpasses, and the capacity of bridges.

A closely related technical factor that helps to explain the existence of increasing returns to size is the indivisibility of inputs. In general, indivisibility means that equipment is available only in minimum sizes or in a specific range of sizes. As the size of the firm's operations increases, the firm's management can switch from using the minimum-sized piece of equipment to larger, more efficient equipment. Thus, the larger the size of the operation, the more the firm will be able to take advantage of large-size equipment that cannot be used profitably in smaller-size operations.

Another technical factor contributing to increasing returns to size comes from the potential benefits from specialization of effort. For example, as the firm hires more labor, it can subdivide tasks and become more efficient.[8] When the firm expands the size of its operations, it can buy specialized pieces of equipment and assign special jobs to standardized types of machinery.

A frequently noted pecuniary factor that helps us explain the existence of increasing returns to size is volume discounts on large purchases of production inputs. Lower input prices paid by larger farming operations could be a major reason why average costs decline as farm size is increased.

Constant Returns to Size.   Increasing returns to size cannot go on indefinitely. Eventually, the firm will enter the phase of constant returns to size, where a doubling of all inputs doubles output. The phase of constant returns to size can be brief before decreasing returns to size sets in. Empirical evidence suggests that the phase of constant or nearly constant returns to size can be fairly long and cover a large range of output levels.

Decreasing Returns to Size.   Can a business keep on doubling its inputs indefinitely and expect its output to double? Most likely, the answer is no. Eventually, there must be a decreasing return to size. The farmer may be the reason for decreasing returns to size. While all other inputs can be increased, his ability to manage larger operations may not. The managerial skills needed to coordinate efforts and resources usually do not increase proportionately with the size of operations.

In Figure 8.8, *B*, the long-run average cost curve is the *envelope* of the set of short-run average cost curves; that is, the long-run average cost curve is tangent to the short-run average cost curve when it is declining. When the long-run average cost curve is rising, it touches the short-run average cost curves to the right of their minimum

---

[8]The benefits gained from specialization are well known. Adam Smith, in his book *The Wealth of Nations*, published in 1776, addressed the gains from the division of labor.

points. The minimum point on the long-run average cost curve is the only point that touches the minimum point on the short-run average cost curve. The declining portion of the long-run average cost curve suggests the existence of increasing returns to size. Beyond the minimum point on this curve, decreasing returns to size exist.

Economists are concerned with the shape of the long-run average and marginal cost curves. The minimum point on the long-run average cost curve represents the most efficient amount in the long run in the sense that the business's average costs of operation are minimized.

The long-run cost curve depicted in Figure 8.8, *B* reflects the conventional shape illustrated in most textbooks. Although the long-run average cost curve no doubt decreases over some range of output before eventually turning up, its shape is not likely to be perfectly U-shaped.

Studies by agricultural economists suggest that there may be some range of output where the long-run average cost curve is relatively flat. In California, Hall and LaVeen (1978) found that the long-run average cost curve becomes relatively flat after initially declining rapidly. They reported that the costs of producing highly mechanized crops generally continued to decline slowly over the entire range of surveyed farm sizes. For vegetables and fruit crops, however, Hall and LaVeen found little or no decline after the initial benefits from expansion were achieved.

Figure 8.8, *C* illustrates the general nature of these findings. Between outputs *OA* and *OB*, the long-run average cost curve is relatively flat. Over this range, all business sizes will have approximately the same costs. Thus, the long-run average cost curve in practice is more L-shaped than U-shaped.

ECONOMICS OF A BUSINESS EXPANSION

In the long run, businesses have time to expand (or contract) the size of their operations. Suppose that the short-run marginal cost and average cost curves for an existing business are represented in Figure 8.9 by SMC_1 and SAC_1, respectively. If the market price for the product is equal to P and the business produced at the point where $P = SMC_1$, the business would sustain a small economic loss on each unit of output produced. At this point, the business would have two options: (1) it could go out of business, or (2) it could expand its existing size, if it could convince its banker of the benefits from this expansion.

If the business expanded to the size represented by SAC_2 and SMC_2, it could produce quantity Q_2 and it would earn an economic profit per unit equal to P minus short-run average cost at quantity Q_2. A profit-maximizing business may want to expand to the size represented by SAC_4 and SMC_4. By producing at Q_4, the business would be operating at the point where P is equal to SMC_4.

While this long-run adjustment for an existing business is taking place, the number of businesses may also increase because of attractive economic profits. Some of these entrants will be newly created businesses. Others may be firms that have shifted out of less profitable enterprises. As these new and modified operations begin to produce output, the market supply of the product will increase. This, in turn, will cause

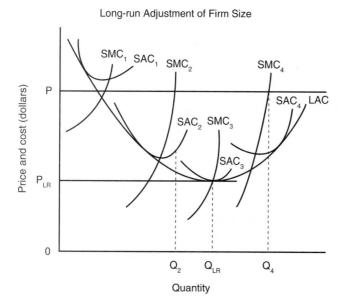

Long-run Adjustment of Firm Size

FIGURE 8.9 A profit-maximizing business may desire to expand the size of its operations to the level corresponding to the SAC_4 and SMC_4 cost curves, given the level of product price P. When others respond to the existence of economic profits, total market supply will expand and the market price will fall. In a free-market setting, businesses will cut back their output and size as best they can. Some may cease producing altogether. The market will be in long-run equilibrium at the point where $P = MC = AC$, which would result in the business producing Q_{LR} units of output at a market price of P_{LR}.

the price of the product to fall.[9] When each business responds to the new lower market price, the output of each will become smaller than before. Those businesses that were just preparing to expand their size in response to the earlier price will be able to adjust their size rapidly. Other businesses that have just completed expansion of their firm will obviously respond more slowly. Those businesses that cannot contract rapidly will lose more money than those that can. This process conceivably will continue until economic profits have been reduced to zero and the incentive for additional firms to enter the sector has been eliminated. Those existing businesses who are losing money will eventually cease producing this product.

The market will be in long-run equilibrium at price P_{LR}. At this price, the firm will be operating at the point at which product price is just equal to the minimum point on the long-run average cost curve (LAC) in Figure 8.9. This figure shows that the optimal size of the business in the long run would see it producing an output equal to Q_{LR}. Businesses expanding this output run the risk of eventually scaling back their operations.

Chapter 9 will further explore the desired expansion of both labor and capital in the long run as the price of inputs fluctuates over time.

Capital Variable in the Long Run

Thus far, the firm has been limited to expanding its use of variable inputs (including rented capital like farmland). The ownership of capital, held constant in the

---

[9]Once we develop the market supply curve for all businesses producing a particular product in Chapter 10, this sequence of events will become more clear.

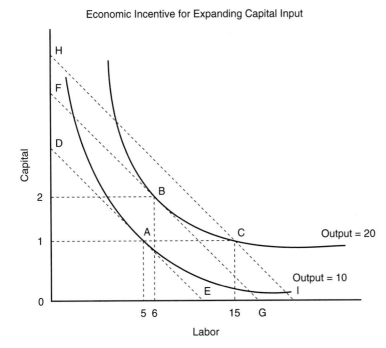

Economic Incentive for Expanding Capital Input

FIGURE 8.10   In the long run, the size of the firm's operations can be expanded if the economic incentive to do so is there. This requires increasing the capital input to its least-cost level (i.e., the point at which the marginal rate of technical substitution of capital for labor is equal to their price ratio). You will recall that this is the same stipulation made for the short-run case. Unlike the short run, however, the firm can now increase its use of capital beyond 1 unit. Producing 20 units of output instead of 10 units can be most efficiently done by operating at point *B*, not *C*.

short run, is allowed to vary in the long run. This input can be thought of as plant size (e.g., the number of manufacturing lines, the capacity of a feedlot, or the number of grain silos owned by an elevator). In the long run, the capital structure of the firm can be adjusted to its least-cost level that occurs when the marginal rate of technical substitution of capital for labor is equal to the input price ratio.

Figure 8.10 shows that at 1 unit of capital, labor is at its least-cost level of use only if output is equal to 10 units (point *A*). At this output level, the least-cost combination of inputs would be 5 units of labor and 1 unit of capital (see point *A*).

If an output of 20 units is desired in the long run, the firm has two options: (1) expand labor use to 15 units and operate at point *C* or (2) expand capital to 2 units and operate at point *B* at which you would employ 6 units of labor. The least-cost combination of inputs to produce 20 units would occur at point *B*. If capital is held constant at 1 unit, the only way this firm can produce 20 units of output would be to employ 15 units of labor (point *C*). The total cost of producing 20 units would be given by the iso-cost line *HI*. Because iso-cost line *FG* lies to the left of iso-cost line *HI*, the total cost of producing 20 units of output at point *B* would be less than producing 20 units of output at point *C*. Suppose the wage rate for labor was $10 per hour and the rental rate for capital was $50 per hour. The total production costs associated with the three iso-cost lines in Figure 8.10 appear in Table 8.1.

Thus, it would cost $200 an hour to operate at point *C*, or to produce 20 units of output with 1 unit of capital and 15 units of labor. It would only cost $160 dollars an hour to produce the same quantity of output with 2 units of capital and 6 units of

TABLE 8.1   Total Production Costs

Iso-Cost Line	Calculation of Total Cost	Total Cost
DE	(1 × $50) + (5 × $10)	$100
FG	(2 × $50) + (6 × $10)	$160
HI	(1 × $50) + (15 × $10)	$200

labor. Therefore, there is economic incentive for the business desiring to produce 20 units of output to expand its capital to the level indicated by point *B*. Using 2 units of capital and 6 units of labor will minimize the cost of producing 20 units of output.

EXPANSION PATH THROUGH ISOQUANTS

In addition to the long-run cost curve, we can examine the expansion of the business by looking at its expansion path in either an input-input or a product-product setting. Beginning with the input-input case, we know that each point on the long-run average cost curve represents the least-cost combination of resources to produce a given level of output. This same phenomenon occurs along the expansion path in Figure 8.11, which is given by the points of tangency between a set of isoquants for two inputs associated with different output levels, and a set of iso-cost lines that have a slope given by the ratio of the prices for these inputs.

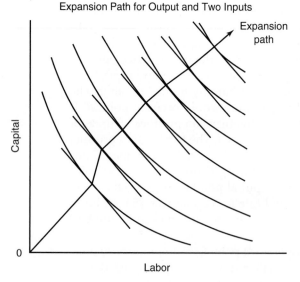

Expansion Path for Output and Two Inputs

Expansion path

Capital

0

Labor

FIGURE 8.11   The long-run average cost curve represents a path of points reflecting the least unit cost of producing specific rates of output. We can also look at the business's expansion path in terms of the least-cost combination of inputs. This figure illustrates that a line connecting a series of points given by the tangency of iso-cost lines and isoquants associated with different levels of output forms an expansion path.

Summary

The purpose of this chapter was to consider several input demand issues facing the business, including the size of the business in the long run and the factors that can influence this size. The major points made in this chapter may be summarized as follows:

1. The least-cost combination of inputs and level of output possible with a given budget is found graphically by looking for the point of tangency between a specific iso-cost line and the highest possible isoquant curve.
2. The cost of production associated with the least-cost combination to produce a given level of output is found graphically by looking for the point of tangency between a specific isoquant and an iso-cost line.
3. The least-cost combination of two inputs can be found numerically by searching for the equality between the ratio of the two input prices (slope of the iso-cost line) and marginal rate of technical substitution or ratio of the two input MPPs (slope of an isoquant).
4. Firms in the long run can expand the size of their operations by using more of all inputs, including forms of capital that were fixed in the short run.
5. The long-run average cost curve, often referred to as the planning curve, illustrates how varying the size of the firm will affect its efficiency.
6. The long-run equilibrium of the firm under conditions of perfect competition will occur at that output level where the product price is equal to both the firm's marginal and average total costs.
7. When businesses expand the size of their operations, they will incur returns to size.
8. If the increase in input is exactly proportional to the increase in input use, the returns to size are constant. If this increase was more (less) than proportional to the increase in input use, returns to size are increasing (decreasing). The business will normally pass through a phase of increasing returns to size or economies of size before encountering constant and then decreasing returns to size, or dis-economies of size.

Definition of Key Terms

Iso-cost line: much like the budget line for consumers, this line reflects the particular level of expenditure for two inputs. The slope of the iso-cost line is the ratio of the prices of the two inputs.

Isoquant: a curve that reflects the combinations of two inputs that will produce a specific level of output.

Marginal rate of technical substitution: the rate of substitution or trade-off between two inputs in the production of a specific product; also represents the slope of an isoquant curve.

Rental rate of capital: the cost of capital broadly defined; the price you would have to pay to rent all the inputs used to produce the business's product.

Technical substitutes: two inputs that can be substituted for one another in the production of a specific product.

ADVANCED TOPIC

The effects of input choice on the business's total costs can be seen in Figure 8.12. Assume a business in Lower Slobovia rents all of its capital and that the rental rate for capital is $100 and the cost of labor is $2.67. The least-cost method of producing 200 units of output would require renting 6 units of capital and hiring 150 hours of labor. Total production costs in this instance would be $1,000. These results are represented by point *A* of Figure 8.12. To increase output to 300 units, the least-cost combination of inputs calls for renting 8 units of capital and hiring 200 units of labor. Total costs would rise to $1,334. These results are represented by point *B* (Figure 8.12).

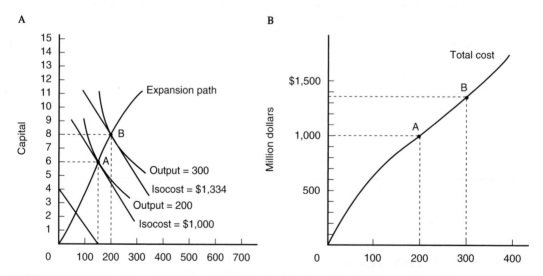

FIGURE 8.12   A, Points *A* and *B* represent the least-cost capital and labor use for producing 200 and 300 units of output, respectively. B, Points *A* and *B* are representations of these input combinations when total cost is graphed against total output.

Repeating this process for all other possible levels of output would result in a curve formed by the tangencies of iso-cost lines and isoquants (Figure 8.12, *A*). The corresponding total cost curve of Figure 8.12, *B* represents the relationship between the total cost curve and output under these conditions and looks much like the total cost curve illustrated in Figure 7.4, *A*. Because more resources are required to produce more output, total cost increases when output increases.

REFERENCE

Hall, FF, and LaVeen, EP, "Farm Size and Economic Efficiency: The Case of California," *American Journal of Agricultural Economics* 6(4): 589–600, 1978.

EXERCISES

1. A firm uses corn and protein supplement to mix a particular type of hog feed. Corn costs $.08 per pound and protein supplement costs $.12 per pound. Let's assume the firm has $3,000 on these two inputs. Plot the iso-cost line suggested by this information in the graph below. What is the value of this curve's slope?

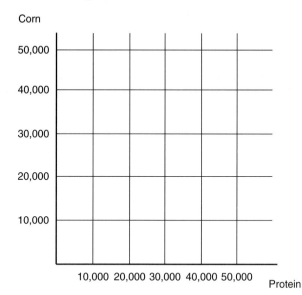

2. Suppose the wage rate for labor is $20 an hour and the rental rate for capital is $50 per hour. Based on this information, please answer the questions appearing below the graph.

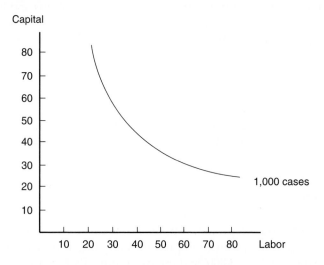

a. Draw the iso-cost line associated with an hourly budget of $1,000.

b. What is the least-cost level of capital and labor this business should utilize when packaging 1,000 cases of fruit juice? How did you arrive at this answer?

c. How much does it cost this business to package 1,000 cases of fruit juice? If the firm can sell the juice for $50 per case, what is its accounting profit?

3. In the boxes appearing above *each* of the following four graphs, please describe what caused the iso-cost line to change *in each box* and the nature of the change *on each line*.

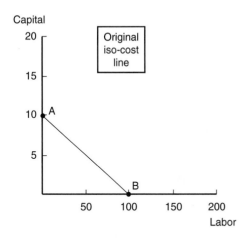

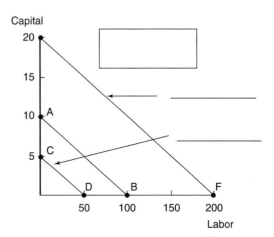

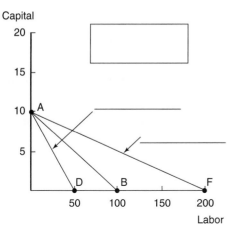

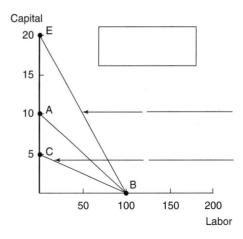

4. Given the following graph, briefly respond to the questions appearing directly
   below this graph.

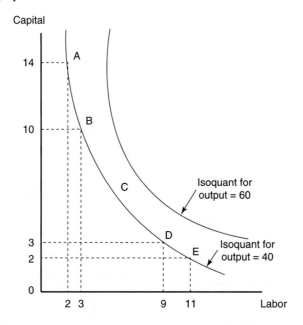

a. For an output level of 40 units, calculate the marginal rate of technical substitution between points *A* and *B*.

b. Also for an output level of 40 units, calculate the marginal rate of technical
   substitution between points *D* and *E*.

c. Your results illustrate the principle of _____ .

9

ECONOMICS OF PRODUCT SUBSTITUTION

Even children learn in growing up that 'both' is not an admissible answer to a choice of 'which one?'

Paul A. Samuelson

TOPICS OF DISCUSSION

CONCEPT AND MEASUREMENT OF THE PRODUCTION POSSIBILITIES FRONTIER

Many large agribusiness firms such as ConAgra Corporation added new product lines to their operations during the merger-mania era of the late 1980s and much of the 1990s. Crop farmers annually choose what crops to plant and how much of each crop to plant. The addition of new products, sometimes through acquisitions and mergers and the phasing out of others, allows a business to maximize its profits and **diversify** its operations.

In Chapter 8, we examined several issues associated with the combination of inputs used by a business. Part of our interest focused on the degree to which one input could be substituted for another in producing a given level of output. The focus of this chapter is on products instead of inputs. It is important to understand the substitution among the different products the business can produce. The purpose of this chapter is to examine the **production possibilities frontier (PPF)** a business faces, both in the short run, when some inputs are fixed, and in the long run, when all inputs are variable.

Production Possibilities Frontier

Table 7.1 introduced the concept of technical efficiency by indicating the minimal number of hours required to produce given levels of output. In the multiproduct case, we can think of technical efficiency in terms of the maximum outputs possible from given levels of inputs.

Suppose SunSpot Canning Company has the option of canning either all fruit, all vegetables, or some combination of these two products as shown in Table 9.1. As fruit (vegetable) canning is increased, vegetable (fruit) canning must be decreased because of the plant's fixed current canning capacity. Thus, a substitution among products occurs in the same sense that inputs were substituted for one another in the preceding chapter.

If SunSpot specialized in fruit canning, it could can 135,000 cases of canned fruit a week. If it specialized in vegetable canning, SunSpot could can 90,000 cases of canned vegetables a week.[1] Column 3 in Table 9.1 reflects the physical trade-off this canning plant faces for these two products, or the **marginal rate of product transformation.** It represents the slope of the production possibilities frontier.

Product Substitution

The marginal rate of product transformation represents the rate at which the canning of fruit must contract (expand) for a one-case increase (decrease) in vegetable canning. The marginal rate of transformation in absolute terms is given by:

---

[1]We will assume that the fruit pack or canning season and the vegetable canning season overlap and thus compete for use of SunSpot's existing resources.

TABLE 9.1   Production Possibilities for SunSpot Canning

	(1) Cases of Canned Fruit	(2) Cases of Canned Vegetables	(3) Marginal Rate of Product Transformation Δ(1) ÷ Δ(2)
A	135,000	0	
B	128,000	10,000	−0.7
C	119,000	20,000	−0.9
D	108,000	30,000	−1.1
E	95,000	40,000	−1.3
F	80,000	50,000	−1.5
G	63,000	60,000	−1.7
H	44,000	70,000	−1.9
I	23,000	80,000	−2.1
J	0	90,000	−2.3

$$\text{marginal rate of product transformation} = \frac{\Delta \text{ canned fruit}}{\Delta \text{ canned vegetables}} \qquad (9.1)$$

In Table 9.1, the marginal rate of product transformation of vegetables for fruit is initially very small (i.e., Δ canned fruit relative to Δ canned vegetables is quite small). In column 3, however, the marginal rate of product transformation becomes much higher (i.e., Δ canned fruit relative to Δ canned vegetables becomes quite large). This increasing marginal rate of product transformation is a widely observed and measured phenomenon and has the same general lawlike acceptance as the declining marginal rate of technical substitution discussed for two inputs in Chapter 8.[2]

The substitution relationship between two products can be illustrated further by plotting the combinations of fruit and vegetables shown in Table 9.1. Points *A* through *J* in Figure 9.1 represent production levels of fruit and vegetables for a canning plant with a given technically efficient use of capital and labor. Point *A*, for example, represents specialization in the canning of fruit. Point *J*, on the other hand, represents specialization in vegetable canning. Point *C* would result in the canning of some of both commodities with the *same* inputs. Point *K* can be ruled to be technically inefficient because a smaller amount of output is being produced with the same quantity of inputs, leaving only points *A* through *J* as the efficient production points.

---

[2]To avoid a units-of-measure problem when examining output-output relationships, one can use the elasticity of product transformation, which measures the marginal rate of product transformation for two products in percentage terms. This elasticity is given by the ratio of the percentage change in the output ratio (i.e., the percentage change in canned fruit divided by the percentage change in canned vegetables) divided by the percentage change in the marginal rate of product transformation.

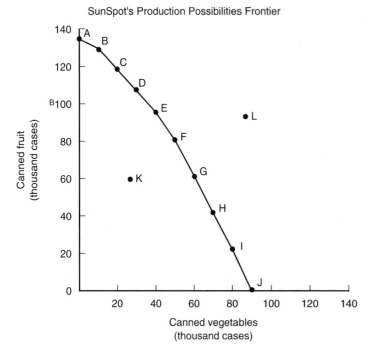

SunSpot's Production Possibilities Frontier

FIGURE 9.1 The downward-sloping production possibilities frontier illustrates the physical trade-offs this business faces in choosing between canning fruit or vegetables as documented in Table 9.1. The concave shape of this curve reflects the less-than-perfect substitutability of input use in switching from canning fruit to canning vegetables.

A curve drawn through these points is called a production possibilities frontier, which gives the product combinations that can be *efficiently* produced using the business's existing resources. Finally, point *L* is impossible to attain with SunSpot's existing resources because it lies outside the production possibilities frontier.

Figure 9.1 suggests that vegetable and fruit canning operations at SunSpot are close—but not perfect—substitutes in competing for the firm's scarce resources in production (i.e., the PPF is neither linear nor has a constant slope of − 1.0). Two enterprises within the firm competing for the firm's scarce resources can also be supplementary or complementary over a range of the PPF. The nature of the PPF in these situations is illustrated in Figure 9.2.

In the case of supplementary products, the production of product one can be increased without taking away from the production of product two. This cannot occur indefinitely, however, and the two products eventually compete for the firm's scarce resources, yielding a segment of the PPF that looks like the downward-sloping PPF curve in Figure 9.1. Two products can also be complements over a range of production possibilities, which results in a PPF curve like the one in Figure 9.2. Here, an increase in the production of product one causes the production of product two to increase also. This will occur only up to a certain point—if at all—after which the PPF again reflects two products competing for the firm's scarce resources and a segment of the PPF again looking like the downward-sloping PPF curve in Figure 9.1.

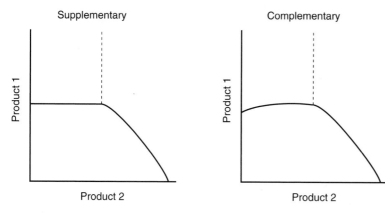

FIGURE 9.2   The nature of SunSpot's PPF.

Finally, increases in the level of one or more inputs used in production, or a technical improvement in the production process, will shift the production possibilities frontier curve outward. This occurs for the same reasons that input changes placed the firm on a higher isoquant in Figure 8.1. Conversely, rising input prices that lower input use will shift the PPF inward. The production possibilities frontier for a larger (smaller) quantity of inputs than that associated with the curve plotted in Figure 9.1 would lie to the right (left) of this curve.

CONCEPT AND MEASUREMENT OF THE ISO-REVENUE LINE

We need to account for the price received by the canning firm for these two products before we may determine what combination maximizes SunSpot's profits. The **iso-revenue line** represents the rate at which the market is willing to exchange one product for another. We may begin to define an iso-revenue line for SunSpot by defining its total revenue, which, for fruit and vegetables, is given by:

$$\text{total revenue} = (\text{price of canned fruit} \times \text{cases of canned fruit}) \\ + (\text{price of canned vegetables} \times \text{cases of canned vegetables}) \quad (9.2)$$

If no canned vegetables are produced by this canning plant, then the number of cases of canned fruit produced for a specific level of revenue is given by the level of revenue divided by the price of canned fruit. Similarly, if no canned fruit is produced, then the number of cases of canned vegetables produced for a specific level of revenue is given by the level of revenue divided by the price of canned vegetables.

The slope of the iso-revenue line is the ratio of the price of the two products, or the price of canned vegetables this business receives by selling a case of canned vegetables in the market, divided by the price of canned fruit. Stated mathematically, the slope is given by:

$$\text{slope of iso-revenue line} = -\frac{\text{price of vegetables}}{\text{price of fruit}} \tag{9.3}$$

For example, if the wholesale price of a case of canned fruit that SunSpot receives is $33.33, and the price of a case of canned vegetables it receives is $25.00, the slope of the iso-revenue line would be −0.75 (i.e., the negative of $25.00 divided by $33.33). One case of canned vegetables is worth three-fourths of a case of canned fruit.

Figure 9.3 illustrates the general nature of the iso-revenue line for these two products. You may wonder why the slope of the line plotted in Figure 9.3 is the negative of the ratio of the price of canned vegetables to the price of canned fruit, when the vertical axis is labeled "cases of fruit" and the horizontal axis is labeled "cases of vegetables."[3] This is entirely consistent with the discussion of the budget constraint in Chapter 3, in which we determined that the slope of the budget line in Equation 3.6 was the negative of the ratio of the price of hamburgers to the price of tacos. This was also the opposite of the labels assigned to the vertical and horizontal axes. (See Figure 3.3.)

The original iso-revenue line associated with a revenue of $1 million, the price of a case of canned fruit of $33.33, and the price of a case of canned vegetables of $25.00 are the basis for iso-revenue line *AB* plotted in Figure 9.3, *A*. The maximum number of cases of canned fruit associated with this level of revenue is 30,000 cases (i.e., $1 million ÷ $33.33), and the maximum number of cases of canned vegetables is 40,000 cases (i.e., $1 million ÷ $25.00). Thus, SunSpot would achieve a revenue of $1 million if it could process 30,000 cases of fruit, 40,000 cases of vegetables, or the specific combinations of these two products that appear on the iso-revenue line.

Figure 9.3, *B* shows that if consumer expenditures for SunSpot's products fell by half, the iso-revenue line would shift in a parallel fashion from line *AB* to line *CD*. Only 15,000 cases of fruit, 20,000 cases of vegetables, or specific combinations of the two products could be sold. A doubling of consumer expenditures for these products would shift the iso-revenue line from line *AB* to line *EF*. Under these conditions, SunSpot would sell 60,000 cases of canned fruit or 80,000 cases of canned vegetables, or specific combinations of these two products. Figure 9.3, *C* shows what would happen to the iso-revenue line if the price of fruit were either doubled (line *BD*) or cut in half (line *BC*). Figure 9.3, *D* shows what would happen if the price of vegetables doubled (line *AC*) or were cut in half (line *AD*). In Figures 9.3, *C* and *D*, the slope of the iso-revenue line became either flatter or steeper as the price of one of the commodities changed.

---

[3]Equation 9.3 can be rearranged algebraically to give the iso-revenue line and its slope:

$$\text{cases of canned fruit} = \frac{\text{total revenue}}{\text{price of fruit}} - \frac{\text{price of vegetables}}{\text{price of fruit}} \times \text{cases of canned vegetables}$$

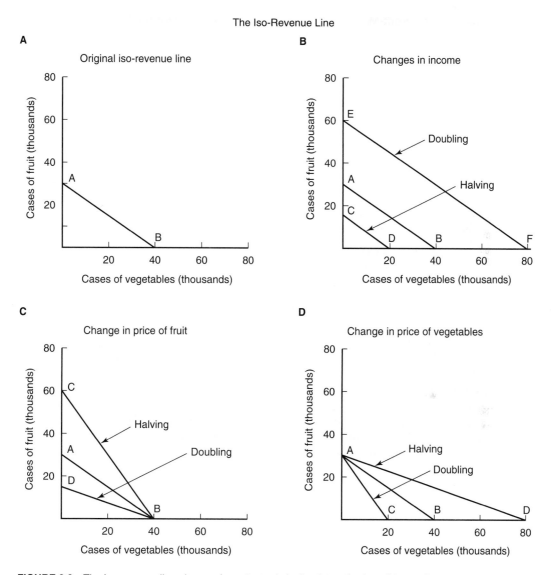

FIGURE 9.3   The iso-revenue line plays an important role in the determination of the profit-maximizing combination of two products. The slope of this line is equal to the negative of the price ratio for the two products.

PROFIT-MAXIMIZING COMBINATION OF PRODUCTS

We can determine the profit-maximizing combination of products under conditions of perfect competition by considering both the physical and economic trade-offs from the alternatives currently available. This requires uniting the concepts of the production possibilities frontier and the iso-revenue line.

TABLE 9.2   Profit-Maximizing Combination of Products for SunSpot

(1) Cases of Canned Fruit	(2) Cases of Canned Vegetables	(3) Revenue $33.33 × (1) + $25.00 × (2)	(4) Marginal Rate of Product Transformation Δ(1) ÷ Δ(2)	(5) Ratio of Price of Vegetables to the Price of Fruit $25.00 ÷ $33.33
135,000	0	$4,499,550		
128,000	10,000	4,516,240	−0.70	0.75
119,000	20,000	4,466,270	−0.90	0.75
108,000	30,000	4,349,640	−1.10	0.75
95,000	40,000	4,166,350	−1.30	0.75
80,000	50,000	3,916,400	−1.50	0.75
63,000	60,000	3,599,790	−1.70	0.75
44,000	70,000	3,216,520	−1.90	0.75
23,000	80,000	2,766,590	−2.10	0.75
0	90,000	2,250,000	−2.30	0.75

Choice of Products in the Short Run

The technical rate of exchange between canned fruit and canned vegetables for SunSpot in the current period is captured by the production possibilities frontier in Figure 9.1. We know from Equation 9.1 that the slope of this curve, called the marginal rate of product transformation, is equal to the ratio of the change in the production of these two products. The slope of this curve is negative, indicating an increasing opportunity cost of product substitution.

The profit-maximizing business seeks to maximize the revenue for the least-cost combination of inputs. In the present context, the business will want to determine the point at which the marginal rate of product transformation is equal to the relative prices of the products being produced. Table 9.2 suggests that the profit-maximizing combination of canned fruit and vegetables for SunSpot in the current period given existing input prices would be between 119,000 and 128,000 cases of canned fruit, and between 10,000 and 20,000 cases of canned vegetables.

The absolute value of the marginal rate of product transformation in column 4 of Table 9.2 will equal the absolute value of the price ratio in column 5 of 0.75 in this range. At this point, the marginal rate of product transformation (slope of the production possibilities curve) for fruits and vegetables equals the ratio of the price of vegetables to the price of fruit (slope of the iso-revenue line). We can, therefore, state the conditions for the profit-maximizing combination of these two products in mathematical terms as:

$$\frac{\Delta \text{ canned fruit}}{\Delta \text{ canned vegetables}} = -\frac{\text{price of vegetables}}{\text{price of fruit}} \tag{9.4}$$

in which both sides of the equation will have a negative value. (Recall the negative values for the marginal rate of product transformation in Table 9.1 and Table 9.2.)

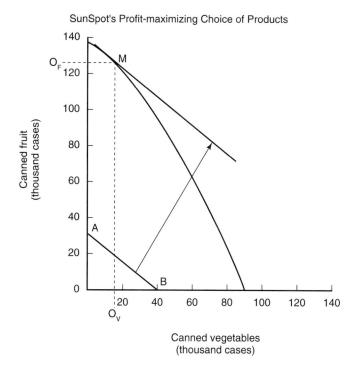

SunSpot's Profit-maximizing Choice of Products

FIGURE 9.4 The profit-maximizing choice of how much fruit and vegetables to can is given by the point at which the iso-revenue line is tangent to this business's current production possibilities frontier (point *M*).

Figure 9.4 suggests the profit-maximizing combination of canned fruit and vegetables for SunSpot would be approximately 126,000 cases of fruit and 13,000 cases of vegetables, which lie in between the ranges discussed in the context of Table 9.2. Total revenue would reach approximately $4,524,580. This combination also represents maximum profits because the business is on the production possibilities curve, which assures maximum technical efficiency.

Effects of Change in Product Prices

Let's assume that we are at point *M* in Figure 9.4, and the wholesale price of fruit suddenly falls to $25. This will alter the slope of the iso-revenue line in Figure 9.5. The new iso-revenue line *CB* is extended out in a parallel fashion until it is tangent to the production possibilities curve at point *N*.

In Figure 9.5, SunSpot's profit-maximizing objective is to *decrease* its fruit canning operations from O_F (Figure 9.4) to O^*_F and *increase* its vegetable canning operations from O_V (Figure 9.4) to O^*_V. This suggests that a business, with a given amount of resources, will alter the allocation of resources between the production of alternative products as their price ratio changes. The quantity of vegetables and fruit SunSpot chooses to can, therefore, will depend on the prices of all its inputs, the stock of its existing fixed inputs, the technology embodied in its labor and capital, and the relative price of fruit and the price of vegetables.

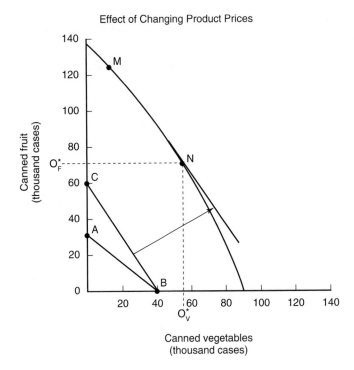

Effect of Changing Product Prices

FIGURE 9.5   A decrease in the price of canned fruit would cause SunSpot to alter the combination of the products it cans.

LONG-RUN EXPANSION OF PRODUCTION

When a business expands its operations, whether it be a farming or ranching operation or an equipment manufacturing operation, it may choose to change its product mix, or the combination it chooses to produce. Under conditions of perfect competition, this may reflect changing trends in the profitability associated with specific products relative to others. A business that 40 years ago manufactured a "one-hoss shay" buggy may today be manufacturing auto parts. Some businesses may also choose to diversify their operations as a means of reducing their exposure to the risk that the profitability of a single product may deteriorate rapidly.

Capital Variable in the Long Run

As indicated by Figures 8.9, 8.10, and 8.11, a business can expand the size of its operations in the long run by increasing its capital stock, such as its land, buildings, machinery, and equipment. For example, we know that each point on the long-run average cost curve in Figure 8.9 represents the least-cost combination of resources to produce a given level of output.

This same phenomenon held along the input expansion path in Figure 8.11 given by the points of tangency between a set of isoquants for two inputs associ-

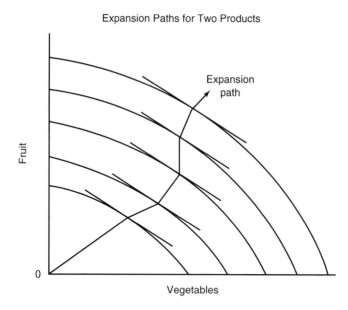

Expansion Paths for Two Products

Expansion
path

Fruit

Vegetables

0

FIGURE 9.6 The long-run average cost curve was said to represent a path of points reflecting the least unit cost of producing specific rates of output. We can also look at the firm's expansion path in terms of the revenue-maximizing combination of products. This figure shows that a line connecting a series of points given by the tangency of iso-revenue lines and production possibilities frontiers associated with different levels of available resources also forms an expansion path.

ated with different output levels, and a set of iso-cost lines in which the slope is given by the ratio of the prices for these inputs. Importantly, when the business moves along this expansion path associated with input use, its choice of specific forms of capital will be influenced by the products in which the business chooses to specialize.

Expansion of Production Possibilities Frontier

In addition to the long-run cost curve and the expansion path in an input-input setting, we can examine the expansion of the business's individual products by determining its expansion path in a product-product setting.

Figure 9.6 illustrates the nature of the expansion path in a product-product setting. Rather than the single production possibilities frontier depicted in Figure 9.1, we see a set of production possibilities frontiers associated with different levels of resource use. A line drawn through the points of profit-maximizing combinations of products also represents an expansion path. All points along this expansion path satisfy the conditions for profit maximization, because the business is maximizing the revenue associated with the least-cost levels of input use. The business, therefore, is economically efficient because it is on the PPF and maximizing profit because it is operating at the point on each PPF where tangency with the iso-revenue line is achieved. Importantly, any deviation from these combinations of products for specific levels of resource use as the business expands will result in a lower level of profit in a particular year.

SUMMARY

The purpose of this chapter was to illustrate the physical and economic relationships associated with the issue of product choice by a business under conditions of perfect competition. The major points made in this chapter may be summarized as follows:

1. The production possibilities frontier in the current period represents the different combinations of two products a business can produce given efficient use of its existing resources. When the business expands its operations in the long run, it will be on higher production possibilities frontiers that reflect the changing nature of its resources.

2. The slope of the production possibilities curve is called the marginal rate of product transformation. This slope reflects the rate at which the business can substitute between the production of two products in the current period. If two products are perfect substitutes, the marginal rate of product transformation will be constant at all points along the production possibilities frontier.

3. The iso-revenue line reflects the rate at which the market is willing to substitute between two products as their prices change. The slope of this line is therefore equal to the ratio of the prices of the two products. The intercept of this line on both axes reflects the maximum quantity of these two products that could be purchased if bought alone, and reflects a given amount of revenue and the prices of the products. Changes in the prices of these products will alter the slope of the iso-revenue line.

4. The profit-maximizing combination of two products to produce is determined by the point of tangency with the business's current production possibilities frontier and the iso-revenue line. At this point of tangency, the marginal rate of product transformation, or slope of the production possibilities frontier, will be equal to the ratio of the two product prices, or the slope of the iso-revenue line.

5. When the business expands its use of capital and labor in the long run, its production possibilities frontier will shift upward. This shift may not be in a parallel fashion if the business's choice of products to produce over time requires specific machinery and equipment that alter the marginal rate of product transformation.

DEFINITION OF KEY TERMS

Diversify: the addition of product lines or enterprises to a business that allows it to guard against the possibility that a downturn in the profitability associated with one product will cause severe financial stress for the business.

Iso-revenue line: the quantity of two products the market is willing to purchase for a given level of expenditure, and the prices of the two products. The slope of this line is the ratio of the two product prices.

Marginal rate of product transformation: the substitution of two products as the business moves along its current production possibilities frontier (i.e., the slope of the PPF).

Production possibilities frontier (PPF): the technically efficient combination of two products a business can produce in the current period given its existing resources and technology.

EXERCISES

1. Suppose a firm can receive $25 for a case of canned vegetables and $30 for a case of canned fruit. Based on this information, please answer the questions appearing below the graph.

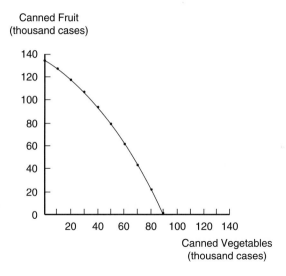

a. Draw the iso-revenue line associated with a target revenue of $1 million.

b. What is the profit maximizing level of vegetables and fruit this business should can? How did you arrive at this answer?

c. What is the firm's level of total revenue given your response to part b?

2. Given the following graph for two products, please respond to the questions below.

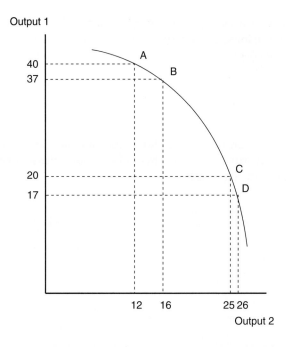

a. Calculate the marginal rate of product transformation between points *A* and *B*.

b. Calculate the marginal rate of product transformation between points *C* and *D*.

c. Your results illustrate the principle of _____ .

3. In the boxes above *each* of the following four graphs, please describe what caused the iso-revenue line to change *in each box* and the nature of the change *on each line.*

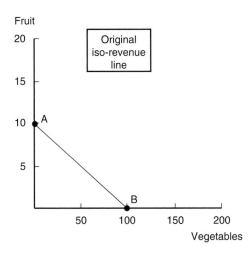

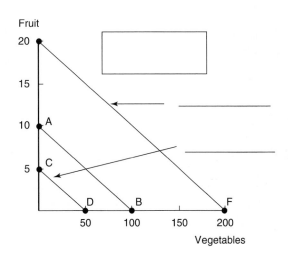

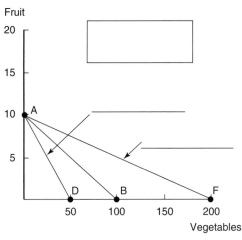

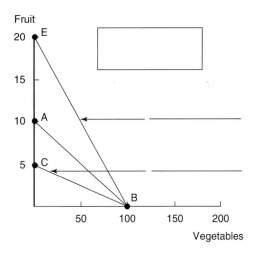

10

MARKET EQUILIBRIUM AND PRODUCT PRICE: PERFECT COMPETITION

The market price of every particular commodity is regulated by the proportion between the quantity which is actually brought to market and the demand of those who are willing to pay the natural price of the commodity.

Adam Smith (1723–1790)

TOPICS OF DISCUSSION

Part II of this book discussed the derivation of the market demand curve, based on the demands of individual consumers, and its elasticity. This represented exactly one-half of the relationships needed to understand changing market conditions, including the market equilibrium price. The other part of the puzzle is the market supply curve. Beginning with Chapter 6, we began explaining the microeconomic factors that influence the behavior of producers in the food and fiber industry.

The purpose of this chapter is to explain how we can derive the market supply curve for a particular product under conditions of perfect competition and interpret what the equilibrium means for consumers and producers. Attention will also be given to the forces that cause changes in the market equilibrium price, and the nature of the adjustment to the new equilibrium.

DERIVATION OF THE MARKET SUPPLY CURVE

The market supply curve for a particular product is based on the decisions of what and how much to produce made by individual businesses in an industry.

Firm Supply Curve

The marginal cost curve and the average variable cost curve help determine the minimum price at which a business can justify operating from an economic perspective. For our hypothetical business TOP-AG, whose costs of production were presented in Table 7.3, the minimum acceptable product price would be approximately $16, which is far below the $45 TOP-AG is currently receiving for its product. If the price of TOP-AG's product fell to $10, should the business continue to operate? Would TOP-AG be covering all of its costs of production at this product price? Would the business even be covering its variable costs of production? In the discussion to follow, we will see that the marginal cost curve lying above the minimum average variable cost represents the business's supply curve in the current period.

Output O_{BE} in Figure 10.1 represents the break-even level of production for TOP-AG, or the point at which the marginal cost curve in Figure 7.4 intersects the average total cost curve. At this level of output, average revenue is just equal to average total cost. This means that at P_{BE}, economic profits are equal to zero. Output O_{SD} in Figure 10.1 is identical to the point in Figure 7.4 at which the marginal cost curve intersected the average variable cost curve. This means that if the price fell to P_{SD}, average revenue would just equal average variable cost. The business could minimize its losses in the current period by continuing to produce if prices were below price P_{BE}. If the product price corresponding to the segment of the marginal cost curve lies between prices P_{BE} and P_{SD}, the firm could cover all of its variable costs and some, though not all, of its fixed costs by continuing to produce.

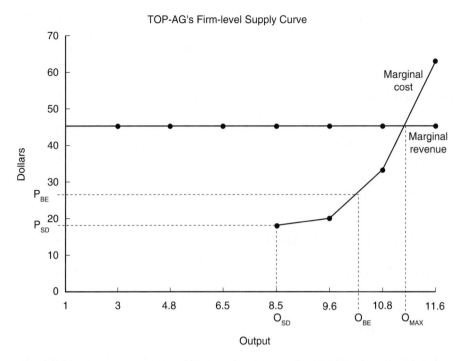

FIGURE 10.1   The portion of this business's marginal cost curve that lies above the average variable cost curve represents a business's current supply curve under conditions of perfect competition.

A rational competitive firm will cease producing in the short run only when the product price falls below the average variable costs of production, which occurs at price P_{SD} in Figure 10.1. Operating when price is below point P_{SD} on the marginal cost curve will only add to the firm's losses because TOP-AG is no longer covering all variable costs. Furthermore, the suppliers of variable inputs (such as fuel and hired labor) will likely cease supplying these services to the business when the checks start to bounce. This is why the level of output associated with price P_{SD} on the marginal cost curve (output O_{SD}) is known as the shutdown point. This is also the reason why we present only the portion of TOP-AG's marginal cost curve appearing above its average variable cost curve when illustrating this business's supply curve in Figure 10.1.

Market Supply Curve

Figure 4.6 illustrated the fact that the market demand curve represents the summation of the quantities desired by all consumers at specific market prices. The market demand curve for the two-consumer example depicted in that figure was found by horizontally summing the individual demands of both consumers. The market supply curve under conditions of perfect competition is determined in a similar manner.

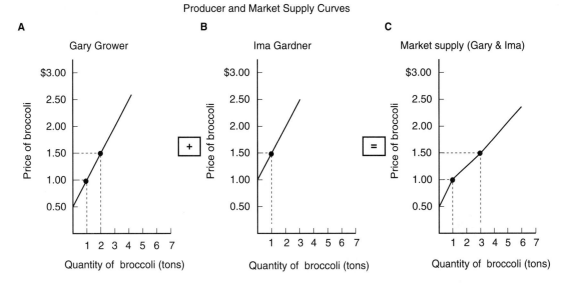

FIGURE 10.2   The market supply curve is found by horizontally summing the quantities supplied by all producers for given levels of market price.

Figure 10.2, *A* suggests that Gary Grower would be willing to supply 1 ton of fresh broccoli if the market price were $1.00 per pound, 2 tons if the price were $1.50 per pound, and so on. Figure 10.2, *B* shows that Ima Gardner would decline to produce at a market price of $1 per pound, but would supply 1 ton of broccoli at a market price of $1.50 per pound, and so on. Figure 10.2, *C* shows that if the market supply were limited to these two producers, the total supply of broccoli forthcoming at a market price of $1 per pound would be 1 ton, 3 tons at a market price of $1.50 per pound, and so on.

Like the demand curve, we can also characterize the properties of the market supply curve by examining the elasticity of this curve.

Own-Price Elasticity of Supply

The market supply curve for a particular product generally has a positive slope because the quantity supplied by businesses increases when the price it receives goes up. It is helpful to think of the behavioral response of producers in the context of their own-price **elasticity of supply.** This elasticity is expressed as:

$$\begin{matrix} \text{own-price} \\ \text{elasticity of supply} \end{matrix} = \frac{(Q_{SA} - Q_{SB})/[(Q_{SA} + Q_{SB})/2]}{(P_A - P_B)/[(P_A + P_B)/2]} \qquad (10.1)$$

in which Q_{SA} is the quantity supply *after* the change in price from P_B to P_A, and in which Q_{SB} is the quantity *before* the change in price. An own-price elasticity of supply exceeding one indicates an elastic supply, and an own-price elasticity of less than one suggests an inelastic supply.

For example, if the own-price elasticity of supply for a product is 1.5, a 1% increase in product price would cause businesses producing this product to increase their production by 1.5%. Because the percentage change in revenue is equal to the percentage change in price plus the percentage change in the quantity supplied, the total revenue of producers would increase by 2.5%.

Finally, the more (less) elastic or flatter (steeper) the market's supply curve is, the greater (lower) the impact a price change will have on total revenue, all other things constant.[1] What would the impact of a 1% increase in product price be on quantity supplied if the market supply curve were perfectly inelastic? What would the change in total revenue be under these conditions?

Producer Surplus

Economic rent, or **producer surplus,** is the economic return above the firm's variable cost of production.[2] When economic profit exists, surpluses are accruing to businesses. This surplus may be measured for an individual business by examining the business's returns above variable costs of production.

A business will supply the first unit of output at a price equal to the marginal cost of producing the first unit. If this marginal cost were $1 and the price of the product were $4, the business would receive a $3 surplus from producing and exchanging the commodity. If the marginal cost of producing the one-hundredth unit were $3, the surplus would be $1 for producing the unit.

By similar reasoning, the area above the market supply curve and below market price represents the producer surplus accruing to businesses participating in the market. This surplus is represented by area *ABC* in Figure 10.3 when the product price is $4. If the product price rises to $6, producer surplus increases to area *CED*. These areas represent economic profit plus fixed costs.[3] Area *AEDB* represents the gain in producer surplus resulting from the rise in the product price from $4 to $6. Hence, producer surplus represents a measure of the gain in economic welfare that businesses receive from producing a particular product in the current period.

MARKET EQUILIBRIUM UNDER PERFECT COMPETITION

One of the conditions for perfect competition presented at the beginning of Chapter 7 is that the business's product is homogeneous, or a perfect substitute for the product sold by the other businesses in this market. Perfect competition enables

---

[1]If firms produce more than one product, and these products are independent of one another, the discussion presented above applies to each product considered separately.

[2]If the rents disappear with entry or exit, they are a short-term phenomenon called quasi-rents.

[3]Because the producer surplus associated with a unit of output represents the marginal economic profit, total producer surplus represents total revenue minus total variable costs, or profit plus fixed costs.

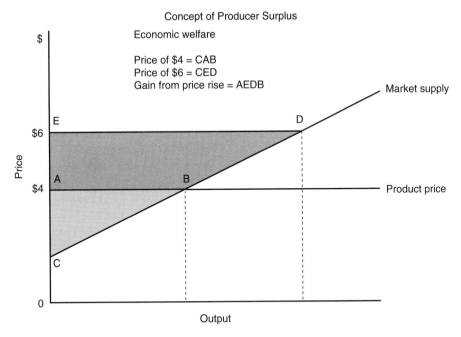

FIGURE 10.3   The change in the economic welfare of businesses can be approximated through the concept of economic rent or producer surplus. The value of this surplus at a price of $4 is given by the darkly shaded area above the market supply curve and below the $4 equilibrium market price for the product. This area reflects the revenue received by a business above the minimum price at which it would have been willing to supply its product.

buyers in the market to choose among a large number of sellers. Another condition enables any business that desires to enter or leave the sector to do so without encountering serious barriers. There must be a large number of sellers and buyers in the market to have perfect competition. No single buyer or seller should have a disproportionate influence on price. Finally, adequate information must exist for all participants regarding prices, quantities, qualities, sources of supply, and so on.

When all four conditions hold, we can say the market's structure is perfectly competitive. Businesses supplying goods in this market are also, by definition, perfectly competitive. Each is a **price taker,** or accepts the price of the product as given. Agriculture probably comes as close as any sector in the economy to satisfying these conditions. There are thousands of corn growers in the United States producing no. 2 yellow corn, each having similar access to market information, and none confront legal barriers to enter or leave the sector.

Market Equilibrium

The equilibrium price in a perfectly competitive market is established by the point of intersection of the market's demand and supply curves. Let D_M represent the market demand schedule for the sector's product, and S_M represent the market supply

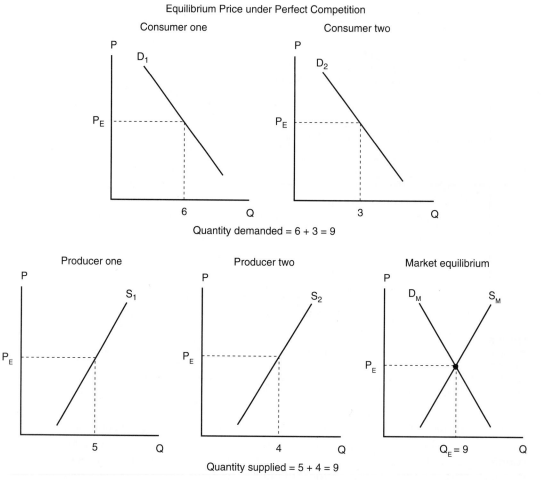

FIGURE 10.4   The equilibrium price in a competitive market is given by the intersection of the market demand and supply curves. As shown in this figure, this would result in a price of P_E and quantity Q_E. At this price, consumers collectively would demand nine units, and producers collectively would supply nine units.

schedule for all businesses in the sector. We have assumed, for ease of presentation, that there are only two buyers (consumers) and two sellers (producers) in this market. Figure 10.4 shows that the equilibrium price in this market would be equal to P_E. At this price, businesses would be willing to supply quantity Q_E (or nine units) and the buyers of this product would desire to purchase quantity Q_E (also nine units). Thus, P_E and only P_E is the price per unit that will "clear the market."

Shifts in either the demand curve or the supply curve will result in a new equilibrium market price. Four possible events can occur that will affect the market equilibrium price and quantity:

1. Demand increases, shifting the demand curve to the right.
2. Demand decreases, shifting the demand curve to the left.
3. Supply increases, shifting the supply curve to the right.
4. Supply decreases, shifting the supply curve to the left.

The effects of each situation on the equilibrium price that clears the market are illustrated in Figure 10.5. In Figure 10.5, *A*, for example, we see that an increase in demand (perhaps consumer disposable income increased) will result in a higher market price (P^*_e). Buyers will now demand, and businesses will supply, a quantity equal to Q^*_e instead of Q_e. The opposite effect occurs when demand decreases (see Figure 10.5, *B*). In both cases there is a change in demand and a change in the quantity supplied.

Turning to supply, let us assume that the supply of the product increases, or that the supply curve shifts to the right. Figure 10.5, *C* illustrates that this shift will lead to a decline in the market-clearing price from P_e to P^*_e. At this new price, firms will supply, and buyers will demand, a quantity equal to Q^*_e. Figure 10.5, *D* shows that the opposite outcome will occur if there is a decrease in supply. In both cases, there is a change in supply and a change in the quantity demanded.

The elasticity of the demand and supply curves plays an important role in determining how much the equilibrium price will change if either demand or supply changes. For example, the more inelastic or steeper the demand curve, the greater the rise (fall) in the market price will be for a given decrease (increase) in supply. The relatively inelastic nature of the demand for farm products, coupled with a volatile supply curve that can shift to the right or the left depending on the vagaries of weather, helps explain the high variability of farm income that we often see from one period to the next. More will be said about this when we discuss the traditional farm problem in Chapter 12 and government programs to support farm prices and incomes in Chapter 13.

Total Economic Surplus

In Chapter 4, we learned that consumer surplus is given by the area below the demand curve and above the equilibrium price (see Figure 4.9). We also learned that producer surplus is given by the area above the supply curve and below the equilibrium price (see Figure 10.3). These areas represent the economic well-being achieved by consumers and producers at the equilibrium or market-clearing price. If we add these two triangular areas together, the newly formed triangle represents the economic well-being achieved by all market participants in this particular market. In Figure 10.6, the summation of consumer surplus (area 1) plus producer surplus (area 2) represents the total area above the supply curve and below the demand curve, and hence the total economic surplus received by all market participants.

Now suppose that because of low yields, the supply curve for this market shifts inward to the left from S to S^* (Figure 10.7). Producer surplus would now be equal to area 4 plus area 6. Thus, if areas 4 plus 6 sum to less than areas 6 plus 7, we may

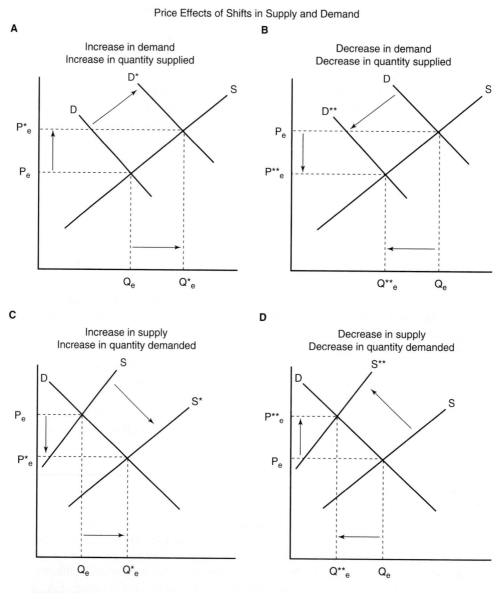

FIGURE 10.5   It is important to distinguish between changes in supply or demand and changes in the quantity supplied or demanded.

conclude that the economic well-being of producers would have declined. We can also conclude that consumer surplus declined because consumers received the equivalent of areas 3, 4, and 5 before but now receive only area 3. Finally, we can conclude that this decrease in supply means that the economic well-being of market participants in general would have fallen by an amount equal to area 5 plus area 7.

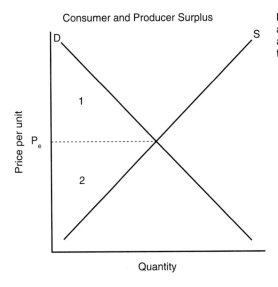

Consumer and Producer Surplus

FIGURE 10.6   Area 1 represents consumer surplus and area 2 represents producer surplus. The sum of both areas represents the economic well-being of society from participating in this market.

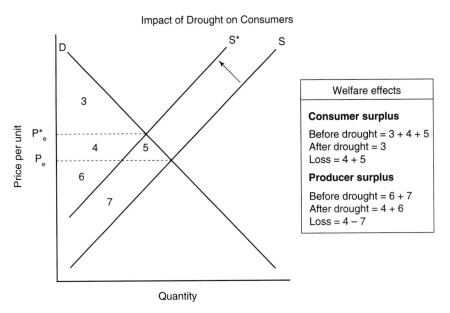

Impact of Drought on Consumers

Welfare effects
Consumer surplus
Before drought = 3 + 4 + 5
After drought = 3
Loss = 4 + 5
Producer surplus
Before drought = 6 + 7
After drought = 4 + 6
Loss = 4 − 7

FIGURE 10.7   This figure shows what would happen to the economic welfare of consumers and producers if a drought caused the aggregate supply curve to shift from *S* to *S**.

Applicability to Policy Analysis

The concept of producer and consumer surplus is an important analytical tool to economists. We may be examining the economic welfare implications of a major drought, such as the drought of 1988, which results in a shift in the current supply curve as discussed previously. Or we may be examining a change in agricultural or macroeconomic policy that causes a shift in either the demand or the supply curve in the current period. The concept of producer and consumer surplus helps in assessing the relative effects of these externalities. It is not unusual for economists, when testifying before Congress on the effects of a drought or a particular policy change, to use the concept of consumer and producer surplus to illustrate the effect this would have on the economic well-being of market participants.

The concept of producer and consumer surplus will be used in the next chapter when we assess the economic welfare implications of imperfect competition. It will also be used extensively in Chapter 13 when we examine the effects of agricultural policy on consumers and farmers.

ADJUSTMENTS TO MARKET EQUILIBRIUM

Markets are not always in equilibrium. In fact, many rarely are. Instead, changing demand and supply conditions across numerous markets result in market disequilibrium. This section describes the symptoms of market disequilibrium and how markets adjust to a new equilibrium.

Market Disequilibrium

At prices above the market-clearing price P_e, there would be an excess quantity supplied by businesses, or a **commodity surplus.** At prices below the market-clearing price, an excess quantity demanded, or **commodity shortage,** would exist. For example, Figure 10.8, *A* shows that at price P_s buyers would wish to purchase Q_d, and sellers would want to supply Q_s. The difference between these two quantities ($Q_s - Q_d$) represents the surplus available on the market at the price P_s. This suggests that the market is in disequilibrium instead of equilibrium because the market has not been cleared at this price. The opposite is illustrated in Figure 10.8, *B*. At a price of P_d buyers would wish to purchase quantity Q_d, and sellers would only want to supply quantity Q_s. Thus, a shortage equal to $Q_d - Q_s$ would exist in the market at a price of P_d.

The existence of these disequilibrium situations will modify over time if prices and quantities are free to seek their equilibrium levels. If a surplus exists, for example, the inventories of unsold production will be unintentionally high. Firms will have incurred costs but received no revenues for this unplanned inventory buildup. As long as these inventories remain unsold, firms will also be incurring storage costs in one form or another. Because they are not maximizing their profits at this point, firms will find it profitable to decrease their level of production and accept a lower price for their inventories.

This adjustment process will stop after prices have fallen from P_s to P_e. If a shortage exists, buyers would compete for available supplies by offering to pay

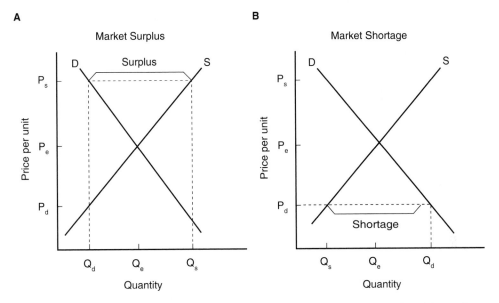

FIGURE 10.8 The equilibrium price in a competitive market is given by the intersection of the market demand and supply curves. If, instead, the price were equal to P_s (A), producers would be willing to supply more than consumers would demand. This phenomenon is referred to as surplus. If the price were instead equal to P_d, the quantity demanded by consumers would be greater than the quantity producers would be willing to supply. This excess demand situation is commonly referred to as a shortage.

higher prices. This will encourage firms to raise and market more of this commodity. This adjustment process will stop after prices have risen from P_d to P_e. At this point, the quantity demanded will be exactly equal to the quantity supplied, and market equilibrium will be restored.

Length of Adjustment Period

The adjustment processes discussed above may suggest that the quantities demanded and supplied are both determined by current prices. In some sectors like agriculture, however, adjustment to market equilibrium takes time. One reason is the biological nature of the production process itself. Once the crop has been planted, for example, little can be done to adjust the supply response of producers until the next production season. Furthermore, when farmers plant their crop, they do not know what the market price will eventually be when they sell their crop several months later.

Let us assume for the moment that farmers base their production plans for this year on last year's price. Price and quantity are now sequentially determined rather than simultaneously determined. The price last year determines this year's production response. This year's quantity marketed, however, will affect this year's price, which will affect next year's production, and so on.[4] If prices were high last year, for example, farmers under free-market conditions would respond by expanding their production activities with the anticipation of eventually marketing more output. The increased level of production will lead to lower prices, all other things constant. This pattern of price and quantity responses forms a pattern like a spider's cobweb over time.

Cobweb Adjustment Cycle

To illustrate cobweb market behavior, let us examine Figure 10.9, *A*. Given the demand and supply curves *D* and *S*, let's suppose that the price of corn last year (year one) is equal to P_1. Because corn farmers base their production for year two on P_1, they will produce Q_2 in year two. This quantity, however, will cause prices in year two to fall to P_2. As shown in Figure 10.9, *B*, corn farmers will respond to this lower price by producing only Q_3 in year three, which will cause market prices to rise to P_3.

   This behavior of prices and quantities over time is referred to as a cobweb pattern, after the cobweb-like nature of the solid lines tracing the movements of prices and quantities shown in Figure 10.9, *C*. This panel illustrates the nature of a converging cobweb. Here, prices and quantities will eventually converge to a market equilibrium at price P_E. This cobweb pattern will occur when the slope of the supply curve is steeper or more inelastic than the slope of the demand curve. A diverging cobweb occurs when the slope of the demand curve is steeper or more inelastic than the slope of the supply curve.[5]

   Events causing changes in demand or supply can cause an interruption to these cycles and lead to a new set of market adjustments over time. As we will discuss later in Chapter 13, federal programs that are designed to modify the booms and busts associated with fluctuating prices and quantities exist for some commodities.

SUMMARY

The purpose of this chapter was to explore the determination of the market equilibrium price under conditions of perfect competition and how the market adjusts to a new market equilibrium when market conditions change. The major points made in this chapter may be summarized as follows:

1. The supply curve for an individual business under conditions of perfect competition is represented by that portion of its marginal cost curve that lies above its average variable cost curve. This suggests that the business will supply this

---

[4]The demand and supply functions in this instance would be given by $P_t = f(Q_t)$ and $Q_t = f(P_{t-1})$, respectively. The response to last period's price in the supply function is thus different from the response to current price assumed thus far.
   [5]A persistent cobweb would occur if the demand and supply curves have identical slopes, which means that the market would continue to oscillate around the market's equilibrium, never converging or diverging.

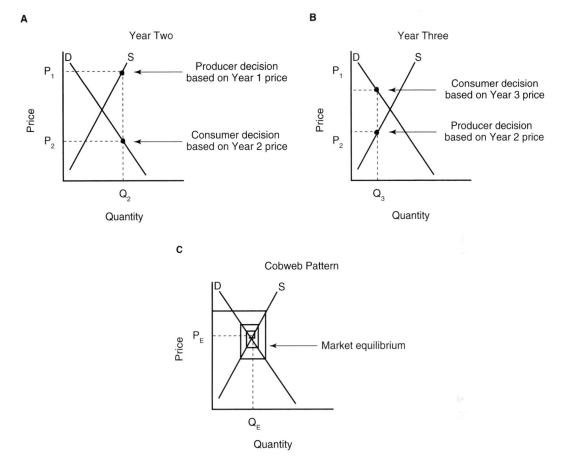

FIGURE 10.9   When producers respond to the previous period's price, markets will adjust to market equilibrium in a cobweb pattern.

product in the current period as long as it is able to cover its variable costs of production.

2. The market supply curve for a particular commodity under conditions of perfect competition represents a horizontal summation of the supply responses of the businesses selling this commodity.

3. The **break-even level of output** occurs for a business when the price of the product is sufficient to just cover its average total costs. Economic profit at this output level would be equal to zero.

4. The **shutdown level of output** occurs when the price of the product is just sufficient to cover the business's average variable costs. If the price of the product were to fall below this level, the firm would cease operations.

5. Producer surplus, the supply-side counterpart to **consumer surplus** discussed in Chapter 4, represents the gain in economic well-being that businesses will

achieve by participating in a particular market during the period. Product surplus is equal to the area above the market supply curve and below the market equilibrium price.

6. Market equilibrium under conditions of perfect competition occurs when the market demand curve intersects the market supply curve. At this price, consumers will purchase the market equilibrium quantity at the market equilibrium price, and businesses will supply the market equilibrium quantity at the market equilibrium price.

7. A shift in the market supply curve will result in a change in supply and a change in the quantity demanded. This is the opposite of a shift in the market demand curve, which results in a change in demand and a change in the quantity supplied.

8. A market disequilibrium occurs when producers respond to an expected market equilibrium price when making their production decisions that turns out to differ from the eventual market equilibrium price. This results in commodity surpluses when expectations are too high and commodity shortages when price expectations are too low. The cyclical pattern of adjustments to a new equilibrium price under these conditions takes on a pattern much like a spider's cobweb.

DEFINITION OF KEY TERMS

Break-even output: level of output at which average total costs equal average revenue or market price.

Commodity shortage: the amount by which the quantity demanded at a given price exceeds the quantity supplied.

Commodity surplus: the amount by which the quantity supplied at a given price exceeds the quantity demanded (not to be confused with producer or consumer surplus).

Consumer surplus: a measure of the savings achieved by consumers at the current market price from the price they would have been willing to pay for a specific quantity of a good or service. Consumer surplus is equal to the area below the market demand curve and above the market equilibrium price.

Elasticity of supply: percent change in quantity supplied with respect to a percent change in the price of the product.

Price taker: a business is said to be a price taker when its actions have absolutely no effect on the price of the input it is buying or the price of the product it is selling.

Producer surplus: a measure of the economic rent or returns above total costs accruing to businesses participating in a market during the current period.

Shutdown output: level of output at which average variable costs equal average revenue or the market price.

Total economic surplus: the sum of producer surplus and consumer surplus.

EXERCISES

1. Given the following graph, please label the curves where asked and answer the questions appearing below the graph.

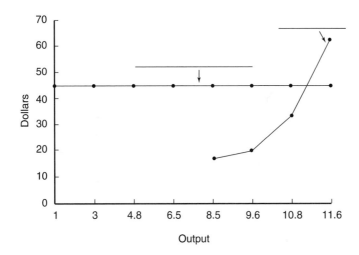

a. What is the profit-maximizing level of output?

b. What is the significance of 8.5 units of output?

2. Please describe what has happened to demand, supply, and market equilibrium in each panel.

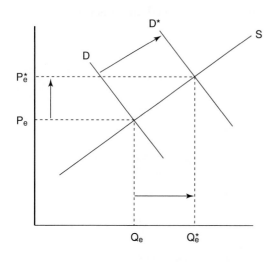

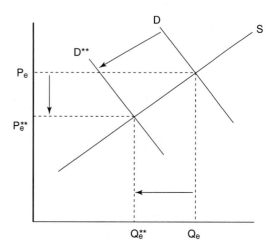

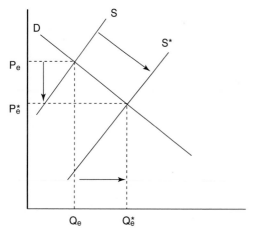

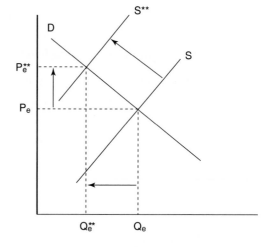

3. Given the following price and quantity-supplied data, graph this firm's supply curve on the axis provided. Also, for each price and quantity-supplied combination, calculate the amount of producer surplus.

Price	Quantity Supplied
0	0
1	1
2	2
3	3
4	4
5	5
6	6
7	7
8	8
9	9
10	10

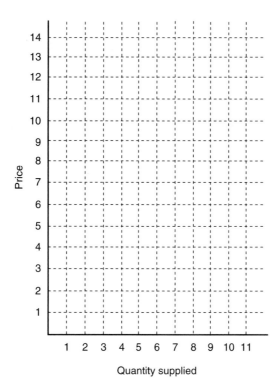

Quantity supplied

4. Assume that the following graph depicts a shift of the worldwide demand curve for beef. Starting in 1999, demonstrate the cobweb adjustment to market equilibrium over time on the following graph. Assume that the market was in equilibrium in 1998 and will eventually converge to the equilibrium depicted by the 1999 demand curve. Assume producers respond to last year's price (the expected price is last year's observed price).

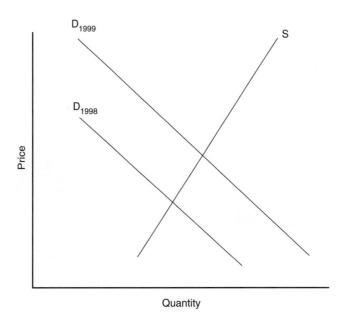

11

MARKET EQUILIBRIUM AND PRODUCT PRICE: IMPERFECT COMPETITION

> *The monopolists, by keeping the market constantly understocked, by never fully supplying the effectual demand, sell their commodities above the natural price.*

Adam Smith (1723–1790)

TOPICS OF DISCUSSION

Food processors, fiber manufacturers, farm input manufacturers, and other non-farm businesses provide an important bridge between farmers and consumers of food and fiber products. These markets send important signals to farmers that help them decide *what* commodities they should produce, *how much* of each commodity they should produce, *how* these commodities should be produced, and *where* these commodities should be distributed.

Up to this point we have assumed that the conditions required for perfect competition exist in the marketplace. In reality, the economy does not consist entirely of perfectly competitive firms. Although farms and ranches generally satisfy the conditions of perfect competition, most of the firms with which they do business in the food and fiber industry do not. In this chapter, we will examine several forms of **imperfect competition** and cite examples of each form in the U.S. food and fiber industry.

Two classic studies of imperfect competition exist in the economic literature—*The Theory of Monopolistic Competition,* by Edward H. Chamberlin, published in 1933, and *The Economics of Imperfect Competition,* by Joan Robinson, also published in 1933. Understanding the theory of imperfect competition is paramount when analyzing the economics of the farm business sector. The economic structure and characteristics of any market define the limits within which market forces are allowed to establish prices. To gain insight into the determination of prices in an industry, we must first understand its structural characteristics and competitive behavior.

MARKET STRUCTURE CHARACTERISTICS

There are four key interrelated structural characteristics used when discussing competitive behavior of a market. These include:

1. the number and size distribution of sellers and buyers,
2. the degree of product differentiation,
3. the extent of barriers to entry, and
4. the economic environment within which the industry operates (i.e., the conditions of supply and demand).

These four characteristics, combined, determine whether an industry or various segments of the industry exhibit behavior conducive to perfect competition or imperfect competition.

Number of Firms and Size Distribution

The number of firms buying and selling in a market is important in determining competitive conduct and price determination processes, because competitive conditions tend to break down when the number of firms involved declines. When competition wanes, prices are less likely to be set by the forces of supply and demand. The distribution of sales or purchases by firms in the industry helps indicate the degree of control that one or a few firms may have over prices. The fewer the number of firms and/or the greater the percentage of product sales accounted for by a few firms in the market, the greater the concentration of the market, and the greater the likelihood that prices are determined by forces other than supply and demand.

Product Differentiation

The degree of product differentiation refers to the extent buyers at each level in an industry perceive a difference in the products offered to them by alternative suppliers. If buyers consider the products to be virtually identical, the product is said to be *undifferentiated,* or *homogeneous.* A seller in an undifferentiated product market has difficulty in setting a higher price than what the forces of supply and demand would dictate because buyers simply shift to other suppliers to obtain the quantities of the product needed. Because the products offered by all suppliers are virtually identical, buyers have little incentive to pay one supplier a higher price than they would another supplier. Thus, sellers of undifferentiated products are price takers, not price makers.

The more differentiated a product becomes in the minds of buyers, the more opportunity the supplier has to raise the price of the product, taking advantage of buyers' perceptions concerning its superior or unique characteristics. It is not necessary that these perceived differences be real. Product differences perceived by consumers may be due entirely to well-planned and executed advertising campaigns.

Finally, products can be either positively or negatively differentiated in the minds of buyers; however, only a positive attitude toward the uniqueness of a product allows its seller to set prices. A negative attitude drives buyers to other products.

Barriers to Entry

Barriers to entry are those forces that make it difficult for firms to enter an industry. Some barriers are created by existing firms or granted by governments; others are simply economic facts of life. If barriers to entry are low, even firms in highly concentrated industries (i.e., small number of firms) have little opportunity to set prices above competitive levels. A firm in an industry with low barriers to entry that is attempting to increase its profits by raising prices would likely encourage new firms to enter the industry eager to share in the profit opportunities. Four common barriers to entry are:

1. absolute unit-cost advantages,
2. economies of scale,
3. capital access and cost, and
4. preferential government policies.

An absolute cost barrier exists if the unit cost of production for an established firm is lower at all levels of output than would be the case for a new entrant. In this case, there is no level of output at which a new firm could operate competitively.

An economies-of-scale barrier generally exists when the smallest efficient-sized plant is large relative to the size of the market, and smaller firms face significantly higher costs of production. New firms entering this industry would have to be large and generate a significant amount of output, causing industry prices to drop drastically and some firms to be driven out of the market.

A **capital cost barrier** exists when the capital investment required for efficient operation is large. Capital cost barriers tend to accompany economies-of-size barriers, but this is not always true. A potential industry entrant may face a significant capital investment cost to begin efficient operation and still only hope to capture a small share of the market.

Finally, the government can administer preferential policies such as patents, copyrights, and import controls that protect one or more sellers from all or some of the potential entrants to a domestic industry.

Economic Environment

Of the many aspects affecting the economic environment in which an industry operates, the one most closely related to the behavior and determination of prices is the existing supply and demand conditions. Three aspects of supply and demand particularly important to determining price in a market are:

1. the level of output in the industry,
2. the responsiveness of supply and demand to price changes and the responsiveness of demand to income changes, and
3. the proportion of consumer food expenditures accounted for by the industry's product.

If an industry's sales volume is relatively large, more profit opportunities will be available for firms to be involved at all levels. This situation makes it difficult for one or a few buyers or sellers to control prices. The opposite is the case if volume is low (a so-called "thin market").

If demand is highly responsive to changes in price and income—which is referred to as an elastic demand—and consumers spend a large portion of their incomes on the product, the industry will grow when income rises and prices fall. Opportunities for new entrants in this industry will be present, increasing the competitiveness at each level. If demand is instead largely unresponsive to price and income changes—which is referred to as an inelastic demand—and consumers, on average, spend only a small proportion of their income on the product, demand will be stable and growth will stagnate.

To summarize, we can classify forms of market structure, in part, by:

- the number of businesses (firms) in the industry,
- the degree of market concentration by the firms,
- the degree of product differentiation,
- the barriers to entry, and
- the economic environment in which firms operate.

Classification of Firms

Four major conditions, combined, define perfect competition and guarantee a market in which the forces of supply and demand determine prices. These conditions are: (1) a large number of firms, (2) a homogeneous product, (3) freely mobile resources, and (4) perfect knowledge of market conditions.

The conditions for perfect competition are extreme, but examples of nearly perfect competition abound in agriculture. In selling activities, numerous small producers of livestock and crops are producing a nearly identical product. In buying activities, numerous consumers are purchasing food and fiber products that are not highly distinguishable. The important result of the conditions of perfect competition, or even near-perfect competition in a market, is that buyers and sellers are price takers, not price makers. The actions of any individual buyer or seller do not affect the market. The interaction of market supply and market demand determines the market price in a perfectly competitive market environment.

Market structure from the seller's perspective can be classified into four types, based on the extent to which market prices are set by supply and demand. These four types of competition in selling are:

1. perfect competition,
2. monopolistic competition (imperfect competition),
3. oligopoly (imperfect competition), and
4. monopoly (imperfect competition).

Although farmers and ranchers largely satisfy the conditions of perfect competition, much of the rest of the food and fiber industry is characterized by one of the three forms of imperfect competition in selling.

IMPERFECT COMPETITION IN SELLING

To fully appreciate the complex nature of the nation's food and fiber system, we must consider various market structures that represent imperfect forms of competition. Three forms of imperfect competition in the selling activities of firms are monopolistic competition, oligopoly, and monopoly.[1] A very important characteristic of imperfect competition in selling is that the demand curve faced by the firm is downward sloping, not perfectly elastic as in the case of perfect competition.

---

[1]Imperfect competition among buyers also takes three general forms: monopsonistic competition, oligopsony, and monopsony. The suffix "poly" refers to sellers, and the suffix "sony" refers to buyers. The prefix "mono" means one, and the prefix "oli" means few. Therefore, imperfect competition may pertain to either buyers or sellers. Imperfect competition prevails among sellers and buyers whenever they individually exert a measure of control over price.

Monopolistic Competition

The conditions of monopolistic competition are identical to those of perfect competition with one important difference—*the products sold are no longer perfectly identical or homogeneous.* If many businesses (a smaller number than exists in the case of perfect competition) are selling a **differentiated product** (which is different from the standardized or homogeneous product in perfect competition), **monopolistic competition** exists. The differentiation of the product gives the seller some flexibility in pricing. If an individual seller can convince buyers that its product is superior to those offered by other sellers, it can charge a higher price for its product. Farm input manufacturers and suppliers differentiating their products with the use of advertising to promote branded feed supplements or branded hybrid seeds exemplify this type of imperfect competition. Like perfect competition, there are no barriers to entry or exit. *Thus, product differentiation is the key difference between monopolistic competition and perfect competition.*

A monopolistic competitor becomes a price maker if it can effectively differentiate its product in the market when other sellers in the market are offering similar products. The opportunity for price-making may be limited because the action of each individual firm may have imperceptible effects on the market and because the product may not be unique. Consequently, attempts by a seller to raise the price too much may drive buyers to other sellers of similar products. How much is "too much" is determined by the degree of differentiation of the seller's product. The marketing strategy of a firm in a monopolistically competitive market must focus on creating superior differences in the products it sells, relative to other firms in the market. Price competition, including price specials and discounts, will be an important aspect of marketing efforts.

Monopolistic competitors face the same average total cost (ATC) and marginal cost (MC) curves as perfect competitors. By differentiating their product, however, individual monopolistic competitors face a downward-sloping demand curve (and a downward-sloping marginal revenue curve that reflects the change in total revenue at each output level) instead of the perfectly elastic (i.e., flat or horizontal) demand curve faced by perfect competitors. The relationship between demand (average revenue), total revenue, and marginal revenue is exhibited in Table 11.1 and Figure 11.1.

The relationship between the demand curve and marginal revenue curve is tied to the elasticity of demand along the demand curve. For example, marginal revenue (MR) is zero at the point of unit elasticity of demand (i.e., -1.0). Up to that point, the imperfectly competitive firm's total revenue is increasing as output increases. In Chapter 7, we learned that the demand curve for a perfect competitor is perfectly elastic (i.e., flat as shown in Figure 7.5 and again in Figure 10.1) and that $P = MR$. Because monopolistic competitors face a downward-sloping demand curve, it is no longer true that $P = MR$. The monopolistic competitor must be aware of the decline in its MR curve and the possibility of negative MR at lower points on its demand curve.

Product differentiation conducted by the monopolistic competitor can be accomplished by modifying the particular product or by advertising and sales

TABLE 11.1   Relationship between Demand (Average Revenue), Total Revenue, and Marginal Revenue

Average Revenue or Demand Curve		Total Revenue	Marginal Revenue
Price	Quantity	(Price × Quantity)	(Change in Total Revenue)
15	0	0]_____	
14	2	28]_____	14
13	4	52]_____	12
12	6	72]_____	10
11	8	88]_____	8
10	10	100]_____	6
9	12	108]_____	4
8	14	112]_____	2
7	16	112]_____	0
6	18	108]_____	−2
5	20	100]_____	−4
4	22	88]_____	−6
3	24	72]_____	−8
2	26	52]_____	−10
1	28	28]_____	−12
0	30	0]_____	−14

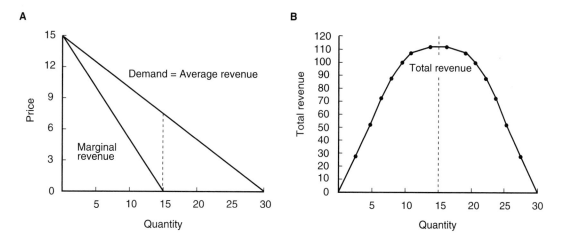

FIGURE 11.1   The relationships between average, marginal, and total revenue in these figures are based on the data in Table 11.1. These concepts determine equilibrium prices and quantities in the case of imperfect competition in selling.

promotion activities. The object is to intensify the demand for the product by distinguishing it from other products in the mind of the buyer. Such activities can be found in most markets in this country. The better the business is at product differentiation, the greater its influence on product price.

Short-Run Equilibrium.   The equilibrium values for price and quantity under monopolistic competition in the short run are determined by the intersection of the marginal cost curve and the marginal revenue curve. In Figure 11.2, this intersection occurs at point E, suggesting a level of output in the short run of quantity Q_{SR}. The business then sets the price it charges for its differentiated product by reading up to the demand curve and over to the price axis. From Figure 11.2, we see the monopolistic competitor will change price P_{SR} in the current period.

The gap between the demand curve (D) and the average total cost curve (ATC) at output Q_{SR} indicates the business achieved either an economic profit or a loss in the current period. If price exceeds average total costs, as it does in Figure 11.2, A, an economic profit exists in the current period. If price is less than average total costs (i.e., the demand curve lies below ATC), as it does in Figure 11.2, B, an economic loss is incurred. The existence of profits (losses) would result in the entry (exit) of additional monopolistic competitors over time.

Long-Run Equilibrium.   Additional entrants to the market when economic profits exist would shift the demand curve downward (decrease average revenue) and lower profits and possibly even create losses. Monopolistic competi-

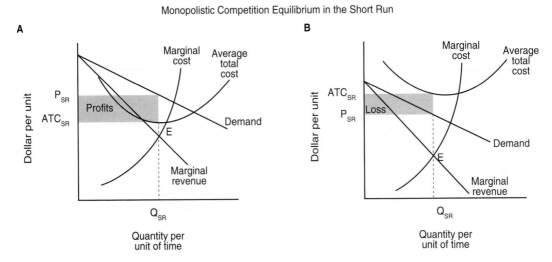

FIGURE 11.2 Monopolistic competitors also equate marginal cost and marginal revenue. The difference between the demand curve, which also represents average revenue, and the average total cost curve represents the average profit or loss per unit. Multiplying this difference by the quantity supplied (Q_{SR}) gives us the economic profit (A) or loss (B) for the business in the short run.

tors exiting the market when economic losses exist will shift the demand curve upward (increase average revenue) and reduce losses and perhaps create economic profits. Figure 11.3 suggests that the monopolistic competitor has reached its long-run equilibrium. How can we tell? There is no gap between the demand curve and the average total cost curve at Q_{LR} in Figure 11.3, which means that there are no economic profits that would attract other monopolistic competitors into the market, and there are no losses that would cause some competitors to leave the market. Note the firm is still equating marginal cost to marginal revenue at point E.

Is the equilibrium quantity here less than what a business would have supplied under perfect competition? The answer is yes. Note in both Figure 11.2 and 11.3 that the monopolistic competitor chooses to operate to the *left* of the minimum point on the average total cost (ATC) curve. You will recall from Figure 8.9 that the minimum point on the (ATC) curve represents the long-run equilibrium output level for a perfectly competitive firm.

Monopolistic competition is therefore less efficient than perfect competition from the viewpoint of consumers. The market price is higher and output is lower

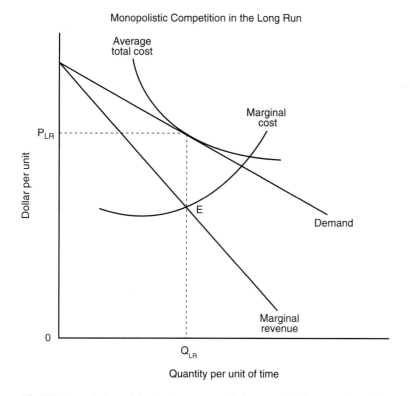

FIGURE 11.3 At the point where the monopolistic competitor is operating at Q_{LR}, average revenue will equal average total cost, indicating that both average profits or average losses have been eliminated.

TABLE 11.2   Top 10 Fast Food Burger Restaurants in 1998

Rank	Brand	Market Share	Advertising Dollars (Million)
1	McDonald's	42.8%	571.7
2	Burger King	20.2	407.5
3	Wendy's	11.5	188.4
4	Hardee's	5.7	50.5
5	Jack in the Box	3.6	51.2
6	Sonic Drive-Ins	3.3	28.1
7	Carl's Jr.	1.9	34.3
8	Whataburger	1.1	6.7
9	White Castle	1.0	10.1
10	Steak n Shake	0.9	5.7
	Total top 10	92.0	1,347.4
	Total market	$42.3 (billion)	1,359.7

than that occurring under perfect competition. Deciding whether society is in a bad position, however, depends on whether we want a marketplace with an undifferentiated or standardized product. Do we all want to wear white shirts and brown shoes? The fashion industry certainly hopes not. Also, do we all want to eat a certain brand of vanilla ice cream and eat a specific brand of white bread? The economy has shown a preference for differentiated products, even if it means paying a higher price. In fact, most of the food and fiber products we purchase are supplied by monopolistic competitors.

Many businesses in the nation's food and fiber industry spend large sums of money annually in an effort to differentiate their product in the eyes of the consumer. For example, Table 11.2 illustrates the top 10 fast food burger restaurants and the amount of dollar advertising associated with the firms. McDonalds, with a market share of 42.8%, spent about $570 million in advertising in 1998. Burger King, with a market share of 20.2%, spent about $408 million in 1998.

Oligopoly

Further removed from the characteristics of perfect competition in selling is the **oligopoly.** The economic conditions that define oligopoly are the same as those of monopolistic competition with one major exception: there are only *a few sellers,* each of which is large enough to have an influence on market volume and price.

Oligopolists are interdependent in their decision making. The actions of an individual oligopolist are seen as a competitive threat to the other oligopolists in the market that may invoke retaliation. It is this interdependence that is the key component in the marketing strategies and pricing behavior of the industry. **Nonprice competition** is the main competitive strategy, including any and all efforts to uniquely differentiate products in the market.

If an oligopolist tries to raise its price, the other oligopolistic firms will not necessarily follow suit. The size of the subsequent drop in the firm's sales will depend on how successful it has been in differentiating its product from those offered by its competitors. Typically the drop in sales will more than offset the price increase, leading to reduced revenues for the price-raising oligopolistic firm.

On the other hand, if an oligopolist attempts to lower its price, competing firms will immediately retaliate by lowering their prices to keep from losing their market shares. The lower price, combined with little or no change in sales, again means that the firm initiating the price decline experiences a loss in sales revenues. Further attempts by the firm to reduce price to gain market share will simply result in further rounds of price cutting with relatively small gains in sales. As a consequence, prices of products sold by firms in an oligopolistic industry tend to be stable. This situation exists today in the soft drink industry and breakfast foods industry.

Once prices in an oligopolistic industry are established, they tend to stay at that level (Purcell, 1979). If there are differences in the prices of the products offered by competing firms in an oligopoly, they are generally the result of successful differentiation of the products by the respective firms. This stable price behavior is not the result of **collusion** but rather of rational economic decision making by each firm in the oligopoly. Also, because oligopolistic firms refrain from competing on a price basis so that prices remain relatively stable, changes in the costs of production, processing, marketing, and so on are not easily passed on and must be absorbed to a large extent.

An oligopolistic market structure in selling often develops in a market because of thin markets or barriers to entry. No matter how freely firms can enter or exit a market, if the market is thin (has a low and stagnant volume), profit opportunities will not exist for more than just a few firms. For any firm to enter a thin market industry, the firm might need to make a sizeable capital investment to capture a large enough share of the market and attain the necessary economies of size. Other barriers to entry can also create the environment in which an oligopolistic market structure will flourish.

Although prices are set in an oligopolistic market in many ways, the most common method is price leadership. In this situation, a particular firm dominates the market either because it controls the largest share of the market or because other firms in the industry view it as more efficient in operation, more proficient in analyzing the market, more experienced, etc. The dominant firm first sets its price so as to maximize its profits. The other firms, with no collusive behavior, simply set their prices at the same level after making any adjustments they feel are justified by the differentiating characteristics of their products. As a consequence of these actions, the price set by the oligopolistic seller is higher than that under perfect competition, making the volume produced and sold lower than would exist under perfect competition. However, the dominant firm may be cost efficient enough to set a low enough price and start a price war that would eventually drive all other firms out of the market. Such a move would establish a monopoly, however, with all the difficult legal problems that would have to be faced.

TABLE 11.3   Food Marketing Mergers and Acquisitions between 1982 and 1996

Year	Processing	Wholesaling	Retailing	Food Service	Total
			Number		
1982	250	38	38	51	377
1983	225	38	45	64	372
1984	242	37	60	78	417
1985	291	64	52	73	480
1986	347	65	91	81	584
1987	301	71	65	77	514
1988	351	71	76	75	573
1989	277	65	53	72	467
1990	208	58	37	47	350
1991	181	35	39	36	291
1992	217	59	29	59	364
1993	266	57	39	71	433
1994	232	62	60	78	432
1995	244	56	42	83	425
1996	210	32	37	120	399

Source: Economic Research Service, USDA.

The automobile industry and aircraft manufacturing industry are two non-agricultural examples of oligopolies. In agriculture, the farm machinery and equipment industry can be classified as an oligopoly. John Deere makes combines that are different from those made by J.I. Case or New Holland. The top four brands sold to farmers and ranchers account for about 80% of all two-wheel tractor sales and 89% of all combine sales in the United States. According to the livestock slaughter summary statistics, the top four-firm cattle slaughter concentration ratios rose from 39.0% in 1985 to around 67% in 1995. Thus, the case can be made of increasing concentration in the cattle slaughter industry.

The pesticide industry and the fertilizer industry also can be classified as oligopolies. In the pesticide industry, three or four firms account for the majority of the total sales to farmers. The products developed in this industry are identified or differentiated by company brand or name.

The climate for mergers and acquisitions has been and continues to be hot. As exhibited in Table 11.3, between 1982 and 1996, nearly 6,500 mergers, divestitures, and leveraged buyouts took place in the food marketing system.

The interdependence of pricing policies among firms and product differentiation makes it difficult to analyze oligopolistic situations. Figure 11.4 illustrates the "kinked" demand curve that economists use to explain the rigid price behavior of oligopolists. Let the demand curve *DD*, which passes through point 1, represent the demand curve when all sellers move prices together and share the total market. And let the demand curve *dd* represent the demand curve when a single oligopolist changes its price. Note that the demand curve *DD* is more inelastic than

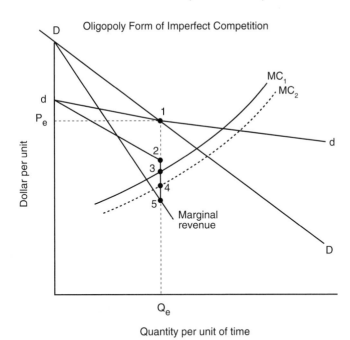

FIGURE 11.4   This figure illustrates the equilibrium price and quantity of an oligopoly. The fact that oligopolists match all decreases but not all increases in price leads to a "kinked" demand curve and discontinuous marginal revenue curve.

the *dd* curve. Because oligopolists take account of the reaction of other oligopolists, there is no single demand curve facing a particular oligopolist. Oligopolies typically match all price decreases by their fellow oligopolists (they do not want to be undersold), but they do not match all price increases (they want to capture a greater market share).

Below point 1, where other oligopolists match the firm's price cut, the demand curve *DD* prevails. Above point 1, the demand curve *dd* will prevail because rival oligopolists will not match the firm's price increase. The kinked demand curve, given by d1D, leads to a break, or vertical discontinuity, in the marginal revenue curve at output Q_e. The segment between points 2 and 5 represents the magnitude of this vertical discontinuity. Shifting marginal cost (*MC*) curves because of technological advances (downward shift from MC_1 to MC_2) intersecting the marginal revenue curve at point 4 rather than point 3 will not change the oligopoly price and quantity.

In meeting demand along the lower segment of the kinked demand curve (i.e., to the right of point 1), the oligopolist will be maintaining its market share, which explains why there is a tendency for prices to remain at P_e. Each oligopolist will earn an economic profit per unit of P_e minus its average total cost.

For a variety of reasons, including the inherent uncertainty of knowing how others will respond, oligopolists facing similar demand and cost conditions may behave in a collusive fashion (i.e., arrange to charge the same price for their output) instead of in the manner depicted in Figure 11.4. Their objective may be to maximize their joint profits. The prices and quantities observed when this situation occurs will be much the same as those charged by a single seller or monopolist. Each oligopolist would charge price P_e, produce its predetermined share of Q_E, and share in the higher level of profits.[2]

Monopoly

At the opposite end of the spectrum from perfect competition in selling is the **monopoly.** Instead of many firms or even a few firms, there is only one seller in the market. A monopoly exists for the same reason that oligopolies exist: barriers to entry. In the case of a monopolist, however, the barriers are sufficiently high to discourage all potential competitors from attempting to enter the industry. The monopoly is similar to an oligopoly, except that the monopolist has no concern for retaliation by competitors in response to changes in pricing. The monopolist sets a price that is higher than would exist under perfect competition to maximize profits. The volume sold also is below that observed under perfect competition. Because the monopolist has no competitors, the price set is usually even higher than would occur under oligopoly.

In practice, however, monopolists tend to hold the price below their profit-maximizing level to discourage both the entry of competitive firms and antitrust litigation. Also, the lack of competitors means that a monopolist can pass on changes in cost to consumers more easily than an oligopolist. Consequently, the prices set by monopolists tend to move closely with movements in input costs that they face in production.

The long-run equilibrium price and quantity under a monopoly are depicted in Figure 11.5. The monopolist will maximize profits by operating where marginal cost equals marginal revenue and then determine the highest price for its product. Figure 11.5 shows that this quantity and price combination occurs at Q_E and price per unit P_E. If the monopolistic firm produced at a lower level, the firm's profits would not be maximized (i.e., it could achieve additional profits by expanding its production to Q_E). If the monopolist produced at a higher level, the costs of producing the additional units would exceed the revenue received from their sale.

Economic profits will exist under a monopoly in the long run because there are likely both economic and legal barriers preventing other firms from entering the sector. The economic barriers might be prohibitive production costs of product prices or outright restraints of trade. The legal barriers might be patents on the design of specific products. Other laws may permit a group of firms to act as a monopoly. Electric power companies and telephone companies are two examples of monopolies. In agriculture, **marketing orders** for specific products such as fluid

---

[2]Firms attempting to collude to reap monopoly profits are referred to as **cartels.** The cartel members jointly establish monopoly prices and quantities and each member's share of total sales. Perhaps the most famous cartel in recent years has been OPEC, or the Organization of Petroleum Exporting Countries. OPEC had a dramatic impact on the world price of oil in the 1970s by restricting the amount of crude oil coming into the world market. Explicit collusion among domestic sellers, however, is in violation of antitrust laws, which we will discuss shortly.

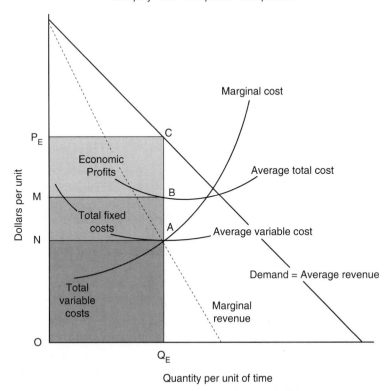

Monopoly Form of Imperfect Competition

FIGURE 11.5   A monopolist also equates marginal costs and revenue. The price charged to consumers at this equilibrium quantity is read off the demand curve. The monopolist differs from other forms of imperfect competitors in that it strives to block others from entering the marketplace. The demand curve faced by the individual firm is the market demand curve. $ONAQ_E$ represents total variable costs, area $NMBA$ represents total fixed costs, area $OMBQ_E$ represents total costs, area OP_ECQ_E corresponds to total revenue, and area MP_ECB depicts economic profits, or total revenue minus total costs.

milk and oranges permit the setting of a higher price by controlling the flow of products into the market. This arrangement enables farmers to receive a higher price than they would have otherwise received.

As with the oligopolist and the monopolistic competitor, the monopolist will choose to sell a smaller quantity than would occur under perfect competition. Each of these imperfectly competitive firms prefers to operate to the left of the minimum point on their average total cost curve, where long-run equilibrium occurs under perfect competition. Thus, there is a cost to society associated with all forms of imperfect competition. The quantity supplied is smaller, the unemployment of resources is higher, and prices are higher than under perfect competition.

TABLE 11.4   Alternative Forms of Market Structure in Selling

Item	Market Structure in Selling			
	Perfect Competition	Monopolistic Competition	Oligopolies	Monopolies
Number of sellers	Numerous	Many	Few	One
Ease of entry or exit	Unrestricted	Unrestricted	Partially restricted	Restricted
Ability to set market price	None	Some	Some	Absolute
Long-run economic profits possible	No	No	Yes	Yes
Product differentiation	None	Yes	Yes	Product unique
Examples in U.S. food and fiber system	Corn producers	Soft drink bottlers	Manufacturers of farm tractors	Marketing orders

Comparison of Alternative Market Structures

Table 11.4 summarizes the basic features of both the perfectly competitive and imperfectly competitive forms of market structure from a selling perspective. For example, Table 11.4 describes the number of sellers, ease of entry and exit, the ability to set market prices, the existence of economic profits in long-run equilibrium, and the extent of product differentiation.

There are numerous corn producers, for example, who sell an undifferentiated product and who individually have no ability to influence market prices. Soft drink bottlers, primarily through advertising, try to differentiate their products. Because the four largest manufacturers of farm tractors account for 80% of total sales, these manufacturers try to differentiate their product (e.g., different colors, model styles, etc.) and have some ability to influence the price of their product. Marketing orders such as that enforced for oranges in the United States are one of the more visible monopolies in agriculture.

Welfare Effects of Imperfect Competition

We can utilize the concept of producer and consumer surplus introduced in earlier chapters to evaluate the economic welfare implications of imperfect competition. We have already learned that imperfect competitors operate to the left of the minimum point on their average total cost curve. This situation, of course, differs from the behavior of the perfectly competitive firm, which operates where the product price is equal to its marginal cost as long as it exceeds the minimum point on its average variable cost curve. This difference occurs because of the downward-sloping nature of the imperfect competitor's demand curve and the desire of these firms to

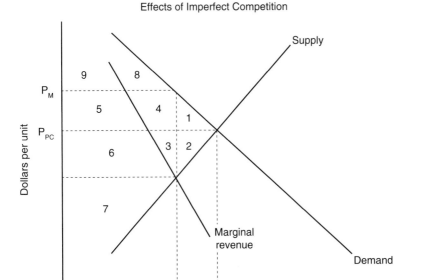

Effects of Imperfect Competition

FIGURE 11.6   Consumer surplus would be lower under a monopoly (areas 8 and 9) than it would be under perfect competition (areas 1, 4, 5, 8, and 9). Producers would have been willing to supply Q_{PC} at a price of P_{PC} under perfect competition. A monopoly, however, would be willing to supply a quantity of Q_M at price P_M. Producer surplus would be higher under a monopoly (areas 3, 4, 5, 6, and 7) than it would be under perfect competition (areas 2, 3, 6, and 7). The economic welfare of society as a whole would decline by the sum of areas 1 and 2.

operate at the point at which their marginal costs are equal to marginal revenue, and to price their product according to the demand curve that lies above the marginal revenue curve. What does this difference in market behavior imply about the economic well-being of producers and consumers?

We can see the maximum impact of imperfect competition on producers and consumers in a particular market if we compare the two extreme forms of market structure: perfect competition and a monopoly. Figure 11.6 indicates that the market equilibrium price under perfect competition would be P_{PC}, and the quantity marketed would be Q_{PC}. Total consumer surplus under these conditions would be equal to the sum of areas 1, 4, 5, 8, and 9. Total producer surplus would be equal to the sum of areas 2, 3, 6, and 7.

If this market exhibited the characteristics of a monopoly instead, the equilibrium price would rise to P_M, and the equilibrium quantity market would fall to Q_M (remember that the supply curve in this case is the monopolist's marginal cost curve). The monopolist would earn economic profits equal to the sum of areas 3, 4, 5, and 6. The total producer surplus would be equal to this economic profit plus area 7. Thus, the producer gained areas 4 and 5 and lost only area 2. Because areas 4 and 5 exceed area 2, we can say that the monopolist would be in a better position economically than all producers were collectively under perfect competition.

Total consumer surplus would fall by the sum of areas 1, 4, and 5 under conditions of pure monopoly, totaling only area 8 plus area 9. So, whereas producers gained areas 4 and 5 and lost only area 2, consumers gained nothing and lost areas 1, 4, and 5. Consumers, therefore, would be in a considerably worse position under a monopoly than they would be under conditions of perfect competition.

Society as a whole would be a net loser if a market's structure were to switch from perfect competition to a monopoly. Society would gain nothing in terms of economic welfare and would lose areas 1 and 2. The sum of areas 1 and 2 is frequently referred to as a **dead-weight loss.** These results support our earlier statements about the relative efficiency of perfect competition from the perspective of consumers and society as a whole.

IMPERFECT COMPETITION IN BUYING

Up to this point, we have considered imperfect competition in selling activities. Imperfect competition in buying activities can influence the market price for resources used in production. Here the supply curve faced by the firm is upward sloping rather than perfectly flat or elastic, as was the case in perfect competition. A meat packing firm can influence market price, for example, by its decisions concerning how many and what kind of livestock of each class and grade to buy. In a given location, this meat packer may be the *only* buyer of beef cattle. In fact, a typical agricultural situation has been one in which a large number of farmers face a single buyer for their product. A textile plant in a small rural "company town" that owns other businesses in town and the apartments that workers live in may have a buyer's monopoly in the local labor market. These situations are termed a **monopsony.**[3]

Monopsony

A buyer in a perfectly competitive input market views the input supply curve as a horizontal line. The perfectly competitive firm's purchases are relatively small and perfectly elastic or do not perceptibly affect market price. A monopsonist, however, is the only buyer in the market and faces an upward-sloping market input supply curve. As a consequence, its buying decisions affect input prices. To increase its input usage, it is necessary for the monopsonist to pay a higher input price. The monopsonist must therefore consider the **marginal input cost** (MIC) of purchasing an additional unit of a resource. Marginal input cost is defined as the change in the cost of a resource used in production as more of this resource is employed.

A numerical example of a set of supply and marginal input cost curves facing a monopsonist is given in Table 11.5. Columns 1 and 2 represent the input sup-

---

[3]Where there are several buyers in a similar situation, oligopsony is the proper term.

TABLE 11.5 Monopsony and Marginal Input Cost

(1) Units of Variable Input	(2) Price per Unit	(3) Total Cost of Input	(4) Marginal Input Cost
1	$3.00	$3	—
2	3.50	7	4
3	4.00	12	5
4	4.50	18	6
5	5.00	25	7
6	5.50	33	8
7	6.00	42	9
8	6.50	52	10
9	7.00	63	11
10	7.50	75	12

ply curve, and column 4 represents the marginal input cost curve. These curves are depicted graphically in Figure 11.7.

The supply curve for a resource facing the monopsonist, which necessarily represents the market supply curve, may also be thought of as the average cost curve for the resource. The marginal input cost curve, which lies above the supply curve, illustrates that the monopsonist must pay higher prices per unit if it wishes to employ greater amounts of the resource. This situation is the mirror image of the monopolistic firm, which must decrease the price of its product if more is to be sold.

A profit-maximizing monopsonist will employ a variable production input such as labor up to the point at which the marginal input cost equals its marginal revenue product. The price of the input is determined by the corresponding point on its supply curve. Marginal revenue product is the addition to total revenue attributed to the addition of one unit of the variable input. Marginal revenue product is equal to the marginal revenue times the marginal physical product, which is the change in output from a change in input use. So long as marginal revenue product exceeds marginal input cost, profit will rise with increasing input use. On the other hand, if marginal input cost exceeds marginal revenue product, profit will rise with decreasing input usage. Therefore, profit is maximized by employing the level of variable input where marginal input cost equals marginal revenue product.

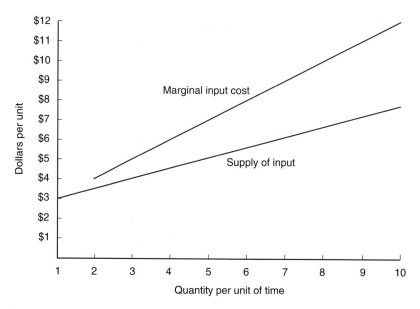

FIGURE 11.7   These curves reflect the cost and input use data presented in Table 11.5.

Equilibrium Conditions under Monopsony and Perfect Competition

FIGURE 11.8   Monopsonists use less of a resource and consequently pay a lower price than they would under conditions of perfect competition.

This situation is depicted graphically in Figure 11.8. The profit-maximizing level of input usage by the monopsonist is given by Q_M, and the price per unit paid is P_M. Here the firm is equating marginal input cost and marginal revenue product at point A, and paying the input price P_M given by the supply curve at point C. This situation differs from the profit-maximizing level of input usage

Equilibrium Conditions under Alternative Combinations of Monopsony, Monopoly, and Perfect Competition

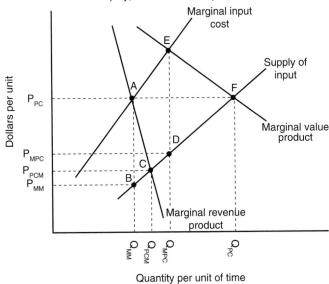

FIGURE 11.9   Each sole buyer or sole seller situation leads to lower levels of production than those given by perfect competition.

under perfect competition given by Q_{PC} and price per unit P_{PC}. Point B is the point at which demand and supply forces jointly determine price. Under perfect competition, the marginal revenue product curve is the same as the marginal value product given by Equation 7.15, because output price P equals average revenue and marginal revenue.

Note that the price paid for the resource is higher under conditions of perfect competition than under a monopsony. The level of resource use also is higher under perfect competition relative to monopsony. The difference between the prices paid for the input under perfect competition and monopsony is termed the **monopsonistic exploitation** of the input.

Finally, consider the situation in which the monopsonist is not only the sole buyer of a resource but also the sole seller of a product. This situation might correspond to that of a meat packer who is the only purchaser of beef cattle and the only seller of wholesale dressed beef in the area. What is the firm's profit-maximizing level of input under this scenario? The answer to this question is illustrated, along with other market structure combinations, in Figure 11.9. The profit-maximizing level of input occurs at point A, where marginal input cost equals marginal revenue product. This point corresponds to input level Q_{MM} and price P_{MM}.

Under conditions of perfect competition in buying but monopoly conditions in selling in a particular market, the profit-maximizing level of input occurs at point C at which marginal revenue product equals the supply of the input. This situation corresponds to input level Q_{PCM} and price P_{PCM}.

Under conditions of perfect competition in selling and monopsony conditions in buying in a particular market, the profit-maximizing level of input is given where marginal input cost equals marginal value product (at point E), which corresponds to Q_{MPC} and P_{MPC} as the price paid for that input.

Under conditions of perfect competition for both selling and buying in a particular market, the profit-maximizing level of input usage is given where marginal value product equals the supply of the input (at point F). Thus, depending on whether imperfect or perfect competition prevails, profit-maximizing input levels and the prices paid for those inputs may vary considerably.

Oligopsony and Monopsonistic Competition

Similar considerations may prevail in buying activities in the case of oligopsony. When three meat packing firms dominate a market, farmers often contend that prices are kept down because the respective firms follow a nonaggressive buying policy.

Oligopsony is defined as a market or industry comprised of relatively few firms engaged in the purchase of resources. **Monopsonistic competition** is defined as an industry composed of many firms buying resources with the capacity of differentiating services. Differentiation of services to producers by buyers of resources may exist in terms of convenience of distribution and location of processing facilities, willingness to provide credit and/or technical assistance, and perhaps personal characteristics of the buyer. The quantity of a resource purchased at various prices by the oligopsonist or by the firm under monopsonistic competition to maximize profit is determined in the same manner as shown for the monopsonist.

The amount of profit earned under conditions of oligopsony or monopsonistic competition will depend upon the elasticity of supply for the resource. The more inelastic or steeper the supply curve for the resources, the larger the amount of profit earned. Under oligopsony, supply would most likely be less elastic than under monopsonistic competition as a result of fewer firms and/or substitutable services. Note the symmetry between the increasing elasticity of supply and lower levels of profit in buying activities, and the increasing elasticity of demand and lower levels of profit in selling activities.

MARKET STRUCTURES IN LIVESTOCK INDUSTRY

To summarize imperfect competition from both a buying and a selling perspective, let's look at the market structures found in the U.S. livestock industry illustrated in Figure 11.10. Producers of feeder cattle, feeder pigs, and feeder lambs represent the original suppliers of the raw ingredient in the industry and closely resemble perfect competitors. These producers are so numerous that the actions of an individual producer have little or no effect on the market. They sell a relatively homogeneous product and have a fairly good knowledge of market opportunities and alternative prices. Because there are some differences in cattle,

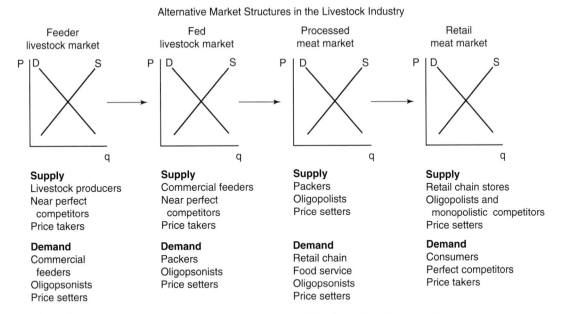

FIGURE 11.10   An examination of the market structures underlying demand and supply in the market channels of the livestock industries suggests the presence of several forms of imperfect competition.

hog, and sheep, and because they do not have perfect knowledge of all possible market conditions, these producers operate within a near-perfect competition market structure. They are basically price takers and are powerless to affect the prices at which they sell their livestock. The consequence is that these livestock producers must accept the price outcomes from the interactions of groups further up the market channel.

A relatively new number of commercial feeders purchase feeder livestock from livestock producers. Consequently, commercial feeders operate as oligopsonists when buying feeder livestock. However, because there are more feeders with less market clout than packers, commercial feeders typically must accept the role of price takers when selling fed animals to packers.

Packers behave as oligopsonists in buying slaughter animals, but behave as oligopolists in attempting to sell their processed meat products. In fact, they sell their product facing the oligopsonistic behavior of both large retail chain stores, large food service buyers, and many small wholesale and retail operations in specific geographical areas. This situation is known as bilateral oligopoly.

Large retail food stores and food service chains operate nationally as oligopolists in product markets, along with a fringe of monopolistic competitors in some local areas. They face a large number of consumers who purchase meat products as near perfect competitors. The retail price is generally set through the price leadership of some dominant retailer or group of retailers in the oligopoly.

At the end of the market channel depicted in Figure 11.10 are consumers, the ultimate users of processed meat products prepared for retail sale. Consumers are price takers in a nearly perfect competition environment. They face the oligopolistic behavior of the large national retail chains and the more price-competitive behavior of local retailers.

GOVERNMENTAL REGULATORY MEASURES

Various measures may be employed to counteract possible adverse effects of imperfect competition in the marketplace. These measures include legislative acts, institution of maximum or ceiling prices received for output, use of lump-sum taxes, and institution of minimum prices paid for resources.

Legislative Acts

Historically, legislative acts have been passed by Congress in an attempt to minimize the social waste of imperfect competition. In 1890, Congress passed the Sherman Antitrust Act, prohibiting monopoly and other restrictive business practices. Section 1 of this historic act makes it illegal to act in restraint of trade by conspiring with other individuals or firms. The act forbids restraining trade through price-fixing arrangements or controlling and sharing industry output by collusive agreement. Section 2 of the act forbids the use of economic power to exclude competitors from any market.

In 1914, two further measures were enacted, namely the Federal Trade Commission Act and the Clayton Act. The Federal Trade Commission was charged with the responsibility of investigating business organizations and practices and with carrying out the provisions of the Clayton Act. Although the Sherman Antitrust Act was general in identifying what actions were illegal, the Clayton Act was quite specific. The Clayton Act thus plugged loopholes and deficiencies in the Sherman Act.

A typical market structure situation in agriculture is one in which a large number of farmers face a single buyer (a monopsonist) for their products. Recall that the monopsonist has sufficient market power to be able to hold down the offered price. To offset this disparity in market strength, farmers historically have organized **agricultural bargaining associations,** or farmers' cooperatives. One example of a cooperative in the marketing and distribution of oranges is Sunkist Growers, Inc. Cooperatives are business organizations that are exempt from federal antitrust legislation and whose chief goal is to improve producer income through either higher prices received for products or lower prices paid for inputs. The growth of farmers' cooperatives in the United States dates from the passage of the Capper-Volstead Act of 1922, which exempted cooperatives from certain restraints imposed by the Sherman Antitrust Act of 1890 and the Clayton Act of 1914. The Capper-Volstead Act of 1922 was the principal legislation exempting cooperatives from antitrust laws.

A number of other acts that are favorable to agricultural organizations have been passed by Congress. These laws, which were enacted to protect agricultural producers and their associations from firms handling their products, are:

- Packers and Stockyards Act of 1921, which reinforced antitrust laws regarding livestock marketing;
- Cooperative Marketing Act of 1926, which permitted farmers or their associations to acquire, exchange, and disseminate a variety of price and market information;
- Robinson-Patman Act of 1936, which primarily covered price discrimination practices; and
- Agricultural Marketing Agreement Act of 1937, which established agricultural marketing orders.

Marketing orders and agreements refer to arrangements among producers and processors of agricultural commodities. The chief goal of a marketing agreement or order is to improve producer income through the orderly marketing of a commodity by a group of producers and to avoid price fluctuations faced by an individual operating alone in the marketplace. Commodity prices may be controlled through negotiations with different groups involved in the marketing process or limitations on the supply placed on the market. Marketing orders have historically been used by the federal government to control the quantity coming into specific markets and, hence, support the market prices and incomes received by farmers.

If an agricultural bargaining association is to be effective in increasing farmer-members' incomes, it must gain control of, or influence, total supply. However, such control or influence is very difficult to achieve and almost impossible to maintain. Not all growers of a commodity are convinced that they should join the bargaining association. A further difficulty is due to the inability to control the supply of substitutes or imported quantities of that good. Finally, success in bargaining brings about a loss in incentive to maintain membership in the association. When product price is increased through the actions of the bargaining association, some of the group members may be tempted to leave the association. Under the scenario of increases in product prices, a strong incentive exists for each producer to expand production and consequently total market supply. Historically, bargaining associations have not been very successful. Nevertheless, they have been a force in improving the degree of competition in markets for agricultural commodities.

To illustrate this situation, consider the article in Box 11.1. This account gives both a historical and current perspective on the creation of a monopoly by the federal government, and what this situation may mean for producers and consumers. This article shows the many sides to the debate over monopolies. They benefit the producer of the product (in this case, oranges); they discourage production through market quotas, which affects input suppliers to these firms and their employees; they lead to higher market prices, which raises incomes of producers; and they lead to higher costs to consumers.

BOX 11.1   Orange Police

It isn't likely that the higher prices for orange products caused by the recent freeze in California are going to shock many people, who are already astounded at the high price they have to pay for products like orange juice, even in normal times. Welcome to the wonderful world of "marketing orders," an archaic federal device that, unlike a freeze, hits consumers every year.

By law, some 35% of California's annual orange crop—three billion oranges—must be exported, juiced or left to rot. Ending citrus marketing orders would spare consumers much of the price hikes for these products.

Federal marketing orders were created in 1937 as a short-term way to tide growers over in the Depression. Like many federal programs, they have long outlived whatever justification they had. By restricting the amount of citrus fruit, many vegetables and most nuts that can be sold, marketing orders keep prices artificially high and ensure the market dominance of such agricultural cooperatives as Sunkist Growers, Inc.

Marketing orders allow Sunkist to use the power of the state to enforce an unregulated monopoly. In 1985, for example, the government went after Carl Pescosolido, a dissent grower from Exeter, Calif., for giving away illegal oranges to charity and failing to fill out quota forms. Mr. Pescosolido is a quieter rebel now, but he is still outraged at the system. "When they say I can sell only 58% of my oranges, it's like someone taking 42% of my land," he told us.

Mr. Pescosolido even has allies inside the government. In October, the Antitrust Division of the Justice Department and the Small Business Administration both called for an end to citrus quotas. "Volume controls impose clear-cut costs to society, whereas any benefits flowing from price-stabilization efforts are speculative," Justice wrote. It noted that marketing orders allow a Sunkist-controlled industry committee "to act as a legalized cartel to set output and hike prices."

The Agriculture Department's own studies show that during the few times quotas have been temporarily lifted, the market has performed well and growers didn't go out of business. Indeed, in recent years the department has ended marketing orders for hops, tart cherries, and some grapefruits at the request of growers who no longer saw any sense in misusing scarce land, capital, labor, and federally subsidized water. But the Agriculture Department is adamant on keeping the citrus quotas intact, a policy that of course ensures that a healthful food remains inordinately expensive in the poorest neighborhoods. "Oranges are not an essential food. People don't need oranges. They can take vitamins," says USDA official Ben Darling in a splendid imitation of Marie Antoinette.

The Agriculture Department is debating whether or not to suspend the current citrus quotas now that so much of the California crop has been lost. The time is ripe for Secretary Yeutter to use his power to make any suspension permanent and end the U.S.'s disastrous 53-year-old legalization of what amounts to a domestic agricultural OPEC.

Source: *Wall Street Journal,* January 2, 1991.

Effect of a Maximum Price Ceiling on a Monopolist

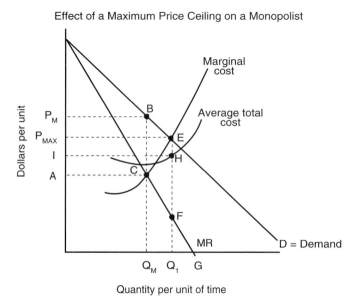

FIGURE 11.11   The profit earned by a monopolist will fall if a price ceiling is instituted.

Countervailing actions as related to the agricultural sector have included the establishment of cooperatives and the development of marketing orders and agreements. However, other measures may also be employed to counteract the possible adverse effects of imperfect competition.

Ceiling Price

A regulatory agency such as the Federal Trade Commission may reduce profit and actually effect an increase in the output of a monopoly by instituting a maximum or ceiling price that can be charged. This countervailing action is illustrated graphically in Figure 11.11. Without assistance from the regulatory agency, the monopolist would produce Q_M, charge P_M per unit, and earn a profit that corresponds to area AP_MBC. If a governmental regulatory agency imposes a price ceiling of

P_{MAX}—which of course is below P_M—the demand curve is given by $P_{MAX}ED$. Over the range in which the demand curve is horizontal, price equals marginal revenue because the monopoly can sell additional units of output without lowering price. Therefore, $P_{MAX}E$ is the marginal revenue curve up to Q_1 units of output. At output levels that exceed Q_1, the original demand curve is unchanged, so the FG segment of the original marginal revenue curve associated with the ED portion of the demand curve is still relevant. The entire marginal revenue curve is then $P_{MAX}EFG$. The marginal revenue curve is discontinuous at Q_1. Therefore, the imposition of the price ceiling of P_{MAX} would cause the monopolist to produce Q_1, charge P_{MAX} per unit, and earn a profit equal to $P_{MAX}EHI$. Note that by instituting a price ceiling, the government would encourage the monopolist to produce more output $(Q_1 - Q_M)$, and charge a lower price $(P_{MAX} - P_M)$. The profit earned by the monopolist falls from AP_MBC to $P_{MAX}EHI$. Thus, this price regulation reduces monopoly profits and lowers price.

Lump-Sum Tax

Alternatively, a governmental regulatory agency may eliminate or reduce profit of a monopoly by assessing a **lump-sum tax** on the firm's operation. This lump-sum tax may be a license fee or one-time charge. In essence, this countervailing action corresponds to a fixed tax, regardless of the level of output. This situation is exhibited in Figure 11.12. Without the lump-sum tax, the monopolist would produce Q_M, given by the intersection of marginal cost and marginal revenue at point F, charge P_M per unit, and earn a profit equal to area AP_MBC. With the imposition of the lump-sum

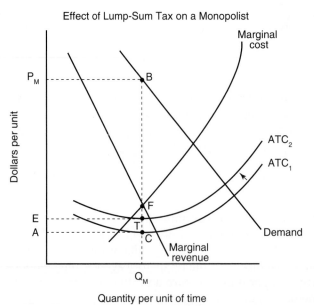

Effect of Lump-Sum Tax on a Monopolist

FIGURE 11.12   Imposition of a lump-sum tax would lower profits but leave the level of output and product price unchanged.

tax, however, the firm's average total cost (ATC) curve shifts upward from ATC_1 to ATC_2. Hence, the monopolist under this arrangement would still produce Q_M and charge P_M per unit. However, the profit now earned by the monopolist is reduced to area EP_MBT, which is less than the profit earned without the tax.

Minimum Price

In monopsony, the government could regulate the price of a resource purchased by imposing a minimum price that must be paid for the resource. In the case of the resource labor, this government regulation is the minimum wage law. This countervailing action is depicted graphically in Figure 11.13.

Without this minimum price regulation, the monopsonist would employ Q_M units of the resource and would pay P_M per unit. With the minimum price regulation, that is, imposing a price P_F that is higher than P_M, the firm's marginal input cost curve becomes P_FDCB with a discontinuous portion equal to the vertical distance CD. If the firm employs a level of input less than Q_F, marginal input cost is less than marginal revenue product, and the firm will be able to expand profits by employing more units of the input. At an input level greater than Q_F, marginal revenue product is less than marginal input cost, so the firm would not employ inputs at this level. This reasoning suggests that the firm will employ an input level of Q_F, to achieve profit maximization. Under this regulation, the monopsonist would employ Q_F units of the resource and pay P_F per unit. Interestingly, the minimum price regulation causes the monopsonist to employ more units ($Q_F - Q_M$) and pay a higher price per unit ($P_F - P_M$) for the resource.

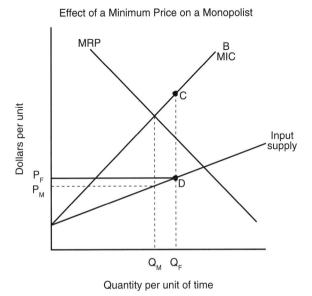

Effect of a Minimum Price on a Monopolist

FIGURE 11.13 The impact of this government regulation is to raise the level of resource use and cost per unit.

SUMMARY

The purpose of this chapter was to discuss the determination of the price and quantity experienced under conditions of imperfect competition. The major points made in this chapter may be summarized as follows:

1. Imperfectly competitive forms of market structure can be classified by:

   - the number of firms and size distribution in the market,
   - the degree of product differentiation,
   - the extent of barriers to entry, and
   - the economic environment within which the industry operates.

2. If there is a relatively large number of firms selling (buying) a differentiated product, monopolistic (monopsonistic) competition is said to exist.

3. If there are just a few firms selling (buying) a standardized product, an oligopoly (oligopsony) is said to exist.

4. If there is only one firm selling (buying) a product, the firm is said to be a monopolist (monopsonist).

5. The profit-maximizing level of output for any firm is determined where marginal revenue equals marginal cost.

6. The profit-maximizing level of input usage for any firm is determined where marginal revenue product equals marginal input cost.

7. Social costs of imperfect competition exist in the marketplace.

8. Several countervailing measures exist for offsetting the adverse effects of imperfect competition, including price regulation and taxation, and marketing orders and agreements.

DEFINITION OF KEY TERMS

Agricultural bargaining associations: associations (e.g., cooperatives) formed by producers to improve the degree of competition in markets for agricultural commodities.

Bilateral oligopoly: a market situation in which there are only a few sellers (oligopoly) facing only a few buyers (oligopsony).

Capital cost barrier: a barrier wherein the capital investment required for efficient operation is sufficiently large.

Cartel: cooperative pool formed by oligopolists to set an artificially high price.

Collusion: a situation designed to increase or maintain profits through price-fixing and/or to restrict entry of new firms in an industry.

Countervailing actions: measures to counteract the possible adverse effects of imperfect competition.

Dead-weight loss: social costs of imperfect competition.

Differentiated product: a product that is made different from others through advertising or quality variation.

Imperfect competition: all those market possibilities that do not meet the conditions of perfect competition.

Lump-sum tax: a fixed tax (e.g., a license fee), irrespective of the level of output, used by regulatory agencies to eliminate or reduce the profit of a monopoly.

Marginal input cost: sometimes referred to as marginal resource cost or marginal factor cost. The term is the addition to total cost of a resource as the result of employing an additional unit of that resource.

Marketing orders: arrangements among producers and processors of agricultural commodities in which the chief goal is to improve producer income through the orderly marketing of a commodity. Historically, marketing orders have been used by the federal government to control the quantity of a commodity coming onto specific markets.

Monopolistic competition: a market structure in which a large number of firms produce a differentiated product. It is relatively easy to enter such a sector.

Monopoly: a market structure that has only one firm selling to buyers.

Monopsonistic competition: a market structure in which there are a large number of buyers of resources, but there exists the capacity of the buyers to differentiate services. Differentiation of services includes convenience of distribution and location of processing facilities as well as willingness to provide credit and/or technical assistance.

Monopsonistic exploitation: the difference between the prices paid for an input under perfect competition and monopsony.

Monopsony: a market structure that has only one firm buying from sellers.

Nonprice competition: attempts to increase the demand for a product or service via product differentiation and to make the demand for the product less price-elastic.

Oligopoly (oligopsony): a market structure in which there are a small number of sellers (buyers). Each seller or oligopolist (buyer or oligopsonist) knows how the other sellers (buyers) will respond to any changes in quantity marketed or prices he or she might initiate.

REFERENCES

Dahl C, Hammond W: *Market and price analysis: the agricultural industries,* New York, 1977, McGraw-Hill.

Ferguson CE, Gould JP: *Microeconomics theory,* 4th ed., Homewood, Ill, 1975, Richard D. Irwin.

Kohler H: *Intermediate microeconomics: theory and applications,* Glenview, Ill, 1982, Scott, Foresman.

Purcell WD: *Agricultural marketing: systems, coordination, cash, and future prices,* Neston, Va, 1979, Neston Publishing.

Robinson J: *The economics of imperfect competition,* London, 1933, Macmillan.

EXERCISES

1. Plot the marginal revenue curve associated with the following demand curve faced by a monopolist or a monopolistic competitor.

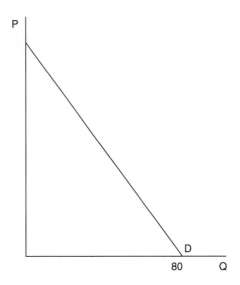

2. Use the graph to the right to answer the following questions:

   a. What price is charged by the monopolist in order to maximize profits?

   b. Calculate the total revenue accruing to the monopolist at the profit-maximizing output.

   c. Calculate the total cost to the monopolist at the profit-maximizing output.

   d. Calculate the profit for the monopolist.

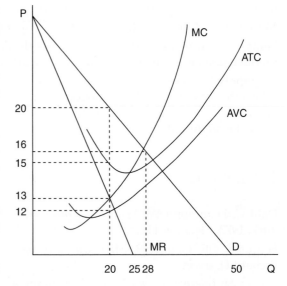

   e. Calculate the total variable and fixed costs of the monopolist at the profit-maximizing output.

   f. Now assume the *MC* curve represents market supply for a perfectly competitive market. What would the equilibrium price and quantity be for perfect competition? Are consumers better off or worse off with perfect competition or monopoly?

3. List differences and similarities among monopolies, oligopolies, and monopolistic competition. Be prepared to give examples of each form of imperfect competition on the selling side.

4.

Units of Variable Input	Price/Unit	Total Cost of Input	Marginal Input Cost
1	2		
2	2.5		
3	3		
5	4.5		
8	6		

   a. Calculate the total input cost and the marginal input cost.

   b. If the marginal value or marginal revenue products were 4, what would be the profit maximizing level of input?

5. a. Find the equilibrium price and quantity for a monopsonist in the graph below.

   b. Find the equilibrium price and quantity under perfect competition in the graph below.

   c. What is the magnitude of monopsonistic exploitation?

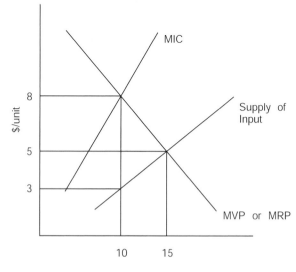

Quantity per unit of time

6. Explain the significance of the following acts:

    a. Clayton Act

    b. Capper-Volstead Act

    c. Packers and Stockyards Act

7. List and explain the various measures that may be employed to counteract possible adverse effects of imperfect competition in the marketplace.

8. On the following graph, show the effect of a lump-sum tax on a monopolist.

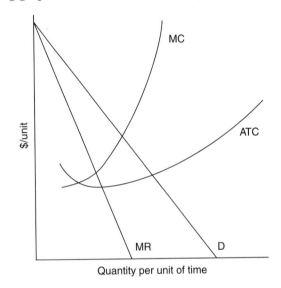

9. Using the graph below, answer questions a through d.

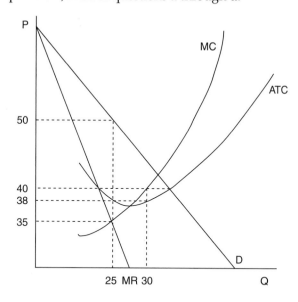

a. What are the profit-maximizing price and quantity levels for the monopolist?

b. Calculate profit.

c. Suppose the government imposes a price ceiling of $40. Now what is the optimal price and quantity combination?

d. Calculate the new level of profit.

PART

IV

GOVERNMENT IN THE FOOD
AND FIBER INDUSTRY

12

GOVERNMENT INTERVENTION IN AGRICULTURE

> *All governments like to interfere; it elevates their position to make out that they can cure the evils of mankind.*

Walter Bagehot (1826–1877)

TOPICS OF DISCUSSION

RATIONALE FOR GOVERNMENT INTERVENTION

In Chapter 1, we made the distinction between capitalism, which is largely free of government intervention, and socialism, which is characterized by massive government intervention. We concluded that the United States exhibits a mixed economic system, or a market economy in which specific sectors are regulated and/or protected by government.

Government provides many important functions, including the provision of law and order, a national defense, an infrastructure (construction and maintenance of highways, sanitation programs, medical programs, education programs, communication facilities, etc.), and economic and social regulations. The latter function includes providing the legal foundation and social environment conducive to an efficient market system, maintaining competition in the marketplace, redistributing income and wealth to accomplish specific social objectives, monitoring the use of resources, and stabilizing the economy.

The rationale typically advanced for government intervention in the agricultural sector of an economy is the need to:

- support/protect an infant industry,
- curb market powers of imperfect competitors when necessary to promote social good,
- provide for food security,
- provide for consumer health and safety, and
- provide for environmental quality.

Tariffs, a form of tax on imported goods that raises their price to domestic consumers, and import quotas are typically used to shield an infant industry from foreign competition. The nation's food and fiber industry in the new millennium clearly no longer qualifies as an infant industry. Although there may be other grounds for justifying government intervention in the farm sector and the food and fiber industry, the justification of the U.S. food and fiber industry being an infant industry is no longer true.

The **food and fiber industry** in general, and the farm sector in particular, are not characterized by small numbers of imperfect competitors. The federal government, for example, has taken action to prohibit what it perceived as imperfect competitor behavior by the nation's breakfast cereal manufacturers. The **Federal Trade Commission (FTC),** an agency of the federal government charged with the responsibility of prohibiting companies from acting in concert to increase their market control and of prohibiting false and deceptive trade practices, charged that Kellogg, General Foods, and General Mills acted in such a way as to maximize their joint profits. These three firms, which together accounted for more than 80% of total cereal sales, were accused of preventing others from entering the sector by saturating consumers with heavy television advertising and a wide variety of brands, which effectively dominated shelf space in grocery stores. Ultimately, however, the FTC dropped the case.

Many nations intervene in their food and fiber industry to ensure an adequate food supply from domestic sources. Although it is an important issue in all countries, some nations drain off resources from other potential uses in an effort to meet their food needs internally.

Sometimes this rationale is based upon the uncertainties associated with imported food supplies and prices, and the drain on the country's foreign exchange reserves required to pay for imports. Although the U.S. Secretary of Agriculture watches stocks of specific commodities when conducting federal commodity programs, food security reasons are rarely cited as a rationale for federal government intervention in the food and fiber industry.

The food and fiber industry is one of the nation's most highly regulated industries. Regulations that govern the use of inputs to produce crops and livestock, regulations that govern the processing and manufacturing of food and fiber products, and regulations that govern the markets in which these products are traded are just a few examples of how the government controls the food and fiber industry. Government regulations cover everything from the regulation of chemical use in crop production to the inspection of crops and livestock products before they reach the retail grocery store shelves.

The **Environmental Protection Agency (EPA)** and other government agencies are charged with the responsibility of monitoring the use of the nation's resources, including land (soil erosion), water (pollution), labor (occupational safety), and air (pollution). Thus, the farm sector and the rest of the nation's food and fiber industry are heavily influenced by government intervention.

The purpose of this chapter is to broadly discuss the issues associated with the production of food and fiber products in the United States. Chapter 13 will go into more detail on specific federal programs aimed at the farm sector and their implications for other sectors in the food and fiber industry.

FARM ECONOMIC ISSUES

Politicians often debate the federal government's practice of making payments to farmers in an effort to support their farm incomes. This debate, which often pits rural politicians with their urban counterparts, has increased in intensity as the farm sector's relative contribution to national output has declined steadily over the post–World War II period. We will examine the nature of the issues involved and introduce the various ways in which the federal government intervenes to alter farm economic conditions.

Defining the Farm Problem

What is the farm problem? Two symptoms of the farm problem illustrated in Figure 12.1 are: (1) output fluctuations from one year to the next due to weather patterns, disease, and technological change, and (2) low net farm incomes. The inelastic nature of the own-price elasticity of demand for agricultural products, the lack of

Effects of Inelastic Demand on Farm Revenue

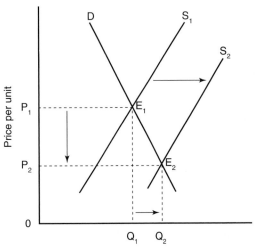

Price per unit

Quantity per unit of time

FIGURE 12.1   This figure illustrates what happens to the total revenue received by farmers facing an inelastic demand for their products if the supply curve were to shift to the right. Total revenue would fall because area $OP_1E_1Q_1$ is greater than $OP_2E_2Q_2$.

market power by farmers and ranchers, the interest sensitivity of the sector, and the fixity of farm assets and chronic excess capacity represent the real roots of the farm problem.

Inelastic Demand and a Bumper Crop.   Any sector facing an inelastic demand for its products will suffer a decrease in its total revenue if market prices fall.

Crop and livestock production is subject to the vagaries of climatological and biological phenomena. Wet springs, dry summers, and early frosts, combined with things such as corn blight and cholera, will shift the sector's supply curve to the left, and bumper crops and technological change will shift supply curves to the right. A record crop, which is more the rule than an exception, can lead to sharp declines in farm product prices and income levels.

Before the shift in the supply curve from S_1 to S_2, farmers were receiving a total revenue equal to the area formed by $OP_1E_1Q_1$. After the shift in the supply curve to the right, however, farmers were receiving a total revenue equal to the area formed by $OP_2E_2Q_2$. Obviously, the second area is smaller than the first, implying that the economic well-being of farmers would be diminished if supply increased. Observed another way, market prices would fall (P_1 to P_2) more than quantities marketed increased (Q_1 to Q_2).

Would farm revenue fall by a smaller amount in years of bumper crops if the demand for the crop were more elastic? Figure 12.2 illustrates the effects of a more elastic demand on farm product prices. In Figure 12.2, we see that the market equilibrium price would decline from P_1 to P_3 if the demand curve facing farmers were represented by curve D_1 and the supply curve were to shift from S_1 to S_2. If the sector were instead confronted by a more elastic or flatter demand curve, rep-

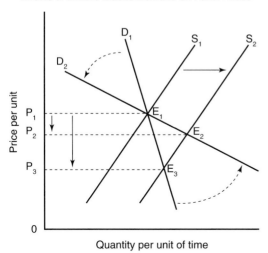

Effects of a More Elastic Demand on Farm Prices

FIGURE 12.2   This figure illustrates an important phenomena of interest to farmers: the more inelastic the demand, the more prices will fall if supply increases.

resented by curve D_2, what would happen to the product price if the supply curve were to shift by the same amount (i.e., from S_1 to S_2)? The market equilibrium would be E_2 instead of E_1, and the equilibrium price would only fall to P_2 instead of all the way to P_3.

We can conclude from Figure 12.2 that the more inelastic or steeper the demand for farm products, the greater the decline (rise) in market prices will be for a given increase (decrease) in supply. A bumper crop, for example, could cause farm product prices to drop sharply under these conditions. Conversely, a drought that shifts the supply curve to the left could lead to a substantial increase in farm product prices.

Lack of Market Power.   In Chapter 10, we learned that farmers and ranchers come close collectively to satisfying the conditions of perfect competition. There are a large number of producers producing a homogeneous product. No one farmer or rancher has sufficient market power to influence the market equilibrium price. Therefore, if Walt Wheaties suffers a sharp decline in wheat yields and does not have crop insurance, his revenue from wheat production will fall markedly, unless a large number of other wheat farmers experience the same decline in yields, and the sector supply curve shifts to the left.

Interest Sensitivity.   The farm sector is one of the most highly capitalized sectors in the U.S. economy. There is more capital invested per worker than in any other sector of the economy. Farmers must borrow substantial amounts of money through short-, intermediate-, and long-term loans to purchase variable and fixed inputs. Thus, an increase in interest rates will increase farmers' interest expenses and, hence, their total production expenses. The high interest rates in the early 1980s caused interest payments to reach almost 20% of total variable production expenses, as compared to approximately 7% in the late 1970s. Higher production expenses will lower net farm income, all other things constant.

In addition to their effects on farm production expenses, higher interest rates in the U.S. economy vis-à-vis the rest of the world will raise the value of the dollar in foreign exchange markets. The higher exchange rate between the dollar and other currencies will make it more expensive for other nations to import U.S. farm products. This will create lower demand for U.S. farm products, which will cause farm prices for products such as wheat and corn to drop, thereby lowering farm revenue.

Asset Fixity and Excess Capacity.   Another major problem facing farmers and ranchers is the notion of asset fixity and excess capacity. **Asset fixity** refers to the difficulty that farmers have in disposing of tractors, plows, and silos when downsizing or shutting down the business. These assets have little or no alternative uses and often sell for pennies on the dollar during hard times in the sector like the early and mid-1980s. They would have experienced a similar fate in the late 1990s were it not for substantial government subsidies. **Excess capacity** refers to the fact that the sector often produces more than it can sell, leading to rising stocks of surplus commodities such as corn, wheat, and cheese. Technological change, which shifts the sector's supply curve to the right, can lead to excess capacity. The combination of excess capacity and asset fixity can lead to a decline in farm asset values in times of surplus stocks. This makes it difficult for farmers and ranchers to scale back their existing operations to desired levels.

Forms of Government Intervention

Government intervention in agriculture required to improve farm economic conditions has taken many forms. We will discuss four forms in this chapter.

Adjusting Production to Market Demand.   One approach to improving farm economic conditions is to reduce the number of resources employed to produce a product plagued by surplus conditions. This is achieved by the government's renting whole farms or by paying farmers not to produce the product by requiring them to set aside part of the land normally used to plant this crop in order to qualify for farm program benefits. With less land planted to this crop, supply will decline and market prices will rise. If the demand curve is inelastic, farm revenue will increase.

Consider the opposite of the bumper crop case illustrated in Figure 12.1. Assume the original supply curve was S_2, and that policies that restrict resource use shift the supply curve back to S_1. Market equilibrium will now occur at E_1 instead of E_2. Because prices rise by a greater percentage (from P_2 to P_1) than quantity declines (from Q_2 to Q_1), total revenue represented by area $OP_1E_1Q_1$ will be greater than the total revenue represented by area $OP_2E_2Q_2$.

Price and Income Support Payments.   Another general approach by government to improve farm economic conditions is to directly support farm prices and incomes, which can be achieved by establishing a price floor supported through government purchases of surplus commodities. If the level of production rises during the year, the government can step in and buy (and store) excess supply at the announced price floor. This would prevent farm revenue from falling below minimum desired levels. An alternative approach is to support farm incomes through direct transfer payments from the government to farmers. The **FAIR Act,** passed in 1996, called for a number of efforts to help producers manage their exposure to risk. This included the establishment of the Risk Management Agency within the USDA and the implementation of pilot programs designed to provide revenue insurance. Thus, the nature of programs employed in the past to support pricing and/or income has taken several different forms. Since most payments under the 1996 FAIR Act are now fixed, farm income can fluctuate more from one year to the next as market conditions change. In response to the low commodity prices in the late 1990s, Congress doubled the fixed $5 billion in scheduled payments with emergency disaster relief in 2000. The federal government also provided marketing assistance loans that guaranteed a minimum price for major commodities, which paid participating farmers $7 billion in 1999. These programs and their implementation over the 1950–2002 period are discussed in Chapter 13.

Foreign Trade Enhancements.   A third approach to improving farm economic conditions involves the link between agriculture and foreign markets. There are essentially two general ways in which such enhancements can improve farm economic conditions in the United States. First, the government can institute **tariffs,** or a tax on imports, which make imported agricultural products more expensive to domestic consumers, or institute **quotas,** which limit the quantity of a particular good that can be imported. Both actions protect producers in the domestic agricultural sector. The Foreign Agricultural Service (FAS) has the responsibility of negotiating and enforcing trade agreements. The FAS in 2000 must address some 650 new trade barriers, up from the 400 barriers in 1993. It is also authorized to spend over $1 billion in 2001 on export activities, including subsidies to U.S. firms facing unfair subsidies by foreign competitors and loan guarantees to foreign buyers of U.S. farm products. Quality restrictions on imports that grade imports or specify their sanitary conditions (pesticide and other chemical residues) can also effectively prohibit imports. These actions have the effect of limiting the supply coming into the market and raising the farm revenues of domestic producers. For example, a higher quota would shift the supply curve to the left and thus, like production controls, have the opposite effect of the bumper crop example illustrated in Figure 12.1. Prices will increase by a greater percentage (from P_2 to P_1) than will the decline in quantity (from Q_2 to Q_1), and farm revenue will increase from area $OP_2E_2Q_2$ to area $OP_1E_1Q_1$.

Second, the government can enhance the attractiveness of U.S. produced agricultural products in foreign markets by subsidizing their purchase, thereby stimulating the export demand for U.S. agricultural products. Export credits help potential buyers finance the purchase of U.S. agricultural products. Subsidies are grants given by the government to private businesses to assist enterprises deemed advantageous to the public. Commodity assistance programs, such as P.L. 480, promote exports under direct food aid or subsidized concessionary sales.[1] These actions have the effect of shifting the demand curve to the right, and they increase farm revenue of domestic farmers.[2]

The specifics of many of these and other foreign trade enhancements, along with the arguments for international agricultural trade, free trade, and multinational trade agreements, will be discussed in depth in Part VI of this book. The current application of farm commodity policies is discussed in depth in Chapter 13.

Other Forms of Intervention.   The government has used a variety of other approaches to enhance economic conditions in agriculture. These approaches include credit subsidies to farm borrowers who cannot otherwise qualify for credit, and loan guarantees for loans made to farmers and ranchers by commercial banks and other nongovernment lenders. The USDA provides about $700 million annually in direct loans to eligible farmers, who are primarily beginning farmers or socially disadvantaged farmers who cannot obtain credit from traditional lenders. These loans are available at subsidized rates from the **Farm Service Agency.** The USDA also guarantees loans made by traditional lenders used to finance operating expenses and buy farmland. In addition, the federal government sponsors the Farm Credit System discussed earlier in Chapter 2, as well as Farmer Mac, which increases the liquidity of commercial banks and Farm Credit System lending associations by purchasing their loans for resale as bundled securities.

The USDA also helps farmers by providing subsidized crop insurance through private companies. Farmers, for example, pay no premiums for coverage against catastrophic production losses (CAT insurance). In addition, the USDA subsidizes their premiums for higher levels of coverage. Crop insurance expenditures by the USDA run about $1.5 billion a year, which includes payments to private insurance companies.

---

[1]Public Law 480 is the name commonly given to the Agricultural Trade Development and Assistance Act of 1954, which seeks to expand foreign markets for U.S. agricultural products, combat hunger, and encourage economic development in developing countries. The act contains three titles: (1) Title I makes U.S. agricultural products available through long-term dollar credit sales at low interest rates for up to 40 years (often referred to as the Food for Peace Program), (2) Title II makes donations for emergency food relief abroad, and (3) Title III authorizes "food for development" grants.

[2]Of particular interest in recent years is the **Export Enhancement Program (EEP)** initiated in 1985 to help U.S. exporters meet competitor nation prices in subsidized markets. Under the EEP, exporters are awarded commodity certificates, which are redeemable for Commodity Credit Corporation (CCC) owned commodities, thereby enabling them to sell certain commodities to specific countries at prices below those found in U.S. markets. Annual quantity and expenditure levels for the Export Enhancement Program must be within the limits set in the Uruguay Round Agreement on Agriculture and enacted in the 1996 FAIR Act. In fact, total EEP expenditures funded by the USDA are almost $2 billion under the limits permitted by the Uruguay Round.

As a part of the $9.1 billion in emergency disaster relief in 2000, about $1.4 billion in crop loss payments were paid to producers to compensate for natural disasters that occurred in 1999. While a portion of these payments went to uninsured producers, those producers had to purchase insurance during the next two years. The USDA is investigating alternative coverage and delivery programs on a pilot basis.

The federal government spends about a half-billion dollars annually on such programs as the Animal and Plant Health Inspection Service (APHIS) to rid cropland of pests and diseases and make these crops more marketable. The Agricultural Marketing Service (AMS) activities include a microbiological surveillance program on domestic fruits and vegetables through a food safety initiative.

Food Safety

Food safety covers a variety of issues, including the conditions under which raw farm products are produced and the conditions under which these products are processed and distributed to consumers. The safety of the nation's food supply is monitored by three federal government agencies: (1) the **Food and Drug Administration (FDA)** in the U.S. Department of Health and Human Services, (2) the **U.S. Department of Agriculture (USDA),** and (3) the Environmental Protection Agency (EPA). The FDA has the broadest authority over monitoring the safety and wholesomeness of food and beverage products. The USDA inspects meat and poultry products. The EPA regulates the use of pesticides when they affect consumer health and safety.

Regulatory issues with respect to food safety often center around the testing that must take place before the product is cleared for public consumption; the tolerance limits associated with the presence of particular substances, such as rodent hair and insect parts in a food product; and product labeling. The FDA spends considerable resources in testing food and beverage products as required by federal legislation such as the Food, Drug, and Cosmetic Act, which originated in the late 1930s. Tolerance limits recognize that it is impossible to achieve a zero tolerance level, and reflect the concept of acceptable risk. Finally, food processors and manufacturers are typically very informative when it comes to labeling ingredients. Government has, in specific instances, stepped in and required health warnings on product labels. Perhaps the most widely known is the warning on cigarette packaging that says smoking is hazardous to your health. A similar health warning is attached to products containing saccharin.

Growing concern over biotechnology embodied in food products in the late 1990s in Europe has had an effect on exports of specific food products to many European countries. Similar concerns have begun to arise in the United States as well, as research focuses on the spin-off effects of genetic engineering, such as the impacts on the Monarch butterfly. Clearly, better information dispelling rumors and research to more fully understand the impacts of biotechnology are needed before broad acceptance is achieved. Finally, the FAIR Act amended the Federal Meat Inspection Act to authorize a review of inspection policies and procedures.

CONSUMER ISSUES

The United States probably has the safest and most abundant supply of food and fiber products of any nation in the world. This is not by happenstance, but rather by design. The government is actively involved at various times during the production, processing, and delivery stages of the marketing channel to ensure a safe and nutritional food supply. There are numerous perspectives one can take to the discussion of consumer issues with respect to food and related products. The perspectives discussed in this chapter include the topics of food safety, an adequate and cheap food supply, nutrition and health, food subsidies, and consumer interests in animal rights.

Adequate and Cheap Food Supply

Consumers in the United States enjoy a relatively stable and cheap food supply. Consumers spend approximately 15% of their disposable income on food and rarely experience an interruption in the supply of food products.

Several studies of hunger in the United States over the post–World War II period indicate that, although progress has been made in reducing hunger in this country, hunger continues to be a serious problem.[3] In the mid-1980s, one study estimated that 20 million persons, or almost one-tenth of the nation's population, went hungry. These studies suggest that hunger is caused by a lack of income, poor knowledge of proper diets, and, in some geographical areas, poor access to nutritional food supplies.

Nutrition and Health

Consumer issues related to nutrition and health have many dimensions. Malnutrition and poor health can occur despite the existence of a safe food supply and an adequate food budget, because some consumers eat the wrong foods (too much fat, too much cholesterol, etc.) despite dietary recommendations to the contrary.

The government has played an important role in nutrition education. The USDA, for example, established the **Human Nutrition Information Service,** which distributes dietary guidelines that have contributed to improvements in nutrition and health.

Food Subsidies

It is hard to imagine that hunger can exist in a land characterized by surplus agricultural production. Yet, the federal government conducts a food assistance program, which has beginnings tracing back to the 1930s. The current **Food Stamp**

---

[3]For example, see National Advisory Commission on Rural Poverty: *The people left behind,* Washington, D.C., 1967, U.S. Government Printing Office; Kotz N: *Hunger in America: the federal response,* New York, 1979, The Field Foundation; and Brown JL: *Poverty and hunger in America,* Washington, D.C., 1986, U.S. House of Representatives, Committee on Ways and Means.

Program (FSP) was established by the Food Stamp Act of 1964 and is currently authorized by the Food Stamp Act of 1977 as amended in 1990 to aid needy households with food purchases. The FSP helps low-income households improve their diets by providing them with coupons to purchase food at authorized retail food stores. This program, along with educational programs, remains the federal government's major means of providing food assistance to those who qualify. Nearly one in ten Americans receives government help to buy groceries.

The 1990 farm bill (FACT Act) provided additional penalties for fraud and misuse of food coupons. Authority to use food stamps in soup kitchens and restaurants was extended. The 1996 FAIR Act provided additional criteria for disqualifying food stores and wholesale food distributors for program violations.

The **National School Lunch Program (NSLP)** is the oldest and largest child-feeding federal food assistance program in the country. The USDA is the largest buyer of prepared foods in the country through the NSLP; schools serve almost 25 million people each day. This program provides financial and community assistance for meal service in public and nonprofit private high schools, intermediate schools, grade schools, and preschools, and also public and private licensed nonprofit residential childcare institutions. All children may participate in the NSLP. Based upon household income poverty guidelines, a child may receive a free, reduced-price, or full-price meal.

Since 1968, the **School Breakfast Program (SBP),** also administered by the USDA, has provided financial and commodity assistance to schools that agree to serve nourishing breakfasts. These meals are also provided free, or at reduced or full prices depending on the household's income.

The **Special Supplemental Food Program for Women, Infants, and Children (WIC)** was created to provide supplementary food assistance benefits to those individuals deemed by local health officials to be at nutritional risk due to their inadequate income and existing nutrition. Categories of individuals eligible for this program include pregnant women, postpartum mothers (up to six months), and infants and children up to five years of age. Participants are given a voucher redeemable for specified foods at participating retail food stores.

There are other programs that also provide food assistance to eligible individuals and households. For example, the Commodity Credit Corporation, a federal corporation within the USDA, can donate commodities to the states under the **Emergency Food Assistance Program,** which is distributed by charitable organizations to eligible recipients.

Rural Communities

Many rural communities were under severe financial stress during the mid-1980s and late 1990s when farmers and ranchers experienced declining incomes and property values. Many communities tied closely to agriculture experienced declines in sales activity and tax revenues. A study by Knutson and Fisher examined options in developing a new national rural policy, and identified the leading rural

Three Most Frequently Identified Rural Development Issues by Region

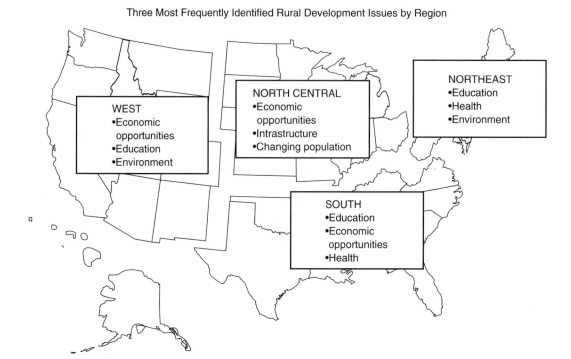

FIGURE 12.3   The nature of the leading rural development issues varies across major geographical regions of the United States. One of the most frequently cited rural development issues was the enhancement of educational opportunities.

development issues. Among other things, this study highlighted the regional differences in development priorities illustrated in Figure 12.3.

A Fund for Rural America was established under the 1996 FAIR Act to augment existing resources for rural development. The Secretary of Agriculture is responsible for coordinating most rural development programs. Under a Rural Community Advancement Program (RCAP), the Secretary of Agriculture is authorized to provide grants, direct and guaranteed loans, and other assistance to meet rural development needs. Funding under RCAP is allocated to rural community facilities, rural utilities, and rural business and cooperative development. The funding authorization for water and waste facilities under the FAIR Act was set at $590 million. Finally, programs for distance learning services and telemedicine were authorized under the FAIR Act. The Secretary of Agriculture can make grants and loans up to $100 million annually to help rural communities build facilities to provide these services.

The **Rural Electrification Administration (REA)** has also played a major role in extending electric power and telephone service to remote rural areas of the country. The REA makes subsidized loans to rural electric and telephone cooperatives. The 1990 Farm Bill established a Rural Development Administration (RDA) in the

USDA. The RDA has been given the authority under Title XXIII to administer the Community and Business Program, which includes business and industry, water and waste disposal systems, and community facility programs. This legislation also created the Rural Investment Partnership Board, which has a revolving loan fund that complements private and other public funds for investment in local rural businesses and guarantees loans to local rural businesses. This program has been targeted for elimination by numerous administrations, including the Clinton Administration.

The success of these programs has direct implications for nonfarm rural residents. By increasing economic activity in rural communities, more jobs are created and the economic base of the community is more diversified and less susceptible to economic downturns in any one sector of the economy. For example, farm commodity policies that temporarily retire land from production and hence reduce purchases of goods and services by farmers will have a less dramatic effect on rural communities if the economic base of these communities is diversified.

RESOURCE ISSUES

Resources in the economy are categorized according to whether they are manufactured resources, human resources, or natural and biological resources. One thing each group of resources has in common is that they are considered scarce, or in limited supply. Government intervenes in agriculture by regulating the use of these scarce resources in the production of food and fiber products. Many of these issues were addressed in previous chapters.

Soil Erosion and Land Use

Soil erosion occurs because the cost of prohibiting soil erosion in the short run exceeds the economic benefit from soil erosion practices. Thus, policies to prevent soil erosion must recognize this imbalance. The **Soil Conservation and Domestic Allotment Act of 1936** recognized this economic fact by providing for soil conservation and soil building payments to participating farmers. Since that time, the USDA has paid farmers more than $25 billion for technical assistance, cost sharing subsidies, loans, and other uses to promote erosion prevention practices.

Both the USDA and the EPA have responsibilities for designing programs to reduce soil erosion. The **National Resource Conservation Service (NRCS)** is the USDA agency that helps provide technical assistance to farmers on preventing soil erosion. Congress mandated the EPA to help control nonpoint sources of water pollution, in which chemicals are carried by eroded soil.

Highly erodible land conservation provisions in current and previous farm legislation protect against erosion by denying program benefits to farmers who do not use approved conservation practices. Under the FAIR Act, farmers were allowed to modify these practices if they could demonstrate that they would provide improved erosion control.

The NRCS is the lead governmental agency in wetlands delineation, or violating provisions by draining wetlands. This agency must make an on-site inspection before imposing any penalties. The NRCS is also charged with implementing regulations on grazing land. The Environmental Conservation Acreage Reserve Program (ECARP) enables the USDA to coordinate various conservation programs related to protection of fragile lands. This program includes the Conservation Reserve Program (CRP) and the Wetlands Reserve Program (WRP) discussed below. The FAIR Act also added an Environmental Quality Incentives Program (EQIP) that provides technical, educational, and cost-sharing assistance to reduce soil, water, and related natural resource problems.

The **Conservation Reserve Program (CRP)** was designed to remove 40 to 45 million acres of highly erodible farmland from production for a period of ten years. This voluntary program called for the government to pay an annual rent to participating farmers plus 50% of the cost of establishing grasses, legumes, or trees on these lands. To encourage participation in the CRP, the rental rate paid by the government must exceed the rate of return the farmer could achieve by farming these lands. To avoid having a major economic impact on the firms providing goods and services to participating farmers and their local community, the CRP prohibited having more than 30% of the land in a single county enrolled in this program.

Another program that conserves the use of soil resources is the **Wetlands Reserve Program (WRP),** which is designed to restore and protect wetlands. The 1990 Farm Bill called for the USDA to enroll up to 1 million acres by 1995 by soliciting bids from farmers. Total enrollment in the CRP and WRP was not to exceed 45 million acres.

Finally, government is actively concerned with the preservation and retention of farmland in agricultural uses. The objective is to ensure sufficient productive capacity and viable rural communities. Some states zone land for agricultural use only. Many states provide for preferential taxation of real estate based upon its agricultural use value rather than its market value. Table 12.1 illustrates the percentage distribution of federal expenditures on resource conservation programs. The CRP represents the largest single expenditure.

Adequacy of Water Supply

Agriculture and the food and fiber industry represent the nation's largest users of water. Many farmers in the more arid regions of the country use large quantities of water to irrigate their crops. And many food manufacturers operate water-intensive processes.

Government involvement in water before the 1970s focused on water availability and the development of new sources of water. Although the focus since then has broadened to include concerns about water pollution, proposals of how to best address water availability and use (ranging from restricting water use by agriculture to proposals to pipe water from Alaska to Southern California) continue to be debated. Projects of this type involve the outlay of billions of dollars.

TABLE 12.1   Resource Conservation and Related Programs Affecting Agriculture

Agency and Program	Percent of Expenditure
U.S. Department of Agriculture (USDA) programs:	
Conservation Reserve Program	26.4
Wetlands programs	1.0
Water Quality Program	2.9
Other conservation	17.1
USDA total	47.5
U.S. Environmental Protection Agency (EPA) programs:	
Water quality programs	7.8
Drinking water programs	2.7
Pesticide programs	1.6
EPA total	12.1
Army Corps of Engineers programs:	
Dredge and Fill Permit Program (wetlands)	1.5
Flood control programs	18.6
Corps total	20.1
U.S. Department of the interior (USDI) programs:	
Range improvement	0.0
Water development and management	14.6
Water resources investigations	2.8
Wetlands conservation	0.0
Endangered species conservation	0.5
Natural resources research	2.2
USDI total	20.3
Federal total	100.0

Withdrawal rates from underground water supplies such as the Ogallala aquifer exceed replenishment rates, causing a drop in the water table that has caused concerns (see Box 12.1). Water, treated as almost a free good until now in many regions of the country, may be priced according to its scarcity in the future. Higher prices will encourage conservation.

Another perspective of the adequacy of the water supply is water quality. Government legislation during the 1970s sought to improve water quality by eliminating point pollution. Federal government policy here was governed by the Clean Water Act passed in 1972. This legislation and subsequent amendments in 1977, 1982, and 1987 made considerable progress in reducing water pollution through the construction of waste treatment plants.

Goals were established to ensure fishable and swimmable water for rivers, lakes, and streams. A regulatory structure was put into place for controlling discharges from factories, sewage treatment plants, and other point sources of water pollution. Such pollution is easy to track and regulate. Nonpoint source pollution, on the other hand, enters the water supply from runoff from cropland, feedlots,

---

BOX 12.1   The Ogallala Aquifer

The Great Plains is one of the richest agricultural production regions in the world. The productivity of Nebraska, South Dakota, Kansas, Oklahoma, and northern Texas is closely linked to the availability of underground water from the Ogallala aquifer, which is estimated to hold a quadrillion gallons of water, or roughly the equivalent of Lake Huron. This aquifer ranges from being 1,000 feet thick in parts of Nebraska to only a few inches thick in parts of Texas.

Today the region's farmers use high-capacity pumps to extract this water for agricultural uses, often with the aid of massive sprinklers. The current annual net drainage (overdraft) is almost equal to the annual flow of the Colorado River. Built up over a period of millions of years, this aquifer will be depleted in a few decades at current rates of use. What does this mean for the type of crops grown? cost of water? water-saving technologies?

---

and pastures. It is caused by rain and melting snow and is influenced by the nature of land use. This form of pollution is difficult and costly to locate, measure, and control. Under the Clean Water Act (CWA), the states took the lead in controlling nonpoint source pollution without specifying the means of controlling it. The USDA and EPA established an action plan in 1997 that laid out three major goals to attain the original mission of the Clean Water Act. These goals were to: (1) enhance protection from public health threats posed by water pollution, (2) develop more effective controls of polluted runoff, and (3) promote water quality protection on a watershed basis.

Nationally, agriculture is the source of pollutants found in 70% of impaired rivers and stream miles and about 50% of impaired lake acres. Well water sampling has revealed widespread evidence of pesticides and nitrogen from agriculture entering groundwater sources. Public concern of animal waste from feedlot operations and intensive spreading of animal waste has also prompted governmental attention, including a joint development by the EPA and USDA of strategies to minimize the environmental risk and public health impacts from animal feeding operations.

Hired Farm Labor

In Chapter 2 we learned that hired labor is a major expense category for farmers and ranchers, but much less than was observed in the early 1950s. Capital has been substituted for labor during the post–World War II period. For example, automated tomato pickers have replaced many laborers who picked tomatoes by hand.

Government intervenes in the labor markets, which provide labor to farming and ranching operations and other businesses in the nation's food and fiber industry in several ways. One labor market intervention is the minimum wage. An amendment to the **Fair Labor Standards Act** in 1966 extended the minimum wage rate to hired farm labor in cases in which the laborer worked more than 50 days in the preceding year.

Of interest from an economic standpoint is whether the increased minimum wage will add further impetus to the substitution of capital for labor over the long run. As you will remember from our discussion in Chapter 8, an increase in the marginal input cost will reduce the demand for that input.

The **National Labor Relations Board,** which oversees the federal legislation that gives workers the right to organize and bargain collectively, is another form of government intervention in labor markets. The National Labor Relations Act outlines forms of unfair labor practices by both unions and businesses. Although meatcutters and many others employed in the food and fiber industry are covered by this federal legislation, hired farm laborers are not at the present.

The **Occupational Safety and Health Administration (OSHA)** created in 1970 oversees the health and safety of laborers by issuing rules covering the workplace. OSHA periodically inspects health and safety conditions at the job site for compliance, cites violations, issues fines, and orders corrective actions. Not all farmers and ranchers view specific OSHA requirements as having much relevance in certain job sites. For example, one requirement is the provision of portable toilet facilities. Senator Wallop of Wyoming gained considerable attention when running for the U.S. Senate in the 1980s by riding off into the sunset on a horse complete with a portable toilet.

A final labor issue that has ramifications for the nation's food and fiber industry is the work of illegal aliens. Many illegal aliens gravitate toward employment by produce growers and other operations in the food and fiber industry. The Immigration Reform and Control Act, which was passed in 1986, made it illegal for farmers and other employers to hire illegal aliens. Repeat violations can result in criminal penalties. This legislation provides an exemption to farm laborers who work for producers of perishable crops (excludes cotton) on a seasonal basis.

Energy

Government intervention in U.S. energy markets has taken many forms since the first OPEC oil market shock in 1973. The Crude Oil Windfall Profit Act of 1980 affected farmers and ranchers who owned land with oil reserves by taxing their "old oil" or windfall profits at a 70% rate, while "new oil" profits were taxed at a 30% tax rate. This legislation proved difficult to administer and was repealed by Congress in 1987.

The federal government has also been involved in the underground storage of imported crude oil, where it is to remain until it is needed to stabilize the effects of world oil supply shocks. Such actions are important to agriculture, which is a heavy user of fuel and fuel-related products such as fertilizers. Conversely, suggestions that a significant tax be placed on imported crude would be detrimental to net incomes received by farmers.

Agriculture's principal link to the nation's energy policy is in its subsidization of ethanol production. Ethanol, derived primarily from corn, is used in manufacturing a blended automotive fuel called gasohol, which is 10% ethanol and 90% gasoline. In the late 1980s, gasohol represented approximately 8% of domestic automotive fuel used. The manufacturing of ethanol is encouraged by several pieces of federal, state, and local government legislation.

The Energy Tax Act of 1978 and subsequent federal legislation exempted fuels containing 10% or more of alcohol produced from renewable resources such as corn from the federal highway excise tax on gasoline, which is five cents per gallon. Similarly, the city of Denver has required drivers in recent years to use gasohol during certain months of the year to combat that city's air pollution problems.

The subsidization of ethanol production expands the demand for corn and increases the revenue of corn farmers, reduces our dependence on imported oil and the nation's merchandise trade deficit, and reduces air pollution.

In Iowa alone, ethanol production accounts for $730 million in value added to the state's corn crop. More than 13,000 Iowa jobs are affected by ethanol, with some 2,500 tied directly to ethanol production. Nationwide, ethanol in 1997 boosted employment by nearly 200,000 jobs, raised net farm income by $4.5 billion, and added $450 million to state and local tax revenue. Approximately one-third of all ethanol produced is manufactured by farmer-owned cooperatives. Figure 12.4 suggests that the relative prices of corn and crude oil influence the attractiveness of ethanol production. To this date, however, ethanol as a significant source of energy has not materialized.

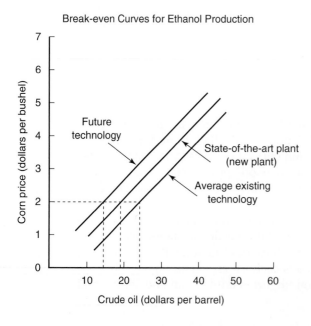

Break-even Curves for Ethanol Production

FIGURE 12.4   Emerging technology of ethanol production will lower the costs of converting corn into ethanol. For example, with corn at $2 per bushel and existing federal subsidies, ethanol is price competitive with crude oil prices at $20 per barrel for state-of-the-art processing plants. Emerging technologies will mean ethanol will be competitive with $15 crude oil rather than $20 crude oil when corn costs $2.

INTERNATIONAL ISSUES

We already addressed several forms of government intervention in international agricultural trade earlier in this chapter when we discussed the effects of export subsidies and import tariffs and quotas on farmers and the food and fiber industry. Two additional international issues are addressed here: (1) the adequacy of the world food supply, which calls for a greater role for government, and (2) the movement toward free trade, which calls for a smaller role for government.

Adequacy of World Food Supply

Thomas Malthus, in the late 1700s, argued that the world would eventually suffer dramatic food shortages because population growth would exceed the growth in the food supply. Specifically, Malthus suggested that population grows at a geometric rate (e.g., doubles every 25 years), and food production, given a fixed amount of tillable land, grows at an arithmetic, or fairly constant, rate.[4]

To illustrate Malthus's point, assume the population of Lower Slobovia was 10 million in 1800 and that it doubles every 25 years. Also assume that this hypothetical country's level of wheat production was 50 million bushels and that it grows at 50 million bushels per every twenty-five years. Based on these assumptions, wheat output per capita would decline as shown in Table 12.2. If this situation did indeed occur, the average citizen in Lower Slobovia would be in a worse position today than was true in 1800.

Now assume that the population of Lower Slobovia did grow as shown in column 1 of Table 12.2, but that output per capita in column 3 remained constant at five bushels over time. How could this be? What Malthus's hypothesis failed to

TABLE 12.2   Wheat Output Per Capita

Year	(1) Population	(2) Wheat	(3) Output per Capita
	(Thousand)	(Mil. Bushels)	
1800	10	50	5.00
1825	20	100	5.00
1850	40	150	3.75
1875	80	200	2.50
1900	160	250	1.56
1925	320	300	0.94
1950	640	350	0.55
1975	1,280	400	0.31
2000	2,560	450	0.18

[4]Malthus T: *An essay on the principle of population,* Homewood, Ill, 1963, Richard D. Irwin.

take into account in Lower Slobovia is technology, which causes the supply curve for wheat in Lower Slobovia (and in the real world) to shift to the right. In the United States, for example, instead of output per capita declining, it has risen dramatically as yield-enhancing technologies have boosted output per capita beyond levels thought possible at the turn of the century.

Government's involvement in preventing Malthus's prediction from coming true in the United States is reflected in the resources it commits to research and development activities, which lead to higher yields. The U.S. Department of Agriculture, along with researchers at land-grant universities around the country and researchers working for private agribusiness firms, have contributed to the yield-enhancing technologies embodied in a wide variety of agricultural inputs such as hybrid seeds and fertilizers that typically result in record crop levels for major food and feed grains in the United States.

Table 12.3 shows that vast differences exist in the expenditure of income on food consumption, the annual population growth rate, the daily caloric intake per capita, and national output per capita between low-income, medium-income, and high-income countries. The percent of total consumption going toward food and the annual population growth rates are generally higher, and the daily caloric intake and national output per capita are generally lower in low-income countries than they are in high-income countries.

Government intervention, including the macroeconomic policies that guide the growth of the general economy and the agricultural policies and government-sponsored research that promote the growth of food and fiber production, are important determinants of the economic trends presented in Table 12.3. Studies by Romer and by Lucas document the significance of the relationship between economic growth and government policy.[5]

Movement toward Free Trade

As indicated earlier in this chapter, international agricultural trade is characterized by a variety of deviations from free trade. Examples cited earlier when discussing government intervention designed to modify farm economic conditions were tariffs, quotas, and export subsidies. The Uruguay round of the General Agreement on Tariffs and Trade (GATT), completed in 1990, gave highest priority to eliminating barriers to free trade. The United States took a free trade stance in these talks, hoping to convince other nations to significantly pare back their subsidies and trade barriers in concert with the United States in an effort to maximize market efficiency (i.e., reduce government costs).

The successfully negotiated United States–Canada Free Trade Agreement and its implementation in 1989 set the stage for the trilateral negotiation of the North

---

[5]See Romer PM: Growth based on increasing returns due to specialization, *American Economic Review*: 701–717, December 1988; and Lucas RE: On the mechanics of economic development, *Journal of Monetary Economics*: 3–42, March 1988.

TABLE 12.3   Selected Indicators of World Food Conditions

Country	National Output per Capita (% of U.S.)	Annual Population Growth Rate	Food Share of Total Consumption	Daily Caloric Intake per Capita
Low income:				
Bangladesh	5.0	2.8	59	1,927
India	4.5	2.2	52	2,238
Sri Lanka	11.2	1.5	43	2,400
Middle income:				
Egypt	15.8	2.6	40	3,342
Tunisia	19.8	2.2	37	2,994
Poland	24.5	0.8	29	3,336
High Income:				
France	69.3	0.4	16	3,336
Canada	92.5	0.9	11	3,462
Japan	71.5	0.6	16	2,864
United States	100.0	1.0	13	3,645

Source: *World development report*, various issues.

American Free Trade Agreement (NAFTA) between the United States, Canada, and Mexico. The gains in aggregate economic welfare for these three nations:

- improve allocation of scarce resources and lower prices to consumers and producers,
- promote economies of scale in manufacturing,
- reduce transactions costs and uncertainty of government policies, and
- promote investment in manufacturing and human resources and technology.

The consensus is that liberalization of international trade and investment results in positive gains for all nations concerned. NAFTA, GATT, and its successor, the World Trade Organization (WTO), promote economic growth in the United States and in other countries. In this context, it is important to identify those individuals, businesses, and regions of the country that may experience a cost associated with trade liberalization. Environmental and food safety concerns, including the notion of identifying the country of origin on food packages, should be carefully studied. The notion of free trade and the benefits and costs of NAFTA and multilateral trade agreements will be addressed more fully in Part VI of this book, beginning with Chapter 19.

WHAT LIES AHEAD?

This chapter introduced some of the ways in which government intervenes in the marketplace for specific resources, goods, and services in agriculture and the food and fiber industry. The reference section of this chapter identifies several sources of additional information on many of these topics. Specific forms of government intervention and their effects on agriculture and the food and fiber industry will be addressed in-depth over the balance of this book.

The specific ways in which the government intervenes in domestic crop and livestock markets to influence economic returns to farmers and ranchers will be addressed in Chapter 13. Macroeconomic policies adopted to combat undesirable economic trends in the general economy and the effects of these policies on agriculture and related sectors in the economy will be discussed in Chapters 14 through 18. Finally, the economics of international agricultural trade, the effects of government intervention, and arguments for free trade are presented in Chapters 19 through 21.

SUMMARY

The purpose of this chapter was to introduce the many ways in which the government intervenes in the nation's farming and ranching operations and other aspects of the food and fiber industry. The major points made in this chapter may be summarized as follows:

1. The rationale for government intervention in a nation's food and fiber industry typically includes: (1) to support/protect an infant industry, (2) to curb market power of imperfect competition when necessary to promote social good, (3) to provide for national food security, (4) to provide for consumer health and safety, and (5) to provide for environmental quality.

2. The U.S. food and fiber industry is one of the most highly regulated industries in the economy. This includes regulations that govern the use of inputs such as agricultural chemicals, regulations that govern the processing and manufacturing of food and fiber products, and regulations that govern the markets in which these products are traded.

3. The inelastic own-price elasticity of demand for farm products, farmers, and ranchers, lack of market power, the interest sensitivity of the farm sector, and the fixity of farm assets and chronic excess capacity represent the roots of the farm problem.

4. A record crop, which is more the rule than the exception, can lead to sharp declines in farm product prices and income levels.

5. The farm sector is one of the most highly capitalized sectors in the U.S. economy. There is more capital invested per worker than in any other sector in the economy.

6. The combination of excess capacity and asset fixity can lead to a decline in farm asset values in times of surplus stocks and makes it difficult for farmers and ranchers to scale back their operations to desired levels.

7. Foreign trade enhancements that promote domestic farm economic conditions include tariffs and quotas on imports from other countries and export credits,

subsidies, and commodity assistance programs to enhance exports of U.S. farm products.

8. Other forms of government intervention in agriculture include subsidized credit to farm borrowers, subsidized crop insurance, and disaster payments to farmers when crop yields fall well below normal levels.

9. Government also intervenes in the food and fiber industry on the behalf of consumers. These efforts are to ensure food safety and to provide an adequate and cheap food supply. Other consumer issues include nutrition and health, the provision of food subsidies, and aid to rural communities.

10. The government also intervenes in the manner in which specific resources are used in the economy. This includes programs to control soil erosion, the adequacy (i.e., quantity and quality) of the nation's water supply, the minimum wages and safety of hired farm labor, and the conservation of energy and production of substitute fuels that use corn and other crops in their manufacturing.

11. Two additional issues that pertain to government's involvement in agriculture are the adequacy of the world food supply and the movement toward free trade.

DEFINITION OF KEY TERMS

Asset fixity: refers to the difficulty that farmers have in disposing of tractors, plows, and silos when downsizing or shutting down their operations.

Conservation Reserve Program (CRP): a major provision of the Food Security Act of 1985 designed to reduce erosion and protect water quality on up to 45 million acres of farmland. Landowners who sign contracts agree to convert environmentally sensitive land to approved permanent conserving uses for 10–15 years. In exchange, the land owner receives an annual rental payment up to 50 percent of the cost of establishing permanent vegetative cover.

Emergency Food Assistance Program: a program that allowed the Commodity Credit Corporation of the U.S. Department of Agriculture to donate commodities distributed by charitable organizations to eligible recipients.

Environmental Protection Agency (EPA): regulates the use of pesticides where they might affect consumer health and safety. This authority was transferred from the US. Department of Agriculture in 1970. Agricultural chemicals must be registered with the EPA before they can be sold.

Excess capacity: the farm sector often produces more than it can sell, leading to rising stocks of surplus commodities such as corn and cheese.

Export Enhancement Program (EEP): a program that targets export assistance to recover lost export markets. The program provides cash subsidies to exporters to help them compete for sales in specific countries. The United States in recent years has agreed to reduce the level of these subsidies in return for concessions made by the European Union.

Fair Labor Standards Act: extended the minimum wage rate to hired laborers in cases in which the laborer worked more than 50 days in the preceding year.

Farm Service Agency (formerly Farmers Home Administration or FmHA): federal agency in the U.S. Department of Agriculture that makes subsidized or guaranteed loans to farmers and loans or grants to rural communities.

Federal Trade Commission (FTC): an agency of the federal government charged with the responsibility of prohibiting companies from acting in concert to increase their market control and of prohibiting false and deceptive trade practices.

Food and Drug Administration (FDA): an agency within the U.S. Department of Health and Human Services charged with the responsibility of monitoring the safety of the nation's food supply; has the broadest authority of ensuring the safety and wholesomeness of food and beverage products.

Food and fiber industry: consists of business entities that are involved in one way or another with the supply of food and fiber products to consumers.

Food Stamp Program (FSP): federal program administered by the Food and Nutrition Service of the U.S. Department of Agriculture that provides food assistance to needy households. This agency also administers the National School Lunch Program, the School Breakfast Program, and the Special Supplemental Food Program for Women, Infants and Children.

Human Nutrition Information Service: an agency within the U.S. Department of Agriculture that distributes dietary guidelines designed to improve nutrition and health of consumers.

National Labor Relations Board: oversees federal legislation that gives workers the right to organize and bargain collectively.

National Resource Conservation Service (NRCS): a federal agency within the U.S. Department of Agriculture that helps provide technical assistance to farmers on preventing soil erosion and related problems.

National School Lunch Program (NSLP): provides financial and community assistance for meal service in public and nonprofit private schools and other specific institutions.

Occupational Safety and Health Administration (OSHA): overseas the health and safety of laborers by issuing rules covering the workplace.

Quota: a ceiling or limit placed on the amount of a particular good that may be imported during the period.

Rural Electrification Administration (REA): a federally subsidized program that provided electrical power and telephone service to remote rural areas.

School Breakfast Program (SBP): initiated in 1968 to provide financial and commodity assistance to schools in providing nourishing breakfasts to children at subsidized prices based upon the household's income.

Special Supplemental Food Program for Women, Infants, and Children (WIC): integrates health care, nutrition education, food distribution and food stamps into a single comprehensive health and nutrition program. It is targeted toward mothers with children who already participate in other welfare programs.

Tariff: a duty or tax placed on imported goods that increases the cost of these goods to domestic consumers.

U.S. Department of Agriculture (USDA): a cabinet level department of the executive branch of the federal government. Its vast programs address the production, safety and marketing of agricultural products as well as the administration of various welfare programs designed to provide food and nutrition services in domestic and foreign markets.

Wetlands Reserve Program (WRP): designed to restore and protect up to 1 million wetland acres.

REFERENCES

Knutson R, Fisher D (eds.): *Proceedings of rural development policy workshops,* Texas Agricultural Extension Service, College Station, Texas, May 1989.

Knutson R, Penn JB, and Flinchbaugh B: *Agricultural and food policy,* 4th ed., Prentice Hall, 1998, Chapter 11.

Lucas RE: On the mechanics of economic development, *Journal of Monetary Economics:* 3–42, March 1988.

Malthus T: *An essay on the principle of population,* Homewood, Ill, 1963, Richard D. Irwin.

Nelson F, Schertz LP: Provision of the Federal Agriculture and Improvement and Reform Act of 1996, Agriculture Information Bulletin No. 729, Economic Research Service, USDA, Washington, D.C., September 1996.

Romer PM: Growth based on increasing returns due to specialization, *American Economic Review:* 701–717, December 1988.

U.S. Department of Agriculture: *Ethanol: economic and policy trade-offs,* Washington, D.C., 1988, U.S. Government Printing Office.

13

SUPPORTING FARM PRICES AND INCOMES

*The government that is big enough to give you
all you want is big enough to take it all away.*

Barry Goldwater

TOPICS OF DISCUSSION

In the early twentieth century, farmers enjoyed continuous prosperity. This period is often referred to as the golden age of American farming. World War I, which occurred during this period, added to the prosperity by increasing the export demand for U.S. farm products when foreign countries sought to compensate for losses in their production. In fact, the economic events from 1910 to 1914 became the basis for the concept of **parity,** which seeks to compare prices in the current period with what they were during this "golden" period.[1]

The depression that occurred in 1920 brought the good times in the U.S. economy to a halt. Unfortunately, when the general economy picked up again in 1921 and continued strong during the "roaring twenties," farmers did not participate in the renewed prosperity. Export demand for U.S. farm products fell when resources were more fully utilized again abroad. The federal government also placed high tariffs on imported goods, and because other countries were unable to export as many goods as before, they reduced their imports also. The farm sector in the United States during the 1920s was therefore plagued with surplus productive capacity, low farm prices, and depressed farm income levels. Little did economists realize in the early 1920s that the problem of excess capacity in the farm sector would continue through today.

By the time the Great Depression hit in 1929, farmers were really hurting. Farm prices and income levels suffered even further declines. In response to this situation, the Federal Farm Board was created and charged with the responsibility to use its budget of $1.5 billion to begin stabilizing farm prices at specific levels, which it did by purchasing crops. Thus began a long history of government involvement in this sector of the economy. Later, other approaches would be instituted to see that the prices of wheat, feed grains, dairy products, cotton, rice, soybeans, sugar, peanuts, and tobacco would not fall below desired levels.

The purpose of this chapter is to illustrate the consequences of an inelastic demand for farm products, some specific approaches the federal government has used to stabilize farm prices and income, and the implications that a more elastic demand for farm products would have on farm revenue. This discussion will explain the major features of the current supply constraint programs and demand expansion programs. The chapter closes with a discussion of proposals to return agriculture to free market forces.

There are essentially two general approaches to altering farm product prices and incomes through government intervention. One approach is to expand demand, thereby shifting the demand curve to the right and raising the equilibrium price. The second approach is to constrain the supply coming onto the market to support farm prices and incomes.

---

[1]The basic idea behind the concept of parity is that if the farmer could exchange a bushel of corn for a pair of overalls from 1910 to 1914, he should be able to make the *same* exchange today. If this occurs, he has achieved 100% parity, with the "golden" period used as the basis for analysis.

PRICE AND INCOME SUPPORT MECHANISMS

The concepts of the low nature of the elasticity of demand for farm products and that an increase in supply causes farm revenue to fall were established in Chapter 12. The historical problem of low returns to resources in agriculture has been dealt with in different ways over time by the federal government. The approach taken has depended in part on the nature of the conditions that existed at the time. The federal government has been involved in altering free market conditions in agriculture for almost 60 years. The roots of the price and income support mechanisms discussed in this section can be traced back to the federal legislation enacted back in the 1930s and 1940s.

There are many excellent summaries of the history of federal government programs for agriculture.[2] Instead of reviewing the labyrinth of farm programs for the last 60 years, we will focus on four key features of the current farm program: (1) loan rate mechanism, (2) set-aside mechanism, (3) target price mechanism, and (4) conservation reserve mechanism. The 1990 Farm Bill sets forth the federal government's farm program for the 1991 to 1995 period. Although the four key features of the current bill reviewed in this section have been around for some time, it is their *current* application in which we are interested.

Loan Rate Mechanism

The U.S. Department of Agriculture has used the **commodity loan rate** mechanism since the early 1930s to support prices for commodities such as wheat, corn, and cotton. The loan rate essentially serves as a floor to farm prices for participating farmers.

Market Level Effects.   To see how this mechanism works at the sector or market level, look at the market demand and supply curves for wheat in Figure 13.1. If competitive market forces were free to work, the market clearing price would be P_F and the quantity marketed would be Q_F. Assume that the federal government wished to support prices at P_G, which lies above P_F. The quantity demanded by consumers at price P_G would be Q_D, and the quantity supplied would be Q_G.

The **Commodity Credit Corporation (CCC),** a corporation operating within the U.S. Department of Agriculture, acts as the purchasing agent for the federal government in this instance. The CCC makes a nonrecourse loan to participating farmers at the loan rate, or a fixed price. The loan, plus interest, can be repaid within 9 to 12 months from the proceeds the farmer receives from selling the commodity on the local market. If it is not profitable for these farmers to repay the loan, the CCC has no recourse but to accept the farmer's pledged collateral (the crop) as payment in full for the loan. The cost of this action to taxpayers in Figure 13.1 would be equal to P_G times $Q_G - Q_D$.

---

[2]For example, see Knutson RD, Penn JB, and Flinchbaugh BL: *Agricultural and food policy,* 4th ed, Englewood Cliffs, N.J., 1998, Prentice Hall.

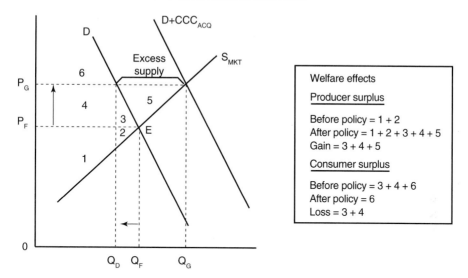

FIGURE 13.1   To achieve support prices of P_G, the federal government would purchase the surplus or excess supply marketed by farmers at the announced support price level. Farmers would be in a better position economically by the amount that area 4 exceeds area 2. Consumers would be in a worse position economically by the sum of areas 3 and 4. They would be paying a higher price (P_G) and consuming less (Q_D).

Participating farmers can either repay their loan if the market price is above the loan rate plus interest charges, or forfeit the commodity pledged as collateral as payment in full for the loan if the market price is below the loan rate. It is the **non-recourse loan** feature that provides the floor to the price received by the farmer.[3] Thus, CCC_{ACQ} in Figure 13.1 represents the additions to CCC stocks acquired by forfeits of commodities pledged by farmers under the nonrecourse loan program.

Firm Level Effects.   As illustrated in Figure 13.2, the individual farm would produce output q_G, where its supply curve (or marginal cost curve) equaled the perceived demand curve (or marginal cost curve) at P_G. The difference between Q_D and Q_G at the sector level represents the excess supply or surplus at P_G that the federal government had to purchase and store for later distribution, if it desired to support prices at this level.

There are several problems associated with the loan rate mechanism when it is used alone. If the support price is higher than the free market price, it encourages higher production. The individual farmer will respond to the higher support price by producing q_G instead of q_F, as shown in Figure 13.2. This leads to greater

---

[3]The term *loan rate* dates back to the time when the purpose of the loan was to extend credit to farmers to help them store some of their production rather than sell it all at harvest time.

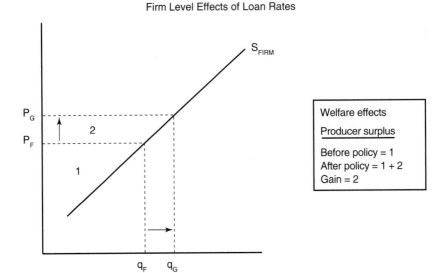

FIGURE 13.2   At the higher announced support price, the individual farmer will produce a higher level of output. The farmer would be in a better position economically by the amount of area 2.

government-held stocks and higher storage costs. Area 2 in Figure 13.2 represents the additional net income earned by the farmer under this farm program.

During the 1950s, the government acquired large quantities of surplus wheat in the form of forfeited nonrecourse loans when market prices were below the loan rate. The government spent more than $1 million a day for storage costs. These charges eventually reached several million dollars a day, which was big money at that time.

In addition, because price supports aid farmers in direct proportion to their level of production, the owners of large farming operations receive the bulk of the program's benefits. The value of expected future benefits from the loan rate mechanism are "capitalized" into the market value of farmland, resulting in an additional benefit to large landowners.

Set-Aside Mechanism

To combat the growing size and cost of government-held stocks, federal government policymakers adopted **set-aside** requirements in the Food, Agriculture, Conservation and Trade (FACT) Act of 1990 to support farm prices. The set-aside mechanism constrains the annual production levels of any single crop and thus avoids accumulating larger surplus stocks, which depress farm prices. Set-aside requirements call for farmers to remove a certain percentage of cropland from production as a condition for receiving farm program benefits. Set-aside requirements were used for most major food and feed grains as a means of reducing production of surplus crops such as corn and wheat. A major problem with the set-aside mechanism is that farmers will idle their poorest land first and crop their remaining acres more

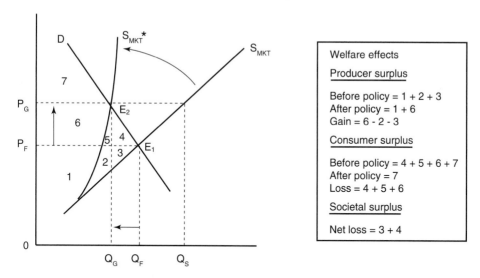

FIGURE 13.3   The set-aside mechanism at market level involves removing the inputs from production that were contributing to the accumulation of surplus stocks. Farmers would be in a better position economically by the amount area 6 exceeds the sum of areas 2 and 3. Consumers, on the other hand, are in a worse position economically by the sum of areas 4, 5, and 6. The net loss to society as a whole due to the reduced level of output would equal the sum of areas 3 and 4. Areas 2 and 5 represent the additional surplus received by firms supplying inputs to agriculture.

intensively, which can result in **slippage** or larger supply and lower prices than desired by policymakers.[4]

The **Acreage Reduction Program (ARP) percentages** spelling out these set-aside requirements were determined in part by the expected ratio of ending stocks to total use. There were individual ARP percentages set for corn, wheat, cotton, and other specific commodities that often varied dramatically from one year to the next depending upon demand and supply factors. The ARP percentage for wheat fell from 27.5% to 5% after the 1988 drought that brought stocks down to pipeline levels. Maximum ARP rates were set for cotton (25%) and rice (35%). The **1996 FAIR Act,** discussed in considerable detail later in this chapter, eliminated the authority for this program, taking the USDA out of the commodity-level supply management game.

Market Level Effects.   To illustrate how set-aside requirements work, study Figure 13.3. D and S_{MKT} represent the market demand and supply curves for a commodity before acreage restrictions are implemented. The market would have cleared

---

[4]Other production control mechanisms used either currently or since the end of World War II include acreage allotments, marketing quotas, land retirement programs, and paid land diversion programs. Farm programs calling for acreage allotments continue today for peanuts and tobacco. Marketing quotas have traditionally been implemented only if two-thirds of the farmers approve them in a referendum. The **Soil Bank Program** utilized in the 1950s led to whole farms being retired from production, eventually retiring approximately 30 million acres.

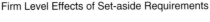

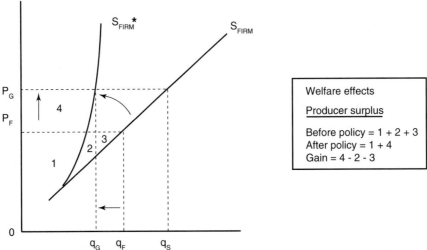

FIGURE 13.4   With the set-aside mechanism at firm level, the individual farmer would produce less than he would under free market conditions and under the loan rate mechanism. Farmers will be in a better position economically as long as area 4 exceeds the sum of areas 2 and 3.

at equilibrium E_1, where price equaled P_F and quantity equaled Q_F. Assume that the government wants to support prices above P_F. At price P_G, the quantity demanded would be equal to Q_G, and the quantity supplied would be Q_S if farmers were free to produce all they desired. We know from the previous discussion of the commodity acquisition approach that this would have resulted in additional surpluses equal to $Q_S - Q_G$. The CCC, in the absence of set-aside requirements, would have had to accumulate these additional stocks if it wished to maintain the market price at P_G.

Firm Level Effects.   The provision for set-aside requirements in the 1990 FACT Act and earlier legislation restricted the amount of land the participating farmer can plant to a particular crop. This would result in a new firm level supply curve S_{FIRM} for this crop that lies to the left of its original supply curve (its true marginal cost curve), as we see in Figure 13.4. The individual farmer operating under the set-aside mechanisms would produce q_G, where the backward-bent firm supply curve for this crop equals marginal revenue (P_G). This quantity is less than the quantity produced under free market conditions q_F and the quantity produced under the commodity acquisition approach (q_S).

The use of set-aside requirements during the 1980s, for example, called for wheat farmers to retire approximately one-fourth of their land from wheat acreage. If a large percentage of these wheat farmers agreed to participate in the federal farm program, the market price would rise from P_F to P_G, as shown in Figure 13.4. The goal is to set price support at above loan rate levels and to avoid having the

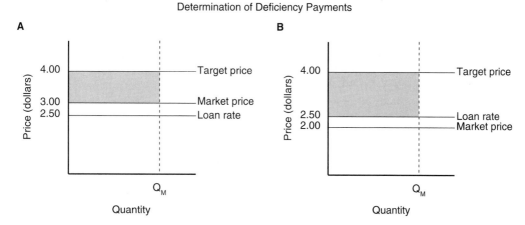

FIGURE 13.5 The target price helps determine the amount of direct payments participating farmers received from the federal government. The amount per bushel is determined by the difference between the target price and the market price or loan rate, whichever is higher.

CCC accumulate large stocks of surplus commodities by farmers forfeiting their output as payment in full for their CCC loan.

To convince farmers to participate in this form of production control program, the federal government offered income support benefits called **deficiency payments,** which are not available to nonparticipants.[5]

Farmers naturally responded to acreage restrictions by idling their worst land first and cropping their remaining acres more intensively, which allowed them to produce output q_G, which may be more than the government expected them to produce.

Target Price Mechanism

A major feature of the 1990 FACT Act, which ended in 1996, was the **target price mechanism.** The concept of target prices was first proposed by Secretary of Agriculture Charles F. Brannan of the Truman administration in 1949. A major fixture since the 1973 Farm Bill, target prices help determine the level of direct payments to farmers. The payment per bushel is equal to the difference between the target price and the market price or loan rate, whichever is higher.

Consider the market conditions depicted in Figure 13.5, in which target prices have been augmented with both the **commodity loan rate** mechanism and the **set-aside** mechanism that limited output to Q_M. If the target price for wheat was $4 per bushel, the market price was $3 per bushel, and the loan rate was $2.50, farmers participating in the program would receive a direct payment from

---

[5]Participating farmers received a **direct payment** from the federal government determined in part by the target price mechanism discussed next in this section.

the federal government of $1 for every bushel of wheat produced, which is illustrated by the shaded area in Figure 13.5, *A*. If the market price fell to $2, the deficiency payment per bushel would have been $1.50, which is illustrated by the shaded area in Figure 13.5, *B*.[6] Thus, the deficiency payment acts as an income support over and above the price support given by the loan rate mechanism.

The target price mechanism used without the set-aside mechanism would stimulate production above free market levels much like the loan rate mechanism would. Farmers would produce to the point at which marginal costs equal the target price rather than the market price, unless otherwise constrained from doing so.

One of the chief challenges of administering a target price program was knowing where to set the target price. Initially, a measure of the national average costs of production for grains and cotton was used to determine what target prices would have to be to cover costs. The 1977 Farm Bill refined this approach by annually adjusting annual target prices based on cost-of-production studies. When the market clearing price fell well below the target price in the late 1970s and early 1980s, this approach to setting target prices came under heavy questioning. Congress instituted its own political judgment in setting target prices instead of studying cost of production. Target prices were specified over the life of the 1990 Farm Bill. The target price of corn was set at $2.75 per bushel, wheat at $4.00 per bushel, oats at $1.45 per bushel, sorghum at $2.61 per bushel, barley at $2.36 per bushel, cotton at 72.9 cents per pound, and rice at $10.71 per hundredweight.

Conservation Reserve Mechanism

The **Food Security Act (FSA) of 1985** enacted the Conservation Reserve Program (CRP) to reduce acreage in production, reduce erosion, and improve water quality. Like the **Soil Bank Program** instituted in the 1950s, CRP is a voluntary long-term land retirement scheme that removes cropland from the production of program commodities such as wheat and corn. The "retired" land must be used for a soil-conserving cover crop such as grass or trees.

The 1985 Farm Bill authorized the retirement of up to 45 million acres of highly erosive land from production, with the caveat that no more than 25% of the land in any one county could enter the program. The 1990 Farm Bill (FACT) expanded the CRP to include lands subject to water quality problems (i.e., a Wetlands Reserve Program [WRP]). Basically, farmers submit sealed bids that are then reviewed to determine which land to retire. If needed, the land can be put back into production.

Under the program, landowners who sign contracts agree to convert environmentally fragile land to approved conserving uses for 10 to 15 years. In exchange, the landowner receives a rental payment annually and cash or payment-in-kind to share up to 50 percent of the cost of establishing permanent vegetative

---

[6]Farmers participating in the program would also likely choose to relinquish title to their crop under the CCC nonrecourse loan program. This means that they would receive the $2.50 loan rate from the CCC rather than only $2 from the spot market. Thus, they receive the full $4, or $1.50 in direct deficiency payments and $2.50 from the CCC nonrecourse loan.

cover. The **FAIR Act** in 1996 set the maximum CRP area at 36.4 million acres. New acreage enrolled in the CRP to replace acreage coming out at the end of their contract must meet stiffer criteria regarding soil erosion, water quality, or wildlife benefits. An early release—60 days' notice—allows farmers to remove land from the program if it has been five years or more.

The Conservation Reserve Program calls for the retirement of additional acres of cropland, which means a further "bending back" of the market supply curve in Figure 13.3, lower levels of production for those crops previously grown on these lands, and higher prices in the short run. Farmers would be put in a better position economically, but consumers would be in a worse position economically. Society as a whole would be in a worse position because of the additional loss of output.

Because we are moving up an inelastic demand curve for crops, total revenue received by producers will be higher with the CRP mechanism than it would have been under either free market conditions or with the set-aside mechanism only.

Commodities Covered by Government Programs

The loan rate, set-aside, target price, and CRP mechanisms played a significant role in the 1990 Farm Bill, which spells out the role the federal government will play in agriculture through 1995. Not all crop and livestock products grown on the nation's farms and ranches were covered by the price and income support mechanisms. In fact, you may be surprised to learn that most crop and livestock products are not covered by federal farm programs.

The CCC makes nonrecourse loans at the established loan rate to farmers for wheat, corn, oats, barley, sorghum, cotton, rice, soybeans, and sugar. Crops such as fruits and vegetables, and livestock such as cattle and hogs, are not covered by the loan rate mechanism. There is a price support program for dairy farmers that calls for the CCC to buy butter, cheese, and nonfat dry milk in bulk containers at announced prices, which provides a floor for milk and dairy product prices.[7]

The 1990 farm bill contained set-aside requirements for wheat, corn, oats, barley, sorghum, cotton, and rice—but not soybeans. Again, crops such as fruits and vegetables are not covered by the loan rate mechanism. No set-aside requirement exists for livestock.[8]

Target prices used to determine the level of direct income support payments to farmers participating in the program from 1991 to 1995 were set forth in the 1990

---

[7]In addition to the CCC loan program, farmers may participate in the Farmer-Owned Reserve (FOR) loan program. Farmers are permitted to enter this program under the 1990 Farm Bill if certain stock-to-use relationships are met and entry trigger prices are met. Only 300 to 450 million bushels of wheat and 600 to 900 million bushels of feed grains (corn, sorghum, oats, and barley) are permitted at any one time. Farmers may sell grain from the FOR program at any time.

[8]The closest we have come to restricting production of livestock is the dairy termination program of 1986, which "bought" reductions in production by paying for the slaughter of dairy cows, heifers, calves, and bulls. The intent was to reduce the nation's milk production capacity and thus avoid the continued growing surplus of manufactured dairy products.

TABLE 13.1   Target Prices in Nominal Terms

Crop	Unit	1985 Farm Bill				1990 Farm Bill	FAIR Act
		86–87	87–88	88–89	89–90	91–95	96–02
Wheat	$/bu	$ 4.38	$ 4.38	$ 4.23	$ 4.10	$ 4.00	$0
Corn	$/bu	3.03	3.03	2.93	2.84	2.75	0
Rice	$/cwt	11.90	11.66	11.15	10.80	10.71	0
Sorghum	$/bu	2.88	2.88	2.78	2.70	2.61	0
Barley	$/bu	2.60	2.60	2.51	2.43	2.36	0
Oats	$/bu	1.60	1.60	1.55	1.50	1.45	0
Cotton	$/lb	.810	.794	.759	.734	.729	0

Farm Bill. Target prices in nominal terms appear in Table 13.1. The 1990 Farm Bill also set payment limitations on the amount of deficiency payments a farmer can receive in an effort to curb federal budget deficits. The maximum regular deficiency payment a farmer can receive in one year was $50,000. There was no limit on nonrecourse loans, however, except for honey.

Farm programs have historically played an important role in determining the level of aggregate net farm income. As Figure 13.6 illustrates, federal government payments to farmers helped to support farm incomes in the early 1970s when real net farm incomes from farming operations were declining. When exports of wheat and other crops rose markedly in 1972 and continued strongly for several years, cash receipts from farm marketings rose and the federal government provided less support.

Even with federal income supports, net farm income expressed in real terms fell in 1983 to a level not seen since the Great Depression. Higher interest rates, a recessionary economy, and declining export markets were all squeezing the profitability of farming operations. Figure 13.6 illustrates that direct payments from the federal government rose beginning in 1983, approaching $17 billion in nominal terms in the following years.

When annual federal budgets began to rise sharply during the late 1980s and early 1990s, annual expenditures for agricultural programs became increasingly criticized. As a consequence, the 1990 Farm Bill was tied specifically to measures to reduce the size of the federal budget deficit. As a result of this and the 1988 drought, which led to higher farm prices and hence less need for the government to support farm incomes, direct payments to farmers declined.

The 1990 Farm Bill also called for a broadening of the CRP to address surface water, groundwater, and wetland issues. This legislation calls for 40 to 45 million acres to be under the control of the CRP, including the 34.6 million acres currently in the program. In addition, 10 million acres of land near well heads, areas inhabited by endangered species, or areas where farm production poses a threat to the quality of ground and surface water supplied were enrolled during the 1991 to 1995 period.

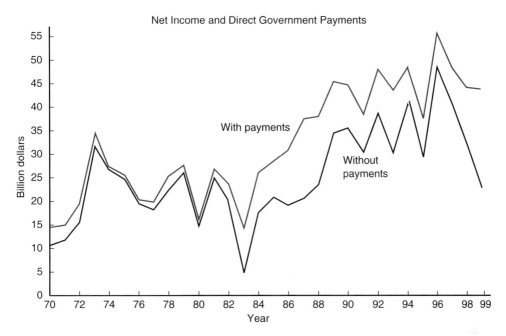

FIGURE 13.6   Annual direct payments to participating farmers support the level of net farm income. The level of government payments rose dramatically as farm commodity prices fell and production costs rose in the 1980s. (*Source:* U.S. Department of Agriculture: *Economic Indicators of Farm Sector.*)

PHASING OUT OF SUPPLY MANAGEMENT

The 1996 Farm Bill, also known as the **Federal Agriculture Improvement and Reform Act** or simply **FAIR Act,** represents a transition to a market-driven agriculture. Under the 1990 Farm Bill, participants in farm programs received deficiency payments reflecting the difference between market prices and target prices. Eligibility for payments was based on the acreage base history the farmer had established for a particular crop. Acreage reduction percentages or ARPs were then used to tell farmers how much of the base acreage they had to set aside in a particular year to receive deficiency payments. The goal from a policy perspective was to manage supply, or the quantity coming onto the market.

The FAIR Act replaced target prices and deficiency payments with annual fixed transition or flexibility contract payments. The receipt of direct government payments was "de-coupled" from planting decisions. Farmers are no longer restrained in their planting decisions by maintaining base acreage and annual ARP levels. They now have the flexibility to plant virtually whatever they want on their base acreage (referred to now as contract acres). Farmers who had planted cotton for generations, for example, were seen switching to corn in response to relative commodity price movements. The receipt transition payments are governed by contract payment rates established in the FAIR Act that decline in value over the 1996–2002 period.

A per-unit **payment rate** for each contract commodity is determined annually by dividing the total annual contract payment rate level for each commodity by the total of all contract farm's program payment production. The annual payment rate for a commodity is then multiplied by each farm's payment quantity, and the sum of these payments across all eligible commodities on that farm represents the farm's annual payment, subject to payment limits.

Exactly what is to happen after 2002 is unclear at this point, although one probable outcome in a largely urban Congress is to freeze these direct payments at 2002 levels.

The FAIR Act also removed the USDA from supply management, eliminating set-aside restrictions or ARP percentages on program crops required to receive deficiency payments, scrapping the **Farmer Owned Reserve** or **FOR program,** and reducing government ownership of stocks. This means that the determination of commodity stocks relative to expected use, and hence annual commodity price movements, is left entirely to private sector forces.

Another significant part of this legislation is the elimination of ad hoc disaster payments to farmers. The assistance made available to farmers following the 1988 drought and the 1993 floods, for example, will no longer be available. It is the responsibility of the farmer to acquire crop insurance to protect against such peril. The FAIR Act was termed "freedom to farm" during its development stages. As some have suggested, this legislation not only provides freedom to succeed in the market place, but *freedom to fail* as well.

The *bottom line* is the strong likelihood of greater price variability and more downside risk associated with net income and asset values as we move into the next decade and a new century. *Gone* are the deficiency payments that made up the difference between announced target prices and spot market prices. *Gone* are the supply management tools designed to stabilize farm prices and incomes. *Gone* are the ad hoc disaster payments used to assist farmers during periods of floods and droughts. U.S. agriculture today is *more reliant* than ever before on a growing world demand for food and fiber products to keep commodity prices at levels that do not result in another financial crisis like that faced by farmers in the 1980s.

The FAIR Act has several questionable features that will no doubt be the focus of debate in the years to come. For example, the flexibility contract payments were made to farmers regardless of the level of market prices and net farm income. The high market prices in 1996, for example, cast the high flexibility contract payments farmers received in that year in an unfavorable light in the eyes of many. The year-to-year instability in agriculture have also increased the risk exposure to farmers and agriculture-related sectors of the economy (farm input suppliers, agricultural lenders, and rural communities).

A Commission on the 21st Century Production Agriculture was established under the 1996 FAIR Act to review the changes in production agriculture under this legislation and recommend the appropriate role of the federal government at the end of 2002, when the FAIR Act expires. The **permanent legislation** under the Agricultural Adjustment Act of 1938 and in the Agricultural Act of 1949 are continued after 2002 if nothing further is done. These provisions authorize marketing quotas, acreage allotments, marketing certificates, and **parity-based price supports** for wheat, feed grains, cotton, and sugar. Congress has traditionally passed legislation like the FAIR Act to avoid reverting to these costly permanent provisions.

DOMESTIC DEMAND EXPANSION PROGRAMS

Raising the level of farm income can also be accomplished through domestic demand expansion, or by shifting the demand curve for farm products in the U.S. economy to the right, which is illustrated at the sector level in Figure 13.7.

When the market demand for a particular crop shifts from D_1 to D_2 and we move up the current market supply curve S, we see that the price of the commodity increases from P_F to P_G. Farmers are better off because the profits generated by their operations are rising (by an amount equal to area $P_1P_2E_2E_1$).

Instead of the level of economic activity on farms and in rural communities, where farmers purchase farm inputs and market their output, being reduced directly or indirectly by production controls, domestic demand causes economic activity to rise. Farm output for this commodity would increase from Q_F to Q_G. The firm level effects of an expanding demand from D_1 to D_2 are much the same as

Market Level Effects of Domestic Demand Expansion

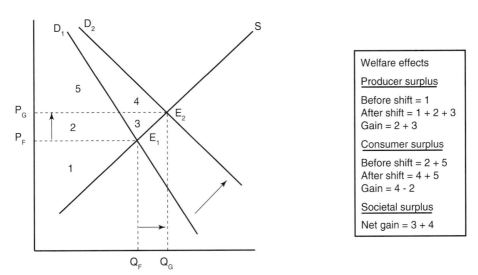

FIGURE 13.7   Under domestic demand expansion, farmers would be producing more than they would under free market conditions.

those illustrated in Figure 13.2 for the commodity acquisition approach. The domestic demand for farm products can be expanded in two ways:

1. The government can institute programs to promote the expansion of domestic demand through school feeding and other nutrition service programs, advertising and promotional programs, and so on.
2. The government can institute programs to subsidize the development of new uses for farm products (e.g., as an intermediate good in the production of other goods). A recent example is the use of state and federal subsidies for the production of ethanol, which is then blended with gasoline to produce gasohol.

The first form of demand expansion program probably holds the least promise of the two in terms of its ability to alleviate the need for price supports in the farm business sector. With a declining rate of population growth in the United States and the low income elasticity for food, it is difficult to foresee how advertising and promotional programs can lead to a significant increase in the domestic demand for food and fiber products over the long run. Furthermore, food and nutrition service programs are already in place. Almost 19 million people participated in the food stamp program in the late 1980s. The federal government's expenditure on this program reached $11.2 billion in 1988. In addition, the federal government annually spends $2 to 3 billion on commodity distribution programs for school feeding, institutions, and needy persons.

The effect of the second form of demand expansion programs is likely to be minor given existing technologies, unless the demand–supply conditions for industrial uses make it more economical for nonfarm businesses to utilize farm products in their production processes. If real energy prices increase, the demand for corn and other sources of starch used to produce liquid energy production will increase. Another factor will be the nature of government policy incentives. Higher subsidies to induce production of ethanol for use in gasohol would also increase the demand for corn and other grains. Increased environmental concerns in cities such as Denver and Los Angeles will also spur the demand for gasohol and other synthetic fuels.

IMPORTANCE OF EXPORT DEMAND

The phasing out of subsidies to producers of export sensitive crops like wheat places increased emphasis on strong export demand to keep prices at or above the old target price levels. Export subsidies represent one potential demand expansion device. The expanded sales of grain to the Soviet Union and other countries during the mid-1970s raised farmers' real gross income to its highest levels during the post–World War II period and greatly reduced the need for government support of farm incomes.[9]

---

[9]Public Law 480 (P.L. 480) is designed to export commodities to foreign countries under several different programs: long-term dollar and foreign currency sales (Title I sales), donations to other governments and various relief agencies to promote better diets and alleviate starvation (Title II donations), barter for strategic materials (Title III), or other foreign assistance programs. Almost 90% of P.L. 480 exports of wheat and flour are Title I sales.

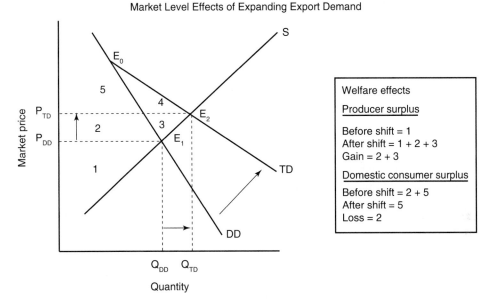

FIGURE 13.8 The domestic demand (*DD*) for farm products in general is highly inelastic. An expanding export demand at lower market prices will result in a more elastic total demand (*TD*) curve and increase the economic well-being of consumers and businesses.

Other programs, such as those that extend export credit to importing nations, target subsidies to specific importing countries, and promote a "buy one, get one free" sale of farm commodities abroad, will be discussed in Part V of this book. The reduction of subsidies under the Uruguay Round of the General Agreement on Tariff and Trade or **GATT** affects this option, however, as discussed in Part V. In later chapters, we will also emphasize two additional factors that have a direct impact on U.S. exports: (1) the purchasing power of national income in importing countries, and (2) the relative cost of the U.S. dollar in foreign currency markets.

Monetary policies designed to combat inflation in this country that result in high interest rates lead to increases in foreign capital flowing into this country (which means that we are exporting recessionary pressures to other countries), and a higher price for U.S. dollars in foreign currency markets (which means that it would now take more units of a foreign currency to buy the dollars necessary to purchase a given quantity of U.S. products). Therefore, expansionary monetary policies will have a positive impact on export demand, and contractionary policies will retard export demand.

The demand curve for raw agricultural products facing U.S. farmers becomes more elastic (less inelastic) with a growing export demand for U.S. farm products, which shifts the total demand curve as illustrated in Figure 13.8.[10] E_0 reflects the

[10]The quantity of U.S. farm products shipped abroad more than tripled during the 1950 to 1990 period.

market equilibrium at a price that almost halts foreign demand for a particular crop. E_1 represents the domestic market equilibrium at the price of P_{DD} if there were no foreign demand. E_2 represents the market equilibrium with foreign demand, which reflects both a higher product price (P_{TD}) and quantity needed (Q_{TD}).

To illustrate the importance of export demand, let us examine the real world case of U.S. wheat farmers and how changing market conditions during the post–World War II period affected their total revenue.

Calculating Total Demand Elasticity

The own-price elasticity of demand for wheat can be partitioned into the contribution associated with the domestic use of the product and the contribution associated with exports as:

$$\begin{matrix} \text{own-price} & & \text{own-price} & & \text{own-price} \\ \text{elasticity of} & = MS_d \times & \text{elasticity of} & + MS_e \times & \text{elasticity of} \\ \text{total demand} & & \text{domestic demand} & & \text{export demand} \end{matrix} \quad (13.1)$$

in which MS_d and MS_e represent the market shares associated with the domestic use and exports of wheat, respectively. By definition, the values for MS_d and MS_e must sum to one.

If all U.S. wheat production is utilized domestically (i.e., $MS_d = 1.00$ and $MS_e = 0.00$) and the own-price elasticity of domestic demand for wheat is -0.20 and the own-price elasticity of export demand for wheat is -1.50, the own-price elasticity of total demand would be:

$$\begin{aligned} \text{own-price elasticity of total demand} &= 1.00 \times (-0.20) + 0.00 \times (-1.50) \\ &= -0.20 \end{aligned} \quad (13.2)$$

The own-price flexibility, or the reciprocal of the own-price elasticity of demand, allows us to focus on the effect that a change in the quantity coming into the market has upon the price of the commodity. In this instance, the own-price flexibility would be -5.00 (i.e., $1.0 \div -0.20$), which suggests that a 1% increase in the quantity of wheat coming into the market would cause the price of wheat to drop by 5%.

Impact of Rising Export Demand

The market share of total wheat output to domestic use and to exports has been changing during the post–World War II period, which has led to an increase in the own-price elasticity of the total demand for wheat. To illustrate, let us continue to use an own-price elasticity of domestic demand for wheat of -0.20 and an own-price elasticity of export demand for wheat of -1.50. The domestic use of wheat output during the 1950 to 1959 period was 62.2% (i.e., $MS_d = 0.622$), and the amount of output going to exports was 37.8% (i.e., $MS_e = 0.378$). The own-price elasticity of total demand for wheat during this period would have been:

$$\text{own-price elasticity of total demand} = 0.622 \times (-0.20) + 0.378 \times (-1.50)$$
$$= -0.69 \qquad (13.3)$$

Thus, the own-price flexibility would be approximately -1.45 (i.e., $1.0 \div -0.69$), which would suggest that a 1% increase in the quantity of wheat coming into the market would cause the price of wheat to drop by 1.45%.

The domestic use of U.S. wheat production during the 1976 to 1982 period fell from 62.2% to 38.3% and export's share rose from 37.8% to 61.7%. Continuing to use the same two elasticities for domestic and export demand, the own-price elasticity of total demand for wheat during this period would have been:

$$\text{own-price elasticity of total demand} = 0.383 \times (-0.20) + 0.617 \times (-1.50)$$
$$= -1.00 \qquad (13.4)$$

The price flexibility associated with an own-price elasticity of unity, or -1.00, is -1.00, suggesting that a 1% increase in the quantity of wheat marketed during the 1976 to 1982 period would have caused the price of wheat that farmers received for their product to decrease by exactly 1%. If the export demand for U.S. wheat exceeded 61.7% of total disappearance and these two elasticities remained constant, then the total own-price elasticity would exceed one, or become elastic.

Impact on Farm Revenue of Wheat Producers

An increase in the own-price elasticity of demand has important implications for total farm revenue.[11] A 1% increase in quantities marketed will result in an increase in total farm revenue only if the own-price elasticity of demand is greater than 1.0.

If the price of wheat is $4 per bushel and the quantity of wheat marketed is 2.5 billion bushels, total revenue would be $10 billion (i.e., 4×2.5 billion bushels). Let's examine what happens to total revenue if the own-price elasticity of demand is -0.69, as calculated using the 1950 to 1959 market shares for domestic use and exports in Equation 13.3, and the quantity of wheat coming into the market increases by 1%. To summarize, we assume the following:

Original price of wheat	$4 per bu.
Original quantity marketed	2.5 billion bu.
Domestic market share	38.3%
Export market share	61.7%
Domestic own-price elasticity	−0.20
Export own-price elasticity	−1.50

---

[11]The volatile nature of the export demand for U.S. farm products and some of the factors causing this variability, including the Asian Crisis in the late 1990s, will be explored later in Chapters 19 to 22.

The first step to determine the impact of a 1% increase in quantity coming into the market is to calculate the price flexibility of demand, or P_{FLEX}. The price flexibility of -1.45 associated with an elasticity of -0.69 (i.e., $1.0 \div -0.69$) tells us the price of wheat would fall by 1.45% if the quantity marketed rose by 1% or:

$$
\begin{aligned}
\text{new price} \quad &= \text{original price} + (\text{original price} \times P_{FLEX} \times \%\Delta Q) \\
&= \$4.00 + (\$4.00 \times (-1.45) \times .01) \\
&= \$3.94 \qquad\qquad\qquad\qquad\qquad\qquad\qquad (13.5)
\end{aligned}
$$

What would the quantity of wheat marketed be under these circumstances? It would be 1% higher than the original 2.5 billion bushels marketed, or:

$$
\begin{aligned}
\text{new quantity} \quad &= \text{original quantity} + (\text{original quantity} \times \%\Delta Q) \\
&= 2.5 \text{ billion} + (2.5 \text{ billion} \times .01) \\
&= 2.525 \text{ billion} \qquad\qquad\qquad\qquad\qquad\quad (13.6)
\end{aligned}
$$

The new level of total revenue received by wheat producers if they increased the quantity of wheat they marketed by 1% would be:

$$
\begin{aligned}
\text{new total revenue} \quad &= \text{new price} \times \text{new quantity} \\
&= \$3.94 \times 2.525 \text{ billion} \\
&= \$9.954 \qquad\qquad\qquad\qquad\qquad\qquad\qquad (13.7)
\end{aligned}
$$

The change in total revenue for wheat producers in general would be equal to:

$$
\begin{aligned}
\text{change in total revenue} \quad &= \text{new total revenue} - \text{old total revenue} \\
&= \$9.954 \text{ billion} - \$10.000 \text{ billion} \\
&= -\$46 \text{ million} \qquad\qquad\qquad\qquad\qquad (13.8)
\end{aligned}
$$

if wheat producers increased the quantities of wheat they marketed by 1%, or 25 million bushels.

What would happen to the total revenue of U.S. wheat producers if they increased the quantity they supplied by 1% and the total own-price elasticity of demand was -0.20, as calculated in Equation 13.2? What would happen with an elasticity of -1.00, as calculated using the 1976 to 1982 market shares in Equation 13.4? Or what would happen if the elasticity was greater than one, which would occur if an increasing share of production was marketed abroad? Table 13.2 illustrates the values calculated in Equations 13.5 through 13.7 for an own-price elasticity of -0.69 and for these additional elasticities.

Column 1 of Table 13.2 gives the alternative elasticity values, column 2 gives the assumed increase in quantity coming onto the market, and columns 3, 6, and 8 report the original quantity marketed, original price, and original total revenue. The values in column 4 are given by Equation 13.6, the values in column 5 represent the reciprocal of the elasticity in column 1, the values in column 7 are given by Equation 13.5, and values in column 9 are given by Equation 13.7.

TABLE 13.2   Importance of Higher Price Elasticity on Demand

(1)	(2)	(3)	(4)	(5)	(6)	(7)	(8)	(9)
Price Elasticity	%ΔQ	Original Quantity (Bil. Bu.)	New Quantity (Bil. Bu.)	%ΔP*	Original Price ($/Bu.)	New Price ($/Bu.)	Original Total Revenue (Bil. $)	New Total Revenue (Bil. $)
−0.20	+1.00	2.500	2.525	−5.00	$4.00	$3.80	$10.000	$9.595
−0.69	+1.00	2.500	2.525	−1.45	4.00	3.94	10.000	9.954
−1.00	+1.00	2.500	2.525	−1.00	4.00	3.96	10.000	10.000
−1.20	+1.00	2.500	2.525	−0.83	4.00	3.97	10.000	10.024

*This amount is equal to the price flexibility for a 1% change in quantity, or the reciprocal of the elasticities reported in column 1.

There are several general conclusions that can be drawn from the values presented in Table 13.2. First, the smaller the own-price elasticity of total demand facing wheat producers, the greater the decline in total revenue as the quantity coming into the market is increased. If the own-price elasticity of total demand is −0.20 and the quantity coming into the market increases by 1%, total revenue will fall by $405 million. Second, if the elasticity was equal to −1.00, a 1% increase in the quantity of wheat coming into the market will leave total revenue unchanged. In this instance, price would fall by 1% when quantity marketed increased by 1%, leaving expenditures of grain buyers and revenue of wheat producers constant. Finally, if the elasticity was equal to −1.20, a 1% increase in the quantity of wheat coming into the market will increase total revenue by $24 million. Obviously, wheat producers would prefer this market elasticity.

The market situations illustrated in Table 13.2 all assumed an increase in quantity coming into the market. What would happen to the value of total revenue in column 9 of this table if the percent changes in quantities marketed in column 2 had been −1.00 rather than +1.00? Try recalculating the new prices, quantities, and revenue in Equations 13.5 through 13.7 for each elasticity and observe what happens to the values in column 9 of Table 13.2. Remember that the signs for both the percent change in quantity in column 2 and percent change in price in column 5 *must be reversed*. You will find the revenue levels are dramatically different.

Implications for Policy

These observations have at least three direct implications for the design of farm programs and conduct of macroeconomic and foreign policy. First, if the long-run, own-price elasticity of wheat were to exceed 1.0, supply constraint programs would restrict the total revenue of wheat farmers rather than support it. Total revenue would decline rather than increase if less wheat output were produced. Supply constraint programs and their attendant costs to the U.S. Treasury would therefore be counterproductive in the long run. For example, if supply constraint programs had limited wheat production in the previous examples to 2.5 billion bushels to support prices at $4 and the own-price elasticity of demand was −1.20, wheat farmers in general would have been penalized by $24 million (i.e., $10 billion versus $10.024 billion). The passage of the FAIR Act in 1996 basically eliminated supply management as a policy tool.

Second, the use of wheat as a foreign policy tool has the potential for reducing the quantity of wheat exports. Embargoes not only have the short-run effect of limiting the quantity shipped, but also the long-run effect of causing export customers to lose confidence in the United States as a trading partner and either seek other trading partners or increase their own production for food security reasons.

Finally, contractionary macroeconomic policies increase the value of the U.S. dollar, which leads to higher foreign currency exchange rates. This would make it more expensive for foreign customers to acquire the U.S. dollars needed to buy U.S. wheat, thus lowering the export demand for wheat. Both policies lower the value of MS_e and increase the value of MS_d in Equation 13.1. This lowers the own-price elasticity of demand and increases the need for supply constraint programs to support the income levels of wheat producers. These policies involve the expenditure of billions of dollars. Their impact on the U.S. farm business sector will be discussed in depth in Chapter 18.

SUMMARY

The purpose of this chapter was to illustrate the economic consequences of the inelastic nature of the demand for farm products and the continuing expansion of the farm business sector's capacity to produce, and to discuss the government programs that have been used to affect the market price farmers receive for their products. The major points made in this chapter may be summarized as follows:

1. Changes in supply, coupled with the highly inelastic demand for raw food and fiber products, can translate into periods of booms and busts in farm income. The ever-expanding nature of annual farm output helps explain the low returns to resources observed historically in agriculture.

2. A short-run price elasticity of demand of −0.20 suggests a price flexibility (the reciprocal of the price elasticity) of −5.00. For farmers, this means that a 1% increase in the quantity they send to market will lower the market price they receive during the year by 5%, all other things constant.

3. One approach to dealing with the historical problem of low returns to resources is the implementation of commodity supply programs, based on federal policy mechanisms such as nonrecourse loans, set-aside requirements, target prices and deficiency payments, and the Conservation Reserve Program.

4. Prior to passage of the FAIR Act in 1996, target prices were used to supplement the price supports under the CCC nonrecourse loan program. The target price concept calls for direct deficiency payments to participating farmers, based on the difference between the target price and the market price when the target

price is higher than the loan rate. This approach was replaced by flexibility contract payments which are fixed annually and decline in value until 2003, when they may disappear altogether.

5. Raising the returns to resources in agriculture can also be accomplished by demand expansion programs. Domestic demand can be expanded by public (e.g., school lunch programs) and private (e.g., advertising) programs. The development of new uses for food and fiber products and expansion of export demand with export subsidies to importing countries also lead to higher domestic prices and farm income.

6. The level of export demand is important because it affects the elasticity of demand for certain farm products. Although farm programs may be needed in the short run to alleviate economic stress, the continuation of farm programs in which the elasticity of demand exceeds one may lower rather than raise farm revenue.

DEFINITION OF KEY TERMS

Acreage Reduction Program (ARP) percentages: spelled out the set-aside requirements facing farmers desiring to participate in the 1990 FACT Act. An ARP percentage of 25%, for example, required the farmer to idle 25 percent of the base acreage established by the farmer over time.

Commodity Credit Corporation (CCC): an agency within the U.S. Department of Agriculture that makes nonrecourse loans to farmers for the purpose of supporting prices at a specific level.

Commodity loan rates: Price per unit (pound, bushel, bale, or cwt) at which the CCC provides nonrecourse loans to farmers to enable them to hold program crops for later sale. Loans can be recourse for dairy farmers and sugar processors.

Deficiency payment: value of payment per unit, equal to the difference between the market price and the target price. This payment is made to participating farmers.

Direct payments: payments in the form of cash or commodity certificates made directly to producers for such purposes as production flexibility contract payments, deficiency payments, annual land diversion, or Conservation Reserve payments.

Farmer Owned Reserve (FOR) Program: mechanism used in past legislation that fostered storage of surplus commodities owned by the producer as opposed to the government. This program was eliminated in 1996 with passage of the FAIR Act.

Federal Agriculture Improvement and Reform (FAIR) Act: Current commodity legislation, which decoupled planting decisions from government payments, giving farmers greater planting flexibility.

Flexibility contract payments: payments to be made to farmers for contract crops through 2002 under the FAIR Act. Payments for each crop are allocated each fiscal year based on the Congressional Budget Office's February 1995 forecast of what deficiency payments would have been under the FACT Act.

Food, Agriculture, Conservation and Trade (FACT) Act of 1990: the omnibus food and agriculture legislation signed into law on November 28, 1990 that provided a 5-year framework for the Secretary of Agriculture to administer various agricultural and food programs.

Food Security Act (FSA) of 1985: the omnibus food and agriculture legislation signed into law on December 23, 1985 that provided a 5-year framework for the Secretary of Agriculture to administer various agricultural and food programs.

General Agreement on Tariffs and Trade (GATT): an agreement originally negotiated in Geneva, Switzerland, in 1947 to increase international trade by reducing tariffs and other trade barriers. The agreement provides a code of conduct for international commerce and a framework for periodic multilateral negotiations on trade liberalization and expansion. The Uruguay Round Agreement established the World Trade Organization (WTO) to replace the GATT. The WTO officially replaced the GATT on January 1, 1996.

Nonrecourse loan: an amount of money equal to support price times the quantity offered as collateral lent by the Commodity Credit Corporation (CCC). The loan is considered paid in full, when turned over to the CCC, when the market price falls below the support price.

Parity: a concept that expresses relative prices (price of farm products relative to input prices) in the current period as a ratio to relative prices in the 1910 to 1914 period. A parity price in the current period (i.e., ratio = 100) would mean that farmers are currently earning the same relative price as what was received during this "golden era."

Parity-based support prices: a measurement of the purchasing power that a unit (e.g., bushel, cwt) of a farm product would have had in the 1910–14 base period. The base prices used in the calculation are the most recent 10-year average prices for commodities. Under "permanent provisions," prices would be supported at 50 to 90 percent of parity through direct government purchases or nonrecourse loans.

Payment rate: the amount paid per unit of production to each participating farmer for eligible payment production under the FAIR Act.

Permanent legislation: legislation that would be in force in the absence of all temporary amendments (farm acts). The Agricultural Adjustment Act of 1938 and the Agricultural Act of 1949 serve as the basic laws authorizing the major commodity programs. Technically, each new farm act amends the permanent legislation for a specified period.

Price support: the minimized price or floor below which the government will not let the price of specific commodities fall. A variety of approaches have been taken to support prices at desired levels.

Set-aside: the amount of land that had to be set aside—either left idle or planted to another crop.

Slippage: extent to which farmers participating in government programs "overproduce" by retiring marginal acres rather than highly productive acres, or decide not to participate at all.

Soil bank: tillable land that farmers take out of production to qualify for price supports.

Target price: price set by the government for selected commodities. This price is achieved by supplementing the market price with a deficiency payment.

REFERENCES

Knutson RD, Penn JB, Flinchbaugh BL: *Agricultural and food policy,* 4th ed., Englewood Cliffs, N.J., 1998, Prentice Hall.

U.S. Department of Agriculture: Agricultural Outlook Supplement, April 1996, Washington, D.C., U.S. Government Printing Office.

EXERCISES

1. Next to the following graph, please place a T or F in the blank appearing by each statement:

   _____ There has been an increase in the quantity demanded.

   _____ Consumers are better off economically if area 3 + area 4 is greater than area 2.

   _____ Producers will be better off by the value of area 2 only.

   _____ This figure illustrates the effects of domestic demand enhancement.

   _____ Society as a whole will be better off by the value of area 3 + area 4.

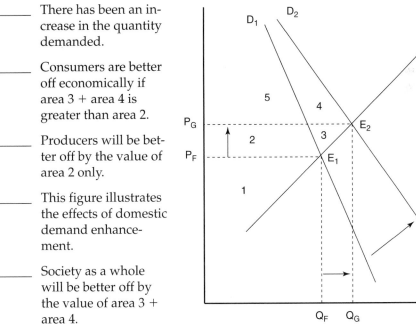

2. Suppose the current market price for wheat is $3 per bushel and that 10 billion bushels are currently being marketed both domestically and abroad. Given the following market shares and own-price elasticities of demand for domestic and export markets, please answer the following questions.

	Market Share	Own-Price Elasticity
Domestic	40%	−0.20
Foreign	60%	−2.00

a. What would happen to U.S. wheat producers' total revenue if the quantity they supplied to the market *increased* by 3% (i.e., how much would their total revenue change from current levels)?

b. Would domestic consumers of wheat products be better off or worse off economically than they were before this increase in supply? Illustrate your answer *graphically*. Why?

3. The following graph depicts the market level effects of using set-aside requirements to support farm prices and income.

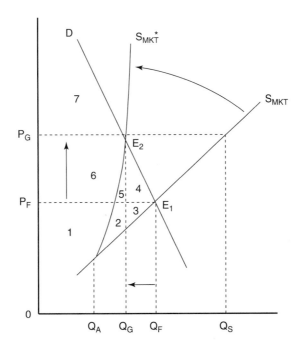

Please enter a T or F in the following blanks to indicate whether the corresponding statement is true or false.

_____ Domestic consumers are made worse off economically in the current period as a result of the government's actions.

_____ The extent to which the government has to enter the market is unaffected by the elasticity of the demand curve.

_____ Q_F represents the quantity sold to foreign governments.

_____ Areas 3 and 4 represent the economic loss to society as a result of this policy action.

_____ Producers gain area 6 from consumer's willingness to pay price P_G for quantity Q_G.

P A R T

V

MACROECONOMICS OF AGRICULTURE

14

PRODUCT MARKETS AND NATIONAL OUTPUT

*If a man marries his housekeeper or his cook,
the national dividend is diminished.*

*A. C. Piqou
(1877–1959)*

The discussion of market outcomes in the preceding chapters was conducted within a **partial equilibrium** framework; that is, we implicitly assumed when studying the events taking place in the market for unprocessed farm products that everything else in the economy would remain constant. We did not recognize the responses registered in all other markets to a new price for farm products.

If we were to analyze the impact of acreage controls placed on the production of wheat in a partial equilibrium framework, we would ignore the effects that this policy action would have on obvious things such as the price of bread and on less obvious things such as the federal budget deficit, national unemployment, and market interest rates. The effects this policy would have on the prices for other crops and livestock would also be ignored. Implicit in an analysis of the equilibrium in a single market is the assumption that the indirect or feedback effects of this action from all other markets are so small that they can be ignored.

The concept of **general equilibrium** regards all markets as being interdependent. Everything depends on everything else—to varying degrees, of course. The impact on the price of wheat resulting from the effect that reduced wheat production has on the prices of other goods and services in the economy would not be ignored. Because the food and fiber industry represents almost one-fifth of the nation's output, a substantial shock to food and fiber markets can have a significant impact on the rest of the economy, and vice versa. In short, general equilibrium captures not only the equilibrium between producers and consumers in the nation's product markets, but also the equilibrium between these products and the nation's money market.

The purpose of this chapter is to illustrate how businesses and households are linked together through resource and product markets, to establish the conditions that must be satisfied for an equilibrium between consumers and producers for a given rate of interest, and to discuss the composition and measurement of national output. The assumption of a fixed interest rate will be relaxed in Chapter 15, when we examine money market equilibrium.

CIRCULAR FLOW OF PAYMENTS

Let us begin our discussion of multimarket relationships in the economy by assuming the existence of a **barter economy,** in which households and businesses exchange goods and services as a means for paying for their purchases. We will then relax this assumption by allowing for the presence of money as a medium of exchange.

Barter Economy

In a barter economy, there is no money to serve as a medium of exchange. Households own all the primary resources (i.e., land, labor, capital, and management), which they supply annually to businesses. Businesses need these resources to produce goods and services. As shown in Figure 14.1, households receive payments in kind; that is, their wages are in the form of the products they helped produce.

Households would have to **barter** among themselves to obtain the mix of goods and services they desire. Can you imagine the difficulties that consumers

FIGURE 14.1   The business and household sectors in a barter economy interact with each other in the economy's resource and product markets. Businesses purchase the services provided by land, labor, capital, and management supplied by households in the resource market. Households buy the goods and services they need in product markets. Because there is no money in this economy, households barter among themselves and with businesses when exchanging goods and services. Bartering occurs even in a monetary economy. A plumber, for example, may do some plumbing work for a dentist in exchange for dental work of equal value. The value of these services would be considered income by the Internal Revenue Service.

and producers would encounter in an economy as complex as ours if bargaining between all parties was necessary to satisfy the demands for goods and services? How would you like to work for a vegetable producer during the summer and be paid in heads of cabbage? This would mean that you would have to barter with the university when paying your tuition, and the university would have to decide how many heads of cabbage they would charge per course.

Monetary Economy

Now let's modify Figure 14.1 to reflect the fact that money is available for use as a medium of exchange. Continuing to assume that there are only households and businesses in our simplified economy, we see in Figure 14.2 that the household sector receives monetary remuneration in the form of rents, wages, salaries, and profits in exchange for providing the business sector with land, labor, capital, and management services. This figure also shows that businesses receive monetary remuneration in the form of consumer expenditures when they supply goods and services to households.[1] The sum of wages, rents, interest, and profits accruing to laborers and capital resource owners in the resource markets represents the economy's national income. The monetary value of the products flowing to households through the product markets represents **national product.**

---

[1]The flow diagrams in Figures 14.1 and 14.2 assume that businesses do not save (i.e., retained earnings are zero). Instead, all profits are paid out in the form of dividends to households that are assumed to own all primary resources in the economy.

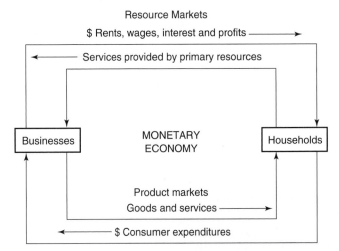

Resource Markets

$ Rents, wages, interest and profits ⟶

⟵ Services provided by primary resources

Businesses

MONETARY
ECONOMY

Households

Product markets

Goods and services ⟶

⟵ $ Consumer expenditures

FIGURE 14.2 The major difference between the monetary economy and the barter economy is that transactions are completed in dollars. When households provide labor and other services to businesses, they are compensated in the form of rents, wages, interest, and profits. When businesses sell goods and services to households, they are compensated by the value of private expenditures. The sum of rents, wages, interest, and profits received by households in the economy's resource markets represents its national income. The value of total expenditures by households represents gross domestic product.

You might have noticed that Figure 14.2 makes no mention of the economic role that government plays in the economy. This simplified flow diagram also ignores the possibility of savings and investment. There is also no recognition of the role financial markets play in the economy. To remedy these deficiencies, let us expand the simple flow diagram of a monetary economy in Figure 14.3 to include government and financial markets.

The expanded flow diagram in Figure 14.3 contains a financial market through which the net savings of households are passed on by depository institutions, the bond market, and the stock market to businesses that must either borrow, issue bonds, or sell stock to finance their expenditures. The term *net saving,* used to describe the flow of money from households to financial markets, and the term *net borrowing,* used to describe the flow of money from financial markets to businesses, signify the fact that money is flowing in both directions (i.e., households also borrow from financial institutions, and businesses also save). Households are classified as net savers because they save more than they borrow, and businesses are classified as net borrowers because they borrow more than they save.

The production of goods and services in the monetary economy depicted in Figure 14.3 also generates income in the form of wages, rents, and interest. This income is captured at the bottom of this figure by the line drawn connecting the business and household sectors.[2] Businesses, of course, can decide to retain part of their profits (i.e., retained earnings) to help finance future expenditures.

The flow diagram in Figure 14.3 also captures the flow of goods and services and the flow of income and expenditures between government and businesses and households. For example, there is a flow of money from households to government in the form of tax payments. Households, in turn, receive money from the government in the form of social security payments, unemployment compensation, and other forms of payments (hence the use of the term *net taxes*).

---

[2]For simplicity, only the money flows between the sectors are reflected in this figure, even though the physical flows of resources, services, and products continue to exist in the economy.

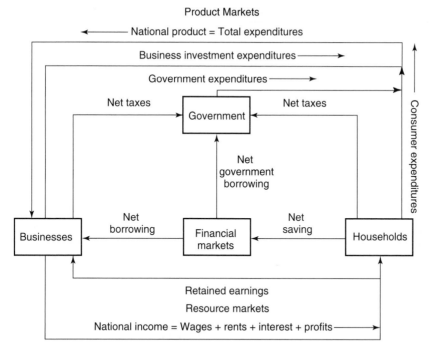

FIGURE 14.3 This figure adds a financial sector and a government sector. For example, households borrow and save. If their saving exceeds their borrowing, they are considered net savers. Businesses also borrow and save in financial markets. The government sector also borrows in the nation's financial markets to finance its budget deficits.

Businesses also pay taxes and receive government payments. The government borrows in the nation's financial markets to finance its budget deficits (i.e., when government expenditures exceed tax revenues). It does this through the sale of government securities by the U.S. Treasury. When the government has a budget surplus, which has not happened at the federal level since 1969, it can buy back some of the securities it has issued. Finally, government expenditures for goods and services also represent a component of the **final demand** for the products produced by businesses. These expenditures for finished goods and services are included along with consumer expenditures and business expenditures when measuring the nation's **gross domestic product (GDP).**

COMPOSITION AND MEASUREMENT OF GROSS DOMESTIC PRODUCT

The nation's annual output is referred to as gross domestic product (GDP). There are two basic approaches to measuring the level of GDP: (1) the expenditures approach and (2) the income approach. The expenditures approach measures activity in the product market, and the income approach measures activity in the resources market. Because both approaches result in the same value of GDP, we will

TABLE 14.1   U.S. Gross Domestic Product, 1999

Component	Amount (Billion Dollars)	% of GDP
Consumption		
Durable goods	761.3	
Nondurable goods	1,845.5	
Services	3,661.9	
Total consumption	6,268.7	67.4
Gross private domestic investment		
Nonresidential structures	285.6	
Producer's durable equipment	917.4	
Residential construction	403.8	
Total fixed	1,606.8	
Change in inventories	43.3	
Total investment	1,650.1	17.7
Government purchases of goods and services		
Federal government	568.6	
State and local government	1,065.8	
Total spending	1,634.4	17.6
Net exports of goods and services		
Exports	990.2	
Imports	1,244.2	
Total net exports	−254.0	−2.7
GROSS DOMESTIC PRODUCT	9,299.2	100.0

Source: *Survey of current business,* August 2000.

focus on the measurement and explanation of activity captured by the expenditures approach.

Table 14.1 gives us an insight into the composition of GDP and the relative magnitudes of its components, and Table 14.2 helps illustrate what is included and not included in GDP.

Approximately two-thirds of the nation's output is represented by expenditures made during the period by consumers. One-half of this total is represented by payments for consumer services. Gross private domestic investment, which represented 17.7% of the nation's output in 1999, captures the expenditures by businesses to purchase buildings and equipment, and the expenditures by households to build new residences.[3]

Government purchases of goods and services typically represent about 18 to 20% of GDP, although net exports in recent years actually reduced GDP (i.e., we have been importing more from other nations than we have been exporting). With the exception of net exports, these percentage shares of total GDP accounted for by

---

[3]The term *gross* here means that the annual expenditures to replace worn-out or obsolete machinery and other selected forms of business assets are included in this total.

TABLE 14.2   GDP: What's In and What's Not

Type of Expenditure	Included in GDP?	Examples
Consumer expenditures:		
Durable goods	Yes	Autos, TVs, VCRs
Nondurable goods	Yes	Food, clothes
Services	Yes	Haircut, airplane ticket
Gross private domestic investment:		
Change in business inventories	Yes	Corn stored on farm
Producers' durable equipment	Yes	Computers, tractors
Structures:		
Nonresidential structures	Yes	Factories, office buildings, shopping malls
Residential structures	Yes	Houses, condominiums
Net exports:		
Exports	Yes	Tractors, computers
Imports	Yes	Coffee, bananas, wine
Government purchases of goods and services:		
Intermediate services	Yes	Fire fighters, police officers
Consumption	Yes	City parks, street cleaners
Other activity:		
Government interest and transfer payments	No	Social security, welfare, unemployment benefits
Private intermediate goods	No	Wheat, iron ore, plastic, crude petroleum
Private purchases of used assets	No	Purchases of used home, used autos
Purchases of farmland	No	Cropland, grazing land

consumption, investment, and government expenditures have held remarkably stable during the post–World War II period.

The value of GDP and its components in Table 14.1 are expressed in current dollars, which is commonly referred to as **nominal GDP.** No adjustments have been made for the effects that inflation has on the purchasing power of annual GDP. **Real GDP** reflects the effects of inflation, which is accomplished by dividing nominal GDP by the implicit GDP price deflator, a price index representing a weighted average of the prices of all the goods and services that are captured in GDP. Figure 14.4 illustrates the historical fluctuations in the real GDP growth rate from 1890 to 1999.

Newspaper articles reporting that the economy grew by 3.2% during the year are referring to the percentage change in real GDP from the previous year. During 1982, for example, nominal GDP grew 3.7%, but real GDP actually declined 2.5%. This particular difference is significant because nominal GDP growth gave a picture of a growing economy, while real GDP growth gave a picture of an economy in a recession. Therefore, here and in all other instances, it would be grossly misleading to discuss economic growth in nominal GDP terms.

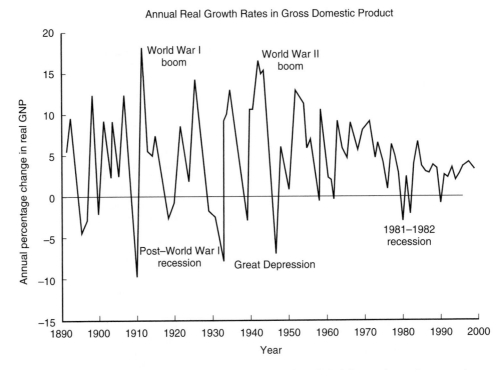

FIGURE 14.4   The percentage change in the real GDP of the United States shows sharp year-to-year variations during the past 100 years. The last 20 years have been more stable than the previous 100-year period. (*Source: Economic report of the president.*)

CONSUMPTION, SAVINGS, AND INVESTMENT

There are two things that households can do with a dollar of **disposable income** (income after taxes): they can "consume" the dollar, or they can save it. If the dollar is used to finance consumer expenditures, it is gone forever. However, if the dollar is saved, it will be available to finance future consumption.

When households make expenditures, they are purchasing what are normally referred to as consumer goods and services. If the good has a life of less than one year, it is called a nondurable good. A service, by definition, is nondurable. If a good is consumed over a longer period of time, however, it is a durable good. Food is an example of a nondurable good, a haircut or an airplane trip are examples of services, and a house or a car are examples of durable goods.

Investment refers to expenditures by businesses on capital goods, such as a new machine or a new building; expenditures by households for new residences; and an increase in the inventories of businesses. We will identify the factors that influence the level of planned consumption expenditures by households and planned investment expenditures by businesses.

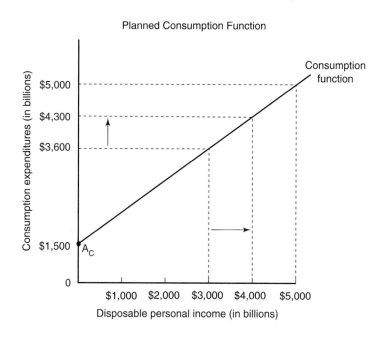

Planned Consumption Function

FIGURE 14.5 The relationship between actual disposable income and planned consumer expenditures by households suggests that households in Lower Slobovia will consume $1,500 billion ($1.5 trillion) per year if their incomes after taxes are actually zero. When disposable personal income increases, consumer expenditures will increase. The rate of this increase will depend upon the slope of this function.

Determinants of Planned Consumption

In 1936, John Keynes identified the major determinant of planned **consumption** by households. Keynes suggested that people "increase their consumption as their income increases, but not by as much as the increase in their income." In other words, there is a relationship—but not a one-to-one correspondence—between the planned consumption expenditures by households and their current level of disposable income.

Figure 14.5 illustrates an annual consumption function and its break-even level of consumption for the hypothetical economy of Lower Slobovia.

Consumption Function.   When the household sector's disposable income goes up in Lower Slobovia, its level of planned consumption goes up also. This relationship represents the household's **consumption function.**[4] As Keynes suggests, planned consumption expenditures rarely change by the same amount as a change in consumer disposable income. The hypothetical annual consumption function in Figure 14.5 suggests that planned consumption would be approximately $1,500 billion even if actual disposable income were zero. This level of consumption expenditures is referred to as autonomous consumption.

---

[4]The term *function* refers to the fact that there is a causal relationship between income and consumption. If income increases, consumption will also increase. If consumption is said to be a function of income, this means that the level of consumption depends on the level of income.

We can express the consumption function depicted in Figure 14.5 in general terms as:

$$C = A_C + MPC\,(DPI) \tag{14.1}$$

in which C represents the level of consumption expenditures for goods and services made by households, A_C represents the level of autonomous consumption expenditures, MPC represents the slope of the consumption function, and DPI represents the level of disposable personal income. Thus, when consumers' disposable personal income increases, consumer expenditures will increase. Assume the level of disposable personal income increased from $3,000 billion to $4,000 billion in Lower Slobovia. Figure 14.5 indicates that consumption expenditures would increase from $3,600 billion to $4,300 billion. Before we can use Equation 14.1 to replicate these graphical results, we must first calculate the slope of this function, or the MPC.

Slope of Consumption Function.   The slope of the consumption function represents consumers' marginal propensity to consume (MPC); that is, it reflects the change in consumption expenditures associated with a change in disposable personal income, or:

$$MPC = \Delta C \div \Delta DPI \tag{14.2}$$

in which Δ represents the change in a variable, C represents the level of planned consumption expenditures, and DPI represents disposable or after-tax personal income.[5]

We can determine the slope or marginal propensity to consume for the economy of Lower Slobovia by analyzing the change in consumption from $3,600 billion to $4,300 billion illustrated in Figure 14.5, when disposable personal income rose from $3,000 billion to $4,000 billion. Using Equation 14.2, we can determine that Lower Slobovia's marginal propensity to consume must be:

$$
\begin{aligned}
MPC &= (\$4{,}300 - \$3{,}600) \div (\$4{,}000 - \$3{,}000) \\
&= \$700 \div \$1{,}000 \\
&= 0.70
\end{aligned}
\tag{14.3}
$$

A marginal propensity to consume of 0.70 means that consumers in Lower Slobovia spend 70 cents of every after-tax dollar on consumer goods and services. Because the consumption function in Figure 14.5 is linear (a straight line), the value

---

[5]An alternative measure of income in the consumption function is given by the permanent income hypothesis. This theory implies that planned consumption expenditures do not depend on current income, but instead on a measure of expected permanent income over the next three to five years. In other words, current planned consumption would not change substantially if current actual disposable income changes, unless this change is expected to continue over time.

of the economy's marginal propensity to consume will be identical over the full range of this curve.

We can verify that this value represents the slope of the consumption function in Figure 14.5 by substituting this marginal propensity to consume into Equation 14.1 and solving for the level of consumption associated with the two levels of disposable personal income used in this example. Given the level of autonomous consumption in this economy of $1,500 billion and an MPC of 0.70, Equation (14.1) can be rewritten to read:

$$C = \$1,500 + 0.70 \text{ (DPI)} \tag{14.4}$$

If the level of disposable income were $3,000 billion, then the level of consumption expenditures in the economy would be:

$$
\begin{aligned}
C &= \$1,500 + 0.70 \ (\$3,000) \\
&= \$1,500 + \$2,100 \\
&= \$3,600 \text{ billion}
\end{aligned}
\tag{14.5}
$$

And if the level of disposable income were instead equal to $4,000 billion, the level of consumption expenditures in this economy would be:

$$
\begin{aligned}
C &= \$1,500 + 0.70 \ (\$4,000) \\
&= \$1,500 + \$2,800 \\
&= \$4,300 \text{ billion}
\end{aligned}
\tag{14.6}
$$

Note these are the same levels of consumption expenditures associated with these two income levels illustrated in Figure 14.5.

Shifts in Consumption Function.
While changes in the level of income correspond to movements along the consumption function illustrated in Figure 14.5, other determinants of consumption expenditures will shift the consumption function.

An increase in wealth of the nation's household sector, for example, would shift its consumption function upward.[6] This implies that the household sector has a greater basis from which to spend for a given level of income than it does with less wealth. As Figure 14.6 illustrates, an increase in the wealth of the household sector in Lower Slobovia would increase consumption expenditures from $4,300 billion to $5,000 billion for a given level of disposable personal income ($4,000 billion). In terms of Equation 14.1, autonomous consumption necessarily must have increased from $1,500 billion to $2,200 billion. A decrease in wealth would have the opposite effect.

---

[6]This measure of wealth should include the current market value of the household's physical and financial assets less debt outstanding, and should be adjusted for inflation.

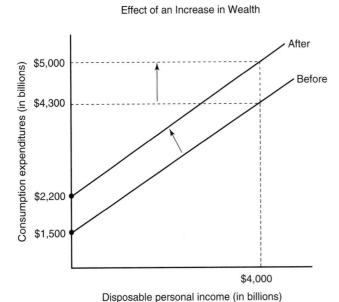

Effect of an Increase in Wealth

FIGURE 14.6   An increase in wealth will shift the consumption function upward, and a decrease in wealth will shift the consumption function downward.

A specific component of the household sector's wealth position—its holding of liquid assets—may be of special importance to the household sector when formulating its spending plans. A liquid asset is an asset that can be converted into money quickly with little or no loss in value. Cash is a perfectly liquid asset. Other liquid assets include investment bonds, shares in savings and loan associations, and savings accounts at commercial banks.

Expectations also may affect how much households with a given level of current disposable income are willing to spend, particularly in the short run. For example, if households expect a higher income in the near future, they may increase their current consumption expenditures. Conversely, the same households may decide to postpone consumption (and increase their savings) if they are pessimistic about the near future.

Determinants of Planned Saving

The current level of planned saving in the economy is equal to the difference between disposable personal income and planned consumption expenditures, or:

$$S = DPI - C \tag{14.7}$$

in which *C* represents the level of consumption expenditures discussed above and *DPI* represents the level of disposable personal income in the economy. This equation simply states that savings by households is equal to their personal income after taxes (or disposable income) and after household spending for goods and services.

The change in saving associated with a change in disposable personal income represents the marginal propensity to save, or:

$$MPS = \Delta S \div \Delta DPI \qquad (14.8)$$

Because the level of savings also represents personal income after taxes not spent on consumption, we know that:

$$MPS = 1.0 - MPC \qquad (14.9)$$

We determined in Equation 14.3 that the marginal propensity to consume in Lower Slobovia was 0.70. The marginal propensity to save in this country must therefore be:

$$MPS = 1.0 - 0.70$$
$$= 0.30 \qquad (14.10)$$

or 30 cents out of every dollar of disposable personal income. This does not mean that the nation's wealth increases by 30 cents for every dollar of disposable personal income. Remember that some consumption, namely autonomous consumption, is carried on even if disposable personal income is equal to zero. We can see if there was saving or dis-saving in the economy during the period by substituting the levels of disposable personal income and consumption into Equation 14.4. For example, if disposable personal income was equal to $3,000 billion in Lower Slobovia, we know from Figure 14.5 and Equation 14.5 that consumption would be equal to $3,600 billion. Thus, the level of saving in Lower Slobovia would be:

$$S = \$3,000 - \$3,600$$
$$= -\$600 \text{ billion} \qquad (14.11)$$

or a dis-savings equal to $600 billion. In other words, consumption exceeded disposable income by $600 billion.

How high would the economy's disposable personal income have to be before dis-saving would be eliminated? A logical extension of this analysis is to determine the break-even level of disposable personal income, or the point at which consumption equals disposable personal income. We can determine this level of disposable personal income by substituting alternative levels of income into Equation 14.3 to determine the level of consumption and then use Equation 14.9 to determine the level of saving. (See Table 14.3.) Thus, a level of disposable personal income of $5,000 billion would be required if the economy's wealth position is not to be less at the end of the year than it was at the start of the year. If disposable personal income is less than $5,000 billion, dis-saving will occur. If disposable income is greater than $5,000 billion, saving will occur.

We can also determine the break-even level of disposable personal income graphically by extending a ray out of the origin in Figure 14.5 at a 45-degree angle.

TABLE 14.3   Level of Disposable Personal Income in Billion Dollars

Disposable Income	Consumption	Saving
3,000	3,600	−600
4,000	4,300	−300
5,000	5,000	0
6,000	5,700	+300
7,000	6,400	+600

This ray indicates all points in this figure at which consumption equals disposable personal income. Draw this line in Figure 14.5 on page 331.

As Figure 14.5 indicates, the break-even level of consumption in Lower Slobovia's economy is $5,000 billion, the point at which the 45-degree ray intersects the economy's consumption function.

As we now turn our attention to the economy's investment function, consider the following:

1. What would happen to the level of consumption expenditures in Lower Slobovia's economy if its federal government increased current income taxes by $10 billion?
2. Can you answer question 1 graphically?
3. Does the consumption function in Figure 14.5 shift downward?
4. Can you answer question 3 using the consumption function equation?

Determinants of Planned Investment

Business investment is defined as business expenditures on new buildings and equipment plus net additions to their production inventories.[7] The level of business investment in the current period is determined in part by the market rate of interest. The level of household investment in new residences in the current period is determined in part by the market rate of interest. Not surprisingly, Keynes stated that the "inducement to invest depends partly on the investment demand schedule and partly on the rate of interest."

Investment Schedule.   The investment demand schedule that Keynes was referring to reflects the investment expenditure plans of businesses and households corresponding with specific rates of interest. Investment expenditures consist of fixed investments in business assets (factories, office buildings, shopping centers, trucks, tractors, silos, apartments, etc.) and changes in business inventories. In-

---

[7]Purchases of land or used capital goods at the aggregate level are not considered as investments because they do not represent capital goods formed in the current period.

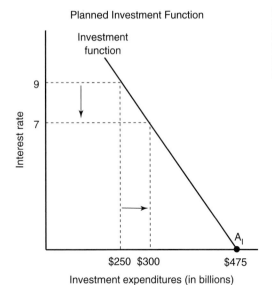

Planned Investment Function

FIGURE 14.7   The relationship between the rate of interest and planned investment expenditures for new structures and equipment is called the planned investment function. The lower (higher) the market rate of interest, the higher (lower) the level of planned investment.

vestment expenditures also include newly constructed houses and condominiums sold to individuals. A homeowner is treated as a business that owns the house as an asset and rents the newly constructed house to itself. The planned investment function for the economy reflects an inverse relationship between the rate of interest and the level of planned investment, or:

$$I = A_I - MIS(R) \tag{14.12}$$

in which *I* represents the level of investment expenditures, A_I represents the autonomous level of investment, or investment that would take place if the market interest rate were equal to zero, *MIS* represents the marginal interest sensitivity of investment or slope of the investment function, and *R* represents the market rate of interest.

The interest sensitivity of investment in Equation 14.12 reflects the impact that a change in interest rates will have upon investment expenditures, and is given by:

$$MIS = \Delta I \div \Delta R \tag{14.13}$$

or the change in investment expenditures associated with a change in the market rate of interest. We would expect the interest sensitivity of investment for businesses to be different from the interest sensitivity of investment for households. In general, the greater the value of *MIS* in Equation 14.12, the more sensitive investment in Lower Slobovia's economy will be to changes in the market rate of interest.

Figure 14.7 presents the annual planned investment function for Lower Slobovia. This figure shows that if the market rate of interest in the current period is 7%, the level of investment expenditures by households and businesses will be $300 billion. But if the market rate of interest rose to 9%, investment expenditures in Lower Slobovia's economy would fall to $250 billion. We can calculate the marginal interest sensitivity of investment for Lower Slobovia based upon this observation using Equation 14.13 as:

$$
\begin{aligned}
MIS &= (\$250 - \$300) \div (9 - 7) \\
&= -\$50 \div 2 \\
&= -\$25.0 \text{ billion}
\end{aligned}
\tag{14.14}
$$

Thus, a 1% increase in interest rates will depress investment expenditures in Lower Slobovia by $25 billion.

If we assume that the level of autonomous investment expenditures in the economy was $475 billion, we can solve Equation 14.12 for the two interest rates depicted in Figure 14.7 and verify the change in investment expenditures suggested by Equation 14.13. Substituting the values of A_I and MIS into Equation 14.12, we see that the level of investment expenditures when the market rate of interest is 7% would be:

$$
\begin{aligned}
I &= \$475 - 25(7) \\
&= \$475 - \$175 \\
&= \$300 \text{ billion}
\end{aligned}
\tag{14.15}
$$

If the market rate of interest in Lower Slobovia increased from 7% to 9%, the investment expenditures economy-wide in Lower Slobovia would be:

$$
\begin{aligned}
I &= \$475 - 25(9) \\
&= \$475 - \$225 \\
&= \$250 \text{ billion}
\end{aligned}
\tag{14.16}
$$

Note that the interest rate and investment expenditure combinations reported in Equations 14.15 and 14.16 represent the values depicted in Figure 14.7.[8]

Shifts in Investment Function.   Other determinants of planned investment cause the investment expenditures function in Figure 14.7 to shift outward to the right or inward to the left. Like shifts in the consumption function, an outward shift in the investment function like that depicted in Figure 14.8 means that more investment will take place at a given market rate of interest (e.g., $350 billion instead of $300 billion at a 7% market rate of interest).

---

[8]An increase in investment expenditures, by inducing additional output and income, "pulls up" the level of consumption expenditures in the economy. The result is an increase in the nation's output that is a multiple of the initial increase in investment expenditures. The size of this multiplier will depend on consumers' marginal propensity to save (MPS). The lower the MPS, the greater the multiplier effect of investment expenditures on output will be.

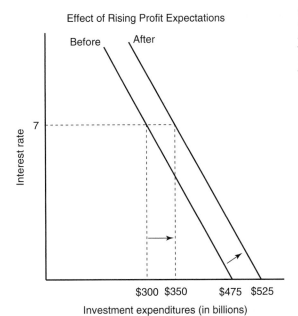

Effect of Rising Profit Expectations

FIGURE 14.8 Favorable developments with respect to expected future incomes of businesses and households, prices of new investment goods, technological change, or taxes will shift the investment expenditures function to the right, and vice versa.

Businesses must project the profitability of an investment project over the life of the investment, and households must project the feasibility of a new home loan before initiating construction. Thus, their expectations about a wide range of factors that affect the future income of producers and consumers will affect decisions as to whether or not they should make an investment in the current period. If businesses are optimistic (pessimistic) about their future profit, the planned investment function will shift to the right (left) for each interest rate, as illustrated in Figure 14.8. If households are optimistic (pessimistic) about their future income, the planned investment function for new residences will shift to the right (left) for each interest rate.

The purchase price of new capital goods also represents a determinant of planned investment expenditures by businesses and households. If the cost of a new plant and equipment were to suddenly decrease (increase), we would expect the planned investment function in Figure 14.7 to shift to the right (left), as shown in Figure 14.8. If the cost of new homes were to suddenly decrease (increase) in Lower Slobovia, we would expect the planned investment function for new residences to shift to the right (left), as illustrated in Figure 14.8.

The technological change embodied in new capital goods also has an effect on planned investment expenditures. Improvements in the productive services, provided by equipment and innovations that affect the functions performed by capital, should shift the planned investment expenditures function to the right, as depicted in Figure 14.8. In other words, planned investment at each interest rate should now be greater than before as firms seek to become more efficient in their operations.

The level of taxes also will affect planned investment expenditures because businesses and households evaluate their expenditure decisions on the basis of their expected after-tax return. If there is a decrease in effective tax rates (i.e., deductions and credits are liberalized), we would expect businesses and households to expand their expenditures for new capital goods. The opposite would be true if tax rates were increased.

What would happen to the level of investment expenditures in Lower Slobovia if the market rate of interest fell by 2%? Can you answer this question graphically? Will the investment function shift to the right? Can you use Equation 14.12 to get a numerical answer to this question?

EQUILIBRIUM NATIONAL INCOME AND OUTPUT

To determine the equilibrium level of national income and output in the economy, we must employ the relationships discussed above for household and business expenditures. The total value of planned consumption, investment, government spending, and net exports in an economy is often referred to as aggregate expenditures. When aggregate expenditures equal aggregate output, the nation's product markets are in equilibrium.

Aggregate Expenditures

Table 14.1 shows that the nation's gross domestic product includes the demand for final goods and services in the economy by consumers, investors, governments, and foreign countries. Assume for ease of exposition that the economy of Lower Slobovia is a closed economy (i.e., it has no imports or exports). Also assume that the market rate of interest is constant. The levels of expenditures in Lower Slobovia can be summarized as:

$$C = \$1,500 + 0.70(\text{DPI})$$
$$I = \$475 - 25(R)$$
$$G = \$880 \qquad\qquad (14.17)$$

in which the consumption expenditure function was developed earlier in Equation 14.4, the investment expenditure function was developed in Equation 14.15, and the government expenditure function above suggests that government expenditures on goods and services in Lower Slobovia is fixed at $880 billion. If we assume that the market rate of interest for the moment is fixed at 7%, then the nation's aggregate expenditures would be given by:

$$AE = C + I + G$$
$$= \$1,500 + 0.70(\text{DPI}) + \$475 - 25(7) + \$880$$
$$= \$2,680 + 0.70(\text{DPI}) \qquad\qquad (14.18)$$

TABLE 14.4 Total Product Market Activity in Lower Slobovia in Billion Dollars

Planned Consumption Expenditures (C)	Planned Investment Expenditures (I)	Planned Government Expenditures (G)	Aggregate Expenditures (C + I + G)	Output (Y)
$5,420	$300	$880	$6,600	$6,000
6,120	300	880	7,300	7,000
6,820	300	880	8,000	8,000
7,520	300	880	8,700	9,000
8,220	300	880	9,400	10,000

in which *AE* represents aggregate expenditures. Table 14.4 summarizes the levels of aggregate expenditures for alternative levels of disposable income and a market interest rate of 7%.[9] The product market will be in equilibrium for a given market interest rate when aggregate expenditures and output are equal to $8,000 billion.

The Keynesian Cross

We can analyze the level of aggregate expenditures in Lower Slobovia in the context of what has come to be known as the Keynesian cross, using the observations for consumption (*C*), investment (*I*) and government (*G*) expenditures presented in Table 14.4. The 45-degree ray coming out of the origin in Figure 14.9 represents all points of equality between aggregate expenditures and aggregate output (gross domestic product also equals national income in the national income and product accounts). Figure 14.9 shows that equilibrium in the product market of the Lower Slobovia economy would occur at $8,000 billion. This is the point at which the aggregate expenditure curve (*C* + *I* + *G*) for our hypothetical economy crosses the 45-degree ray. Table 14.4 also shows that aggregate spending in Lower Slobovia equals the nation's output at $8,000 billion.

Below this equilibrium, planned aggregate expenditures exceeded aggregate output, which should draw down unsold business inventories and increase pressures to expand output. After $8,000 billion, aggregate output exceeded planned aggregate expenditures, which should lead to increases in unsold business inventories and put downward pressure on future aggregate output levels.

Deriving Aggregate Demand Curve

The slope of Lower Slobovia's aggregate demand curve will depend upon the response of consumption expenditures and hence total expenditures to changes in

---

[9]Chapter 17, which discusses the nation's taxation and spending policies, will explain the determination of disposable personal income (DPI). Assume here that taxes in the current period are based upon last year's income, which is $400.

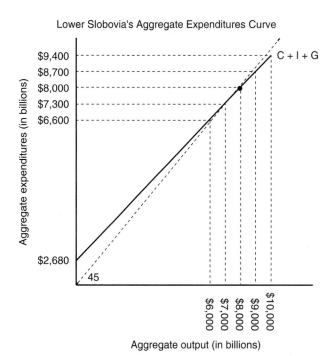

Lower Slobovia's Aggregate Expenditures Curve

FIGURE 14.9   The equilibrium level of aggregate output in the nation's product market occurs where $Y = C + I + G$. This occurs at a level of aggregate output and aggregate expenditures of $8,000 billion. This equilibrium assumes a given market interest rate and general price level.

real wealth (current wealth adjusted for changes in the general price level, which affects the purchasing power of wealth). The three aggregate expenditure curves in Figure 14.10, *A* are associated with three different general price levels. Let us assume equilibrium E_1 is associated with the level of aggregate expenditures illustrated in Figure 14.9.

The general price level P_1 is also associated with an aggregate output or real GDP equal to Y_1. **Equilibrium Output** E_3 would represent the level of aggregate output Y_3 that we would observe at a higher general price level P_3, and equilibrium E_2 would reflect a lower general price level P_2 and higher aggregate output level Y_2. Figure 14.10, *B* presents a plot of the price-quantity combinations associated with E_1, E_2, and E_3. A line drawn through these price-quantity coordinates gives the economy's aggregate demand curve (AD) a traditional downward-sloping demand curve like those seen earlier in this book.

Aggregate Supply and Full Employment

Aggregate supply represents the nation's aggregate output supplied to consumers, businesses, governments, and foreign countries. The aggregate supply curve, therefore, must necessarily represent the nation's aggregate output supplied to these groups for a full range of price levels. When aggregate demand increases, farm and nonfarm businesses in the economy will respond to the higher prices by increasing their output and/or reducing their current unsold business inventories.

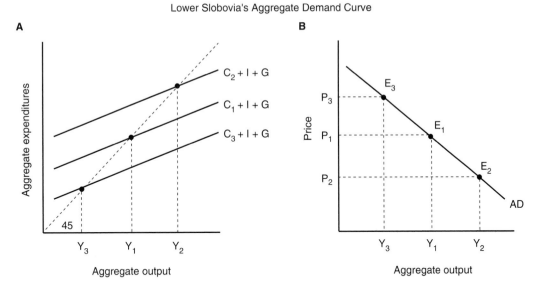

FIGURE 14.10   The aggregate demand curve for Lower Slobovia's economy is derived by observing the level of consumption spending and aggregate spending as we vary the **general price level** and hence real wealth of its consumers, and then observing what this means for the level of aggregate real output (real GDP).

This can be seen as a movement up the economy's aggregate supply curve in the short run, because it takes time for current investment expenditures by businesses in buildings and other productive assets to expand productive capacity, which shifts the supply curve to the right.

If aggregate demand decreases and prices fall, businesses will respond by cutting production schedules, and business inventories typically rise. This will result in a movement down the economy's aggregate supply curve.

The aggregate supply curve can take on a number of slopes depending on where the economy is on the aggregate supply curve. The curve has three distinct ranges: (1) the Keynesian depression range, (2) the normal range, and (3) the classical range. Each of these ranges is illustrated in Figure 14.11.

The depression range of the aggregate supply curve, sometimes referred to as the Keynesian range, after the British economist John Maynard Keynes, indicates that there is a perfectly elastic range of output over which increases in demand result in increases in supply unaccompanied by rising prices. The normal range of the aggregate supply curve takes the form of the firm and market-level supply curves. The general price level will begin to rise as demand increases, more slowly at first and then more sharply when the curve becomes more inelastic. When the economy reaches its capacity to supply goods and services in the current period, the aggregate supply curve becomes perfectly inelastic. This range is referred to as the classical range because it takes on the properties of the aggregate supply curve found in the writings of classical economists who were prominent before Keynes' theories surfaced in the 1930s.

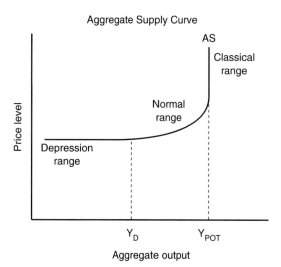

Aggregate Supply Curve

FIGURE 14.11   The economy's aggregate supply curve shows the depression, normal, and classical ranges.

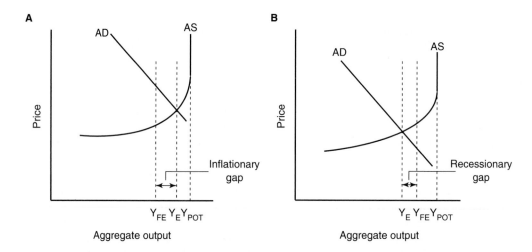

FIGURE 14.12   Let Y_{FE} represent the level of aggregate output at which the economy achieves **full employment,** the point at which labor and capital resources are employed at a noninflationary or natural rate of employment. If the level of aggregate demand is less than what is necessary to be operating at Y_{FE}, the economy is suffering from a recessionary gap. If aggregate demand exceeds Y_{FE} and approaches Y_{POT}, the economy is suffering from an inflationary gap.

The equilibrium between spending and output illustrated in both Figure 14.9 and Figure 14.10, *B* represents one view of equilibrium in an economy's product market for a given market rate of interest. We can also illustrate the equilibrium in the product market by examining the intersection of the economy's aggregate demand curve (*AD*) and the aggregate supply curve (*AS*). Assume that the aggregate supply curve for Lower Slobovia takes the form illustrated above in Figure 14.11 and repeated in Figure 14.12, *B*.

The intersection of the *AD* and *AS* curves in Figure 14.12, *B* identifies the equilibrium output level of the nation's product market, or Y_E. This corresponds with the point at which the aggregate spending curve $(C + I + G)$ intersects a 45-degree ray from the origin, indicating that aggregate expenditures equals aggregate output. Figure 14.12 suggests that equilibrium output in the economy is less than full employment output (Y_{FE}), or the economy's maximum noninflationary level of goods and services. Manufacturing plants and the labor force in the economy at Y_{FE} would be operating at their natural rate of employment. Although further output would be possible in the current period if the economy were at Y_{FE}, it would require businesses to adopt abnormal operating practices, such as putting on a graveyard shift, and laborers to give up some of their current leisure.

The aggregate supply curve becomes even more inelastic at Y_{POT} or the economy's potential GDP. This portion of the aggregate supply curve suggests that aggregate output would be completely unresponsive to further expansion of aggregate demand in the current period, given its existing technology and normal production practices. Further demand expansion in the economy will be totally inflationary.

Recessionary and Inflationary Gaps

The amount by which equilibrium output Y_E falls short of the full employment output level Y_{FE} is called a **recessionary gap.** The rationale is that the economy could have produced output Y_{FE} without creating much inflation. But for one reason or another, the economy only produced output Y_E. The lower level of output also means that fewer people are employed (i.e., the unemployment rate for labor in the economy would be higher) than would be true at Y_{FE}. Figure 14.12 indicates that the recessionary gap in this instance would be equal to $Y_{FE} - Y_E$.

An **inflationary gap** will occur when planned aggregate expenditures are greater than the economy's full employment output. This economic situation can also be seen by reconsidering Figure 14.12, *A*. An inflationary gap will occur when the aggregate demand curve *AD* intersects the aggregate supply curve *AS* to the right of Y_{FE} The size of the inflationary gap would be the difference between the equilibrium level of output and Y_{FE}. This would correspond to the situation in which the aggregate expenditures curve in Figure 14.12, *A* intersects the 45-degree ray from the origin to the right of Y_{FE}. Although not shown, Figure 14.12, *A* and *B* would reflect a departure from the full employment level of output, Y_{FE}.

To summarize, departures from full employment output (Y_{FE}) are classified as either an inflationary gap or a recessionary gap. These gaps exist when the following conditions hold:

$$\text{Recessionary gap} = Y_{FE} - Y_E \text{ when } Y_E < Y_{FE}$$
$$\text{Inflationary gap} = Y_E - Y_{FE} \text{ when } Y_E > Y_{FE}$$

One final point worth noting when looking for the existence of inflationary and recessionary gaps is that, when the aggregate demand curve moves up the aggregate supply curve and approaches Y_{FE}, further increases in aggregate demand will increase the general price level in the economy more than it will increase aggregate output.

WHAT LIES AHEAD?

The coming chapters will show how the aggregate demand and supply curves can be adjusted to eliminate inflationary or recessionary gaps. Chapter 15 focuses on the second condition for general equilibrium; namely, equilibrium in the money market, and the monetary policy options available to the Federal Reserve System. Chapter 16 focuses on taxes, government spending, and budget deficits. Chapter 17 addresses the nature of fluctuations in the economy and what the government can do about them. Chapter 18 will illustrate what this all means for the price of food and the economic performance of farmers and ranchers along with other segments of the nation's food and fiber industry.

SUMMARY

The purpose of this chapter was to introduce the concept of general equilibrium, the relationship between sector output and GDP, and the physical and financial links between the farm business sector and the rest of the economy. The major points made in this chapter may be summarized as follows:

1. A partial equilibrium assumes other things remain unchanged. In many situations, however, it is necessary to take into account events taking place simultaneously in other markets. When we do this, we are focusing on a general equilibrium.
2. The circular flow of payments in a barter economy consists of a two-way flow of goods and services between businesses and households. A grower of cabbage selling his product to a plumber would have to negotiate an exchange of goods and services that is equally fair to both parties (e.g., 25 cabbages per hour of plumbing services). The circular flow of payments in a monetary economy has the benefit of permitting the cabbage grower to receive money in exchange for his product, which he can then allocate across his purchases of other goods and services.
3. When measured in current prices, economists refer to these aggregate expenditures as nominal gross domestic product (GDP). If deflated by the implicit GDP price deflator, however, economists instead use the term real gross domestic product (GDP), when referring to aggregate output of the economy.
4. Consumer expenditures amount to more than two-thirds of total aggregate expenditures in the economy. These expenditures can be disaggregated into the following categories:

- Expenditures on food and related products
- Expenditures on nonfood, nondurable goods
- Expenditures on durable goods
- Expenditures on services

Durable goods are products that are consumed over a period of time, such as cars, television sets, and a good pair of shoes. Nondurable goods are products such as food, gasoline, and newspapers. Services include items such as haircuts and air travel.

5. The slope of the aggregate consumption function is called the marginal propensity to consume (MPC). This coefficient suggests what the change in consumption expenditures will be if disposable personal income changes. The marginal propensity to save (MPS) suggests what the change in saving would be if disposable personal income changes. Thus, MPS = 1.0 − MPC.

6. Other factors influencing the level of consumption expenditures, besides the level of income, are the level of taxes paid by consumers, the level of their real wealth, and their expectations about their future financial position.

7. The slope of the aggregate investment function is known as the interest sensitivity of investment, or the change in investment expenditures given a change in the market rate of interest.

8. The determinants of planned investment expenditures in new business assets by businesses, or new residences by households include factors such as the purchase price of new capital goods, the expectation of future income by businesses and households, technological change, and the level of taxes.

9. The Keynesian cross refers to the determination of the equilibrium level of output in the nation's product market by using a 45-degree ray out of the origin in a graph with aggregate expenditures on the vertical axis and aggregate output on the horizontal axis. The point at which this ray crosses the economy's aggregate expenditures curve indicates the equilibrium level of output.

10. The economy's aggregate supply curve reflects its willingness to supply goods and services at alternative levels of product prices. This curve has the following three ranges:

   a. A depression range, sometimes called the *Keynesian range,* after British economist John Maynard Keynes

   b. A normal range that looks like market supply curves

   c. A classical range, which is named after the classical school of thought that suggests that the aggregate supply curve is completely unresponsive to active demand-oriented macroeconomic policies.

11. An inflationary gap occurs in an economy when planned aggregate expenditures exceed its full employment level of output. A recessionary gap occurs when the opposite is true; namely, aggregate expenditures are less than the economy's level of full employment output.

DEFINITION OF KEY TERMS

Barter: the direct exchange of one good or service for another, without the use of money.

Barter economy: an economy in which money is not used as the medium of exchange. Instead, households and businesses "swap" goods and services in satisfying their needs.

Consumption: expenditures by consumers for food, nonfood, nondurable goods, durable goods, and services.

Consumption function: a mathematical expression of the relationship between the level of consumption and variables such as the level of disposable personal income.

Disposable personal income: income of consumers after taxes have been deducted; personal income minus personal taxes.

Equilibrium Output: the level of aggregate output at which aggregate demand equals aggregate supply.

Final demand: expenditures by consumers, investments by businesses, government spending, and net exports. Final demand is equal to total output intermediate demand.

Full employment GDP: the level of aggregate output at which labor and capital in the economy are employed at their natural or noninflationary rate.

General equilibrium analysis: regards all sectors of the economy as being interdependent. The events taking place in all markets are considered in the analysis.

General price level: a weighted average price of all goods and services in the economy.

Gross domestic income: given by the income approach to measuring economic activity, national income is equal to the sum of wages, rents, interest, and profits.

Gross domestic product (GDP): referred to as the nation's output, GDP is equal to consumer expenditures, business investment, government spending, and net exports (exports minus imports).

Inflationary gap: amount by which equilibrium GDP exceeds full employment GDP.

Investment: expenditures on new plant and equipment by businesses, the increase in business inventories, and expenditures on new residences by households.

Monetary economy: an economy in which money is used as the principal medium of exchange. Households receive remuneration in the form of wages, rents, interest, and profits. Businesses receive remuneration in the form of expenditures by households, businesses, government, and foreign countries.

National product: see gross domestic product.

Nominal GDP: value of aggregate output in the economy measured in that period's prices, or current prices.

Partial equilibrium analysis: assumes the events taking place outside the market under analysis remain constant.

Real GDP: value of aggregate output in the economy measured in the prices of another (i.e., base) period, or constant prices.

Recessionary gap: amount by which full employment GDP exceeds equilibrium GDP.

Savings: that part of disposable personal income not spent on current consumption.

REFERENCES

Barro RJ: *Macroeconomics*, 5th ed., 1997, MIT Press.

Economic report of the president, Washington, D.C., various issues, U.S. Government Printing Office.

Keynes JM: *A general theory of employment, interest and money*, London, 1936, MacMillan Publishing.

Mankiw NG: *Principles of macroeconomics*, New York, 1997, Dryden Press.

EXERCISES

1. Referring to the graph located to the right, please place a T or F in the blank appearing next to each statement:

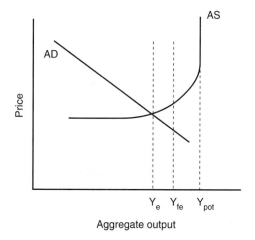

_____ Y_{fe} represents the economy's maximum output in the short run.
_____ This economy is experiencing a recessionary gap.
_____ Supply expansion policies are recommended to eliminate this gap in the short run.
_____ Demand expansion strategies could be used to eliminate this gap.

2. Assume that disposable income increases in the economy from $4,000 to $5,000 and causes consumer expenditures to increase from $4,300 to $5,200.

   a. What is the economy's marginal propensity to consume?

   b. Are households in this economy saving or dis-saving?

   c. What is the marginal propensity to save in the economy?

   d. If taxes were to increase by $500, what effect would this have on consumer expenditures?

3. Suppose the aggregate investment function for the economy is given by:

$$I = \$400 - 40(R)$$

where *I* represents the level of investment expenditures and *R* represents the market rate of interest. Further assume the current rate of interest is 10% ($R = 10.0$)

   a. How much will investment expenditures in the economy change if interest rates fall by two percentage points?

   b. Would the level of autonomous investment change? Why?

   c. What factors might cause the investment curve given by the investment function above to shift to the left?

   d. What factors might cause the interest rate in this economy to decrease by two percentage points?

15

CONSEQUENCES OF BUSINESS FLUCTUATIONS

The modern world regards business cycles much as the ancient Egyptians regarded the overflowing of the Nile. The phenomenon recurs at intervals; it is of great importance to everyone, and natural causes of it are not in sight.

John Bates Clark
(1847–1938)

TOPICS OF DISCUSSION

The goals of macroeconomic policy are to promote employment, price stability, and economic growth. The general economy, as we all know, experiences rising unemployment, inflation, and economic stagnation from time to time. Macroeconomic policymakers must use the monetary and fiscal policy tools available to them to cure these economic ills.

The purpose of this chapter is to examine the nature of fluctuations in business activity, outline the macroeconomic policy actions the federal government can take to achieve specific objectives, and illustrate the effects of these actions on macroeconomic variables such as the market rate of interest, the rate of inflation, the unemployment of labor and capital, and rate of growth in real GDP. Chapter 19 will address what this all means for the U.S. food and fiber system.

FLUCTUATIONS IN BUSINESS ACTIVITY

The nation's economy traditionally goes through periods of ups and downs in the level of its business activity. These business fluctuations, which are often referred to as **business cycles,** are typically thought of in terms of movements in the economy's GDP, interest rates, or unemployment rate. We will discuss the nature of business fluctuations in the general economy, the major indicators of this activity, and the policy actions normally taken to modify these fluctuations.

Nature of Business Fluctuations

Figure 15.1 illustrates the general nature of business fluctuations in the economy, which has four distinct phases. Periods of recession and expansion occur between peaks of these cycles. As the cycle reaches a trough, or "bottoms out," and the recovery begins, the economy enters another **expansionary period,** which is often referred to as a "boom" in business activity. Not every expansionary period reaches a new high. An expansionary period may end prematurely for one or more reasons, and then a new **recessionary period** will begin. If the level of business activity falls sharply, it may be classified as a depression rather than a recession. Of course, nothing in modern U.S. economic history rivals the depths of the Great Depression of the 1930s.

Four of the last seven periods of expansion in the economy coincided with this country's participation in war activities, with the period of expansion being from 1929 to 1945, beginning with the Great Depression and ending with the boom associated with World War II business activity. The U.S. economy also experienced an unprecedented period of expansion during much of the 50-year period beginning in 1950.

There are different definitions of what constitutes a recession in the U.S. economy. The U.S. Department of Commerce takes the position that the economy is in a recession when it experiences two consecutive quarters of negative growth in real

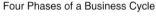

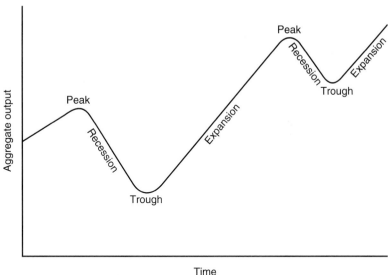

FIGURE 15.1   The traditional business cycle (of fluctuation) has four phases: (1) an expansionary phase, (2) a peak that marks the end of the expansionary phase, (3) a recessionary phase, and (4) a trough that marks the end of the recessionary phase. Most cycles result in a higher peak in economic activity during a period of prolonged economic growth in the economy; however, this need not always be true. During the 1930s, the recessionary phase became so pronounced that it was labeled the Great Depression.

gross domestic product.[1] The National Bureau of Economic Research (NBER) has devoted a considerable amount of its resources to defining and measuring business fluctuations in the economy. After studying business activity "after the fact," the NBER defines the "official" timing of the recession. The NBER's identification of recessions does not necessarily have to contain two consecutive quarters of negative real economic activity.

Indicators of Economic Activity

Of course, businesses and policymakers cannot wait for the NBER to tell them that the economy experienced a recession 12 months ago. Some follow indicators of business activity that either coincide with, lag behind, or lead business fluctuations in the economy.

The U.S. Department of Commerce regularly publishes a series of lagging, coincident, and leading indicators that serve to tell businesses and policymakers what is happening in the economy and what is likely to happen in the future.

---

[1]We continue to be concerned with the real, as opposed to nominal, growth of the economy for reasons outlined in Chapter 14. We have deducted the effects of inflation to focus on the growth of the economy's purchasing power (i.e., its real gross national product).

Coincident Indicators.   **Coincident indicators** move concurrently with business activity in the economy. Examples of coincident indicators are information on current industrial production, the number of employees on payrolls, personal disposable income, and manufacturing sales. These indicators help explain current business activity.

Lagging Indicators.   **Lagging indicators** of business activity usually indicate a change in economic activity about one quarter (i.e., three months) after the fact. Examples of lagging indicators are business inventories, labor cost per unit of output, the average duration of employment, and the average interest rate that banks charge their best customers (i.e., the prime rate).

Leading Indicators.   The **leading indicators** of business activity are defined so as to indicate business fluctuations before they occur; that is, they are expected to peak approximately one to two quarters before business activity actually peaks in the economy. (See Figure 15.2.) Examples of leading economic indicators are new orders for consumer goods, new building permits, new investment in plant and equipment, and changes in selected prices and the money supply.[2]

Forecasting Models.   As an alternative to using leading indicators to make forecasts of what will happen to the economy in the future, many businesses and policymakers employ the services of sophisticated computer models. These models reflect the relationships between past economic behavior of producers and consumers, and selected prices, interest rates, and other variables thought to explain economic behavior.

There are numerous commercial and government-sponsored models that are capable of projecting events in the economy in general and the farm business sector in particular.[3] Their major advantage is their ability to examine what would happen if conditions were to take on alternative values. This gives businesses and policymakers a feeling for the range of outcomes that are likely to occur under specific sets of conditions.

CONSEQUENCES OF BUSINESS FLUCTUATIONS

Two consequences of fluctuating business activity are **unemployment** and **inflation.** Unemployment will rise in periods when the economy is experiencing a recessionary gap, as illustrated in Figure 14.12. Inflation generally rises when the

---

[2]The U.S. Department of Commerce regularly publishes a set of indexes capturing these and other indicators in the publication *Business conditions digest*. These include an index of 12 leading indicators, an index of 4 coincident indicators, and an index of 6 lagging indicators.

[3]Examples of commercial models of the U.S. economy that contain a farm sector are the models developed by Wharton Econometric Forecasting Associates (WEFA). An example of a university-sponsored model of the U.S. economy that contains a farm business sector is the AG-GEM model maintained by Texas A & M University.

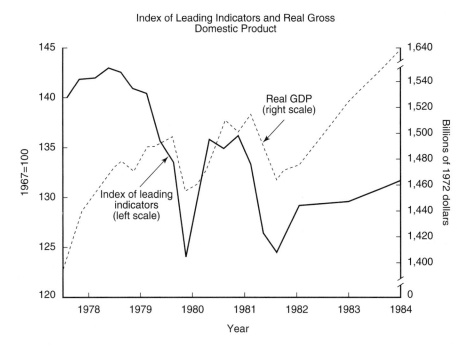

Index of Leading Indicators and Real Gross
Domestic Product

FIGURE 15.2   The index of leading indicators, depicted by the solid line, indicates business
fluctuations before they actually occur. The 1978 to 1984 period was chosen because it
illustrates how well this index can forecast the direction of fluctuations in real GDP in
subsequent quarters. Note in particular the doubled peak, or M-shaped activity forecasted
during the 1980–1982 period.

economy is experiencing an inflationary gap. It is often said that the goal of monetary and fiscal policy is to eliminate inflation (inflationary gaps) and unemployment (recessionary gaps) in the economy. Yet the terms *unemployment* and *inflation*
are frequently used without the benefit of a precise definition of their meaning or
measurement.

Unemployment

Unemployment, broadly defined, refers to the idling of part of the civilian labor
force and the idling of business plants and equipment. Unemployment of part of
the nation's scarce resources results in the loss of output and savings, both of which
affect the potential future growth of the economy and the economic suffering en-
dured by those workers and businesses whose resources are unemployed.

Unemployment of Labor.   The unemployment rate for labor is measured by
subtracting the total number of people not in the labor force (homemakers, students, those physically unable to work, and others who do not wish to work)

from the total noninstitutional population (this total excludes individuals in prisons and persons under 16 years of age). This difference represents the total labor force, which totaled 143.8 million people in 2000. From this figure, we subtract the members of the armed services (1.7 million) to determine the total **civilian labor force** (142.1 million). Unemployment (5.9 million) is found by subtracting the number of employed persons in the economy (136.2 million) from the total civilian labor force.

The **unemployment rate** is then found by dividing the number of unemployed persons by the size of the total civilian labor force, or:

$$\text{annual unemployment rate} = \frac{\text{number of unemployed persons}}{\text{size of total civilian labor force}} \quad (15.1)$$

In 1999, for example, the unemployment rate was 4.2% (i.e., 0.042 = 5.9 ÷ 142.1). Figure 15.3 shows what has happened to the unemployment rate since 1940. This figure shows that the unemployment rate was approximately 15% in 1940, which is high by today's standards but well below the unemployment rate of almost 25% during the Great Depression.[4]

Unemployment for labor falls into several categories. **Frictional unemployment** refers to the continuous flow of people who are changing jobs and, therefore, are currently unemployed. **Cyclical unemployment** refers to unemployment associated with business fluctuations. **Seasonal unemployment** refers to unemployment associated with changes in business conditions that are seasonal in nature. Workers in fruit- and vegetable-processing plants and construction workers in northern states are two examples of workers who may be seasonally unemployed. **Structural unemployment** refers to those workers who are unemployed because of structural changes in the economy brought about by technological change that does away with their jobs. Farm laborers whose jobs have been replaced by tomato pickers are examples of structurally unemployed workers.

Unemployment of Capital.   The equivalent concept for capital (i.e., plant and equipment) is the **manufacturing capacity utilization rate,** established by the U.S. Department of Commerce, or actual output divided by current manufacturing capacity. The lower the capacity utilization rate, the higher the unemployment of capital. **Full employment** refers to the employment of labor and capital when the economy is producing at its maximum noninflationary level of output.

In recent years, full-employment GDP has been thought to occur at an unemployment rate for labor of about 5% and a capacity utilization rate for business capital of 86%.

---

[4]Some economists argue that this unemployment rate understates the *true* unemployment rate because it does not account for discouraged workers who are no longer seeking employment and therefore are not included in the total civilian labor force (the denominator in the unemployment rate). This phenomenon is referred to as hidden unemployment. The labor force participation rate reflects the percentage of available people of working age who are actually in the civilian labor force.

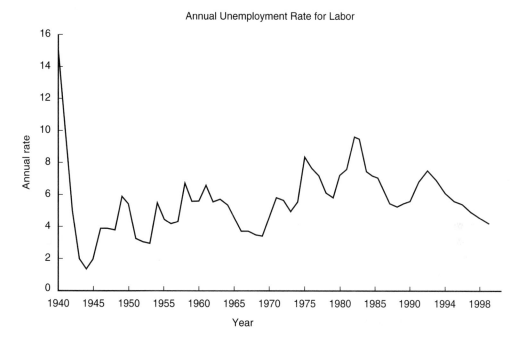

FIGURE 15.3   Compared with the high unemployment rates observed during the early 1940s, today's unemployment rate seems rather small. The unemployment rate is determined by dividing the number of unemployed persons by the total civilian labor force. Some economists argue that during a recession hidden unemployment occurs because workers are discouraged from seeking work. Therefore, the true unemployment rate would be understated.

Inflation

Inflation is generally defined as a sustained rise in the general price level (i.e., weighted average of all prices). Two key terms in this definition require special emphasis.

1. The term *sustained* rules out any temporary rise in prices.
2. The term *general price level* rules out the use of the term *corn price inflation* or *fuel price inflation*.

A temporary rise or spike in the general price level that returns to its initial level is not inflationary. The rise in the price of a single commodity also need not be inflationary, particularly if the prices of substitute goods or services have fallen. Instead, only a sustained rise in the general price level, as measured by the **consumer price index** (CPI), which accounts for changes in the prices of all goods and services purchased by consumers, represents their inflation.

The CPI is one of the most closely watched statistics coming out of Washington. The CPI reflects changes in the cost of living of the typical American family. The cost-of-living statistic is based upon more than 100,000 individual price quotations obtained monthly from approximately 25,000 retail stores, 20,000 households, and 20,000 tenants in 85 cities across the country. These prices include the price of food, housing, clothing, transportation, medical care, and other goods and services. This statistic is also based upon the relative importance of these goods and services in the "market basket" of the typical American family. This information is combined to produce an index that reflects the cost of the goods and services purchased by a typical household. The value of this index for any one year is given by:

$$CPI = \frac{\text{cost of standard market basket in current year}}{\text{cost of standard market basket in base year}} \times 100 \qquad (15.2)$$

Thus, if the standard market basket of goods and services cost $35,000 in 1999 versus $21,084 in the base period, the value of the CPI index for 1999 would be 166.6 (i.e., 166.6 = [35,000 ÷ 21,084] × 100).

The impact of a price change for a specific group of goods on the CPI is calculated based upon the relative importance of the group in the typical market basket of goods and services purchased by consumers. For example, assume that the cost of food eaten away from home has gone up by 10% during the period, and that the relative importance of these expenditures is 5.712%. The percentage change in the overall CPI would be:

$$W_{FAFH} \times \%\Delta P_{FAFH} = \%\Delta\, CPI$$
$$0.05712 \times 10.0 = .5712\% \qquad (15.3)$$

or approximately six-tenths of 1%. The variable W_{FAFH} represents a weight reflecting the relative importance of food consumed away from home to all goods and services purchased by consumers. Since 1987, the relative importance weights used for food consumed away from home and other goods and services have been based upon expenditure patterns observed from 1982 to 1984, scheduled to be updated to 1999–2000 in January 2002. As shown in Table 15.1, housing carries the greatest weight (39.636%), followed by transportation (17.45%) and food (16.302%).

Annual percent changes in the CPI are of more than idle interest for many consumers. Beyond reflecting what happens to the purchasing power of our income, or real disposable personal income, these changes are used to automatically adjust the nominal level of income received by specific groups of consumers as a cost-of-living adjustment (COLA). Steelworkers' wages, for example, are directly tied to the CPI. As a part of their union contract, steelworkers get a raise of $0.01 per hour for every 0.3 points in the CPI index (1982–84 = 100). In 1980, when the CPI rose from 72.6 to 82.4, reflecting a 13.5% inflation rate faced by consumers, this meant that steelworkers' nominal hourly wages rose by almost $0.33 to offset the effects of inflation. Social security benefits also go up automatically whenever the rate of inflation (i.e., the percent change in the CPI) exceeds 3% annually. Veterans and retired federal workers receive similar protection. As a result of these COLAs, it is estimated that a 1% increase in the CPI triggers more than $2 billion in addi-

TABLE 15.1   Consumer Price Index

Expenditure Category		Relative Importance Weights
Food and beverages		16.302
Food at home	9.603	
Food away from home	5.712	
Alcoholic beverages	0.987	
Housing		39.636
Residential rent	7.036	
Homeowners rent	20.470	
Other housing	12.13	
Apparel and upkeep		4.634
Transportation		17.450
New vehicles	4.835	
Used vehicles	1.888	
Motor fuel	3.160	
Public transportation	1.400	
Other transportation	6.167	
Medical care		5.768
Entertainment		6.008
Other goods and services		10.203
All items		100.00

Source: Bureau of Labor Statistics, August 2000.

tional federal government expenditures. According to Bureau of Labor Statistics estimates, the nominal income of more than one-half of all American households is affected by movements in the CPI.

Different age groups of consumers will have different weights. Young unmarried consumers will spend a different percentage of their consumption dollar on food and nondurable goods than a family of four will. Do the relative importance weights in Table 15.1 describe the relative importance of these goods and services to you?

The annual rate of inflation for consumers can be measured by computing the percentage change in the CPI from one period to the next, or:

$$\text{annual inflation rate} = \frac{\text{current CPI} - \text{previous CPI}}{\text{previous CPI}} \qquad (15.4)$$

For example, if the CPI in 1999 was 166.6 and in 2000 was 172.3, the annual rate of inflation for the year 2000 would be:

$$\text{annual inflation rate} = \frac{172.3 - 166.6}{166.6}$$

$$= .0342 \text{ or } 3.42\% \qquad (15.5)$$

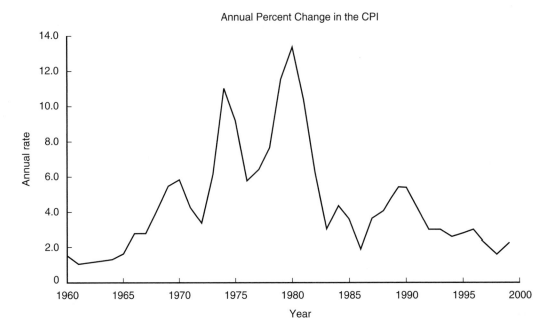

FIGURE 15.4   Inflation is defined as a sustained rise in the general price level. The annual percentage change in the consumer price index (CPI) measures the U.S. consumers' inflation rate. Not all consumers are confronted by the same rate of inflation, however. Young consumers typically purchase a different basket of goods and services than do retired consumers. Inflation also varies between rural and urban areas and across major geographical regions of the country.

Figure 15.4 shows what happened to the rate of inflation from 1960 to 1999. We see the upward trend in inflation that occurred during the late 1970s.

The U.S. Department of Commerce also prepares a **producer price index (PPI),** which accounts for changes in the prices of goods and services purchased by producers. The rate of inflation for producers, as measured by the percentage change in the PPI from one period to the next, is calculated in a manner similar to the CPI; that is, it uses base-year quantities at current prices and compares them with base-year quantities at base-year prices.

The cost of unemployment is clear—particularly to those who are unemployed. Unlike unemployment, however, inflation affects all of us in a direct and immediate way. It reduces the purchasing power of consumers' disposable income and the purchasing power of producers' profits. Persons particularly hurt by inflation include:

1. workers whose salary does not increase enough to offset the effects of inflation (e.g., those on a fixed income),
2. lenders who make loans at fixed interest rates and who are unfavorably surprised by the extent of inflation,

TABLE 15.2   Salary Needed to Break Even with a 5% and 10% Annual Inflation Rate

Year	Salary Required	
	5%	10%
1990	$20,000	$20,000
1991	21,000	22,000
1992	23,160	24,200
1993	24,320	26,620
1994	25,520	29,282
1995	26,800	32,210
1996	28,140	35,431
1997	29,540	38,974
1998	31,020	42,872
1999	32,580	47,159
2000	34,200	51,875

3. individuals or businesses who sign contracts that do not account for inflation, and

4. individuals or businesses who hold money rather than assets that go up in value with inflation.

Consider the example presented in Table 15.2, where we determined what you would have to earn in the year 2000 to be able to afford the same basket of goods and services you purchased with $20,000 in 1990, if the annual rate of inflation during this period was either 5% or 10%.

Table 15.2 suggests that your salary would have to be $51,875 in the year 2000 if you want to be able to buy the same bundle of goods and services that $20,000 bought in 1990, if the annual rate of inflation were 10%. You would only have to earn $34,200 by the year 2000 to break even if the annual rate of inflation were only 5%. This underscores the problems faced by those on fixed incomes or those who receive annual pay raises that are less than the rate of inflation. Table 15.3 shows how much more formidable the task of staying even with inflation is in other parts of the world. There are considerable differences among the industrial and developing countries.

In Chapter 14, we defined nominal and real GDP. Real GDP is found by dividing nominal GDP by the implicit GDP price deflator. The implicit GDP price deflator is nothing more than the broadest price index one can use when accounting for inflation when assessing the magnitude of the nation's output. This price index includes all the goods and services consumers purchase and purchases of other final goods by businesses, governments, and foreign buyers. Real GDP is thus a measure of output that attempts to highlight changes in physical levels of production at the economy level between two different periods in time.

Let's examine an example of an economy producing two products, apples and oranges (Table 15.4). We cannot assess the aggregate output of this economy

TABLE 15.3   Average Annual Inflation Rate for Selected Countries

Country	Inflation Rate (% per Year)	Doubling Time (Years)
Argentina	346.0	0.5
Israel	238.5	0.6
Brazil	158.6	0.7
Mexico	72.9	1.3
Italy	11.5	6.4
Spain	10.7	6.8
France	7.7	9.4
Canada	6.5	11.0
United Kingdom	6.2	11.5
United States	4.7	15.1
Switzerland	3.4	20.8
Germany	3.0	23.5
Japan	2.1	33.3
Industrial countries	8.6	8.4
Developing countries	37.1	2.2
World	15.5	4.8

Source: *International financial statistics.*

TABLE 15.4   Nominal and Real Gross Domestic Product

1992 Nominal GDP	1999 Nominal GDP	1999 Real GDP
100 oranges @ 20¢ = $20 80 apples @ 50¢ = $40 Total spending = $60	120 oranges @ 27¢ = $32 90 apples @ 60¢ = $54 Total spending = $86	120 oranges @ 20¢ = $24 90 apples @ 50¢ = $45 Total spending = $69

in a single statistic that includes market values, because we cannot add apples and oranges. The implicit GDP price deflator for this two-product economy in 1999 would be 124.63 (i.e., [$86/$69] 100). This example shows that real GDP in 1999 expressed in constant or 1992 prices would be $69, which reflects a 15% real growth in this hypothetical economy since 1992. A comparison of nominal GDP over this period would have suggested a 24.6% growth rate, due in part to the growth in physical consumption of apples and oranges and in part to the increase in the prices of these two commodities. The latter does not represent a real expansion of the economy.

Types of Inflation.   Like unemployment, there are also several types of inflation: demand-pull inflation, cost-push inflation, and stagflation.

When the aggregate demand for goods and services is rising and the economy is approaching full employment, **demand-pull inflation** will occur. The result

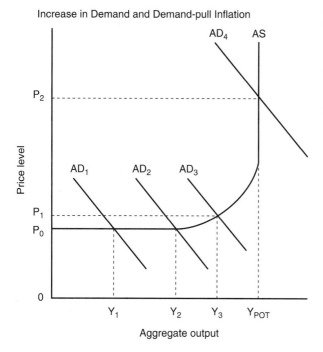

Increase in Demand and Demand-pull Inflation

FIGURE 15.5   Demand-pull inflation occurs when changes in demand result in a sustained rise in the general price level. The increase in aggregate demand from AD_1 to AD_2 was not inflationary. The increase in aggregate demand from AD_2 to AD_3, however was mildly inflationary because the general price level rose from P_0 to P_1. Finally, the increase in aggregate demand from AD_3 to AD_4 would be highly inflationary. The general price level rose from P_1 to P_2.

is a rise in the general price level.[5] This phenomenon is illustrated in Figure 15.5. In this figure, we assume that the AS curve represents the economy's current aggregate supply curve and AD_1, AD_2, and so on, represent a series of alternative aggregate demand curves.[6]

Figure 15.5 suggests that if the aggregate demand curve were to shift from AD_1 to AD_2 there would be virtually no change in the general price level. If the aggregate demand curve instead shifted from AD_1, to AD_3, the general price index would increase from P_0 to P_1. This increase in demand, therefore, is inflationary, although mildly so.[7] The inflation rate in this instance would be equal to $(P_0 - P_1) \div P_1$. Finally, if the aggregate demand curve were to shift from AD_1 to AD_4, the economy would reach its capacity to supply additional goods and services in the current period, Y_{POT}. This desired level of expenditures would be highly inflationary because the general price index would increase from P_0 to P_2. In the last two situations, increases in demand would be "pulling up" the general price level.

---

[5]This also corresponds to the inflationary gap, in which desired spending is greater than the economy's full-employment output, which was introduced in Chapter 14 and will be discussed in considerable detail in Chapter 18.

[6]This aggregate supply curve contains the full range of potential outcomes in the economy. The portion of the AS curve associated with outputs between 0 and Y_2 is called the Keynesian or depression range. The segment associated with Y_2 through Y_{POT} was illustrated earlier in Figure 14.11 and is referred to as the normal range. The perfectly inelastic segment of AS is known as the classical range.

[7]Demand-pull inflation begins to occur before full employment of the nation's resources is achieved. Some economists refer to this as **premature inflation.**

A second form of inflation, which occurs when the economy is not at full employment, is **cost-push inflation.** This form of inflation occurred from 1969 to 1970 and from 1973 to 1975. In both instances, prices were rising even though the economy was nowhere near full employment. Cost-push inflation, which is sometimes referred to as market-power inflation, can arise for at least two reasons: (1) union monopoly power, and (2) business monopoly power.

Some unions may have enough bargaining power in the labor market to impose wage increases on employers, who in turn raise their prices to maintain current profit margins. This wage-price spiral results in a rise in the general price level. Businesses with monopoly powers can also raise their prices if they desire higher profits. Workers will then demand higher wages to compensate for losses in their standard of living (they no longer can buy the same bundle of goods as a result of this price increase). This, in turn, causes the monopolist to raise his prices further to secure his desired profit margin. This price-wage spiral also results in a rise in general price levels.

A phenomenon first experienced in the late 1970s and early 1980s is **stagflation,** which refers to the existence of increasing inflation during a period when the economy is experiencing rising unemployment. This term originates from the combination of economic *stag*nation and in*flation*. Obviously, different macroeconomic policies are needed in each of the foregoing situations if policymakers desire both to promote the growth of the economy and to stabilize prices (i.e., eliminate inflation).

Short-Run Phillips Curve

We would all agree that rising unemployment and rising inflation are bad. Rising unemployment hurts those who are unemployed *and* those who would have sold goods and services to them. Inflation hurts practically everyone by reducing the purchasing power of income and wealth. Are these two problems interrelated? Yes, says British economist A. W. Phillips, who first noted the interrelationship between these two consequences of business fluctuations. He observed that wages increased more rapidly when England's unemployment rate was low than when the unemployment rate was high. This suggested that inflation of wages and other prices does not wait until full-employment GDP is reached, which explains the "normal" range of the aggregate supply curve in Figure 14.11. The Phillips curve seems to hold for specific periods of time, but structural changes and other factors shift the aggregate Phillips curve to the right or to the left from time to time. Few would question the short-run nature of this curve.

Figure 15.6 presents the short-run Phillips curve for the hypothetical country of Lower Slobovia. It suggests that an inflation rate of 10% is associated with an unemployment rate of 4%, and an inflation rate of 4% is associated with an unemployment rate of 9%. Thus, policies that fight inflation will—at least in the short run—lead to higher unemployment. And policies to stimulate employment will lead to higher inflation in the short run. This has obvious implications for the United States' choice of macroeconomic policies.

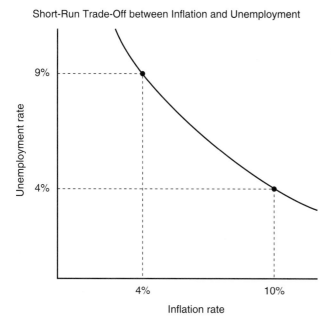

Short-Run Trade-Off between Inflation and Unemployment

FIGURE 15.6   The relationship between the rate of inflation and rate of unemployment in Lower Slobovia is captured by its Phillips curve.

MACROECONOMIC POLICY OPTIONS

Policymakers adopt several approaches toward trends in market activity in the economy. Classical economists such as Adam Smith long argued that macroeconomic policymakers should do nothing, and that the economy will somehow self-correct for undesirable trends in unemployment, inflation, and recessionary GDP gaps.

During the Great Depression of the 1930s, British economist John Maynard Keynes advanced the notion that federal governments should take an activist position in dealing with undesirable trends in the economy. He recommended using demand-oriented monetary and fiscal policies either to stimulate or retard economic activity, depending on whether it was experiencing a recessionary or inflationary gap.

The Reagan era of the 1980s saw the advancement of supply-side arguments designed to stimulate productivity as a means of expanding aggregate output while lowering inflation. The general features of each of these macroeconomic policy options are discussed in the following sections.

Laissez-Faire Macroeconomic Policy

A **laissez-faire,** or "hands off," **macroeconomic policy** assumes that markets in the economy, left to their own devices, will keep the economy near full employment output. If the economy were to drop below full employment, lower general price levels and, hence, higher real wealth of consumers and lower interest rates would automatically stimulate the economy back to full employment. Conversely, if the economy were expanding beyond full employment, higher general price levels and, hence, lower real wealth of consumers and higher interest rates would automatically pull the economy back to full employment.

In summary, a philosophy of laissez-faire reflects the advice classical economists gave to governments in the early 1900s: macroeconomic policymakers should resist the temptation to tinker and instead keep their "hands off" the economy.

Demand-Oriented Macroeconomic Policy

The Great Depression caused real GDP in the United States to fall 30% from 1929 to 1933, forcing both economists and policymakers to reconsider the wisdom of a laissez-faire macroeconomic policy. Keynes' book, titled *A General Theory of Employment, Interest and Money,* discussed the importance of consumer disposable income as the engine for macroeconomic growth along with interest rates. Keynes prescribed that an economy should pursue aggressive macroeconomic policies in its money and product markets to boost aggregate demand when recessionary trends exist and slow aggregate demand when inflationary pressures exist.

In 1934, Keynes actually published an open letter to President Franklin D. Roosevelt in the *New York Times* that outlined how the United States could lift the economy out of the Great Depression by following aggressive **demand-oriented macroeconomic policies,** shifting the aggregate demand curve in Figure 14.10 to the right. The use of demand-oriented macroeconomic policies became so popular in the post–World War II period that President John F. Kennedy announced in an economic policy speech that he was a "Keynesian."

The role of demand-oriented macroeconomic policy can be likened to being the host of a party; that is, policymakers should know when to remove the punch bowl before the party gets out of hand and know when to bring the punch bowl back before the party gets too dull. In economic jargon, policymakers attempt to promote the economic growth of the economy without stimulating inflation. If inflation begins to rise, demand-oriented policy actions can be taken to dampen aggregate demand for goods and services, which reduces inflationary pressures but slows economic growth if the economy is not at full employment. If unemployment begins to rise, demand-oriented policy actions can be taken to stimulate aggregate demand. These actions will reduce unemployment but may be inflationary if the economy is approaching full employment.

Figure 15.7 shows that expansionary demand-oriented macroeconomic policies will increase aggregate output from Y_1 to Y_2. The general price, however, would rise from P_1 to P_2, signaling an increase in inflation. The extent to which general price levels will rise depends on the elasticity of that segment of the *AS* curve intersecting the *AD* curve. The more inelastic or steeper the *AS* curve, the greater the increase in the general price level for a given increase in demand.

The reverse is also true. Contractionary demand-oriented policies will decrease the general price level, but when aggregate output declines, the unemploy-

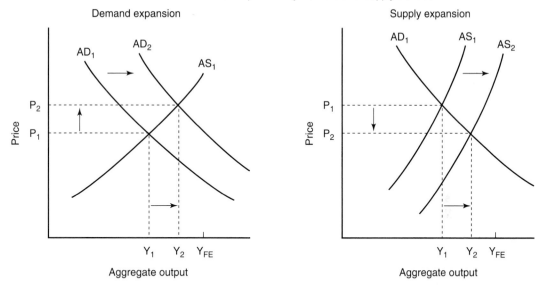

FIGURE 15.7   The implementation of demand-oriented and supply-oriented macroeconomic policies can have different impacts on the macroeconomy. An expansionary demand-oriented policy will shift the aggregate demand curve from AD_1 to AD_2, causing aggregate output to rise from Y_1 to Y_2 and the general price level to rise from P_1 to P_2. An expansionary supply-oriented policy will shift the aggregate supply curve from AS_1 to AS_2, causing aggregate output to rise from Y_1 to Y_2 and the general price level to fall from P_1 to P_2.

ment rate will rise.[8] And the less inelastic or flatter the *AS* curve, the smaller the increase in the general price level for a given increase in demand.

Keynes' proposal to actively use monetary and fiscal policies to stabilize the economy at full employment was "the law of the land" until the 1960s, when monetarists such as Milton Friedman began to question attempts to "fine tune" the economy. Monetarists and other critics of aggressive monetary and fiscal policies suggested that a steady rate of growth in the money supply alone, using the monetary policy instruments discussed in Chapter 14, was a more attractive macroeconomic policy alternative to the prescriptions of Keynes.

Many economists are critical of the demand-oriented, activist macroeconomic policies because they allow politicians to use fiscal policy to manipulate the economy for political short-term goals that may not be desirable in the long run. They argue that lags in the adoption and full implementation of macroeconomic policies may begin to take effect after the economy has largely corrected itself, and thus either cause

[8]This inverse relationship between inflation and unemployment was first popularized by British economist A.W. Phillips. He advanced a graphical relationship between these two important macroeconomic variables that came to be known as the Phillips curve. More recent research has shown that, although this relationship may hold in the current period, changes in technology and productivity distort this relationship over the long run.

an unnecessary recession or inflationary pressures. The economy, they argue, is more like a supertanker than a row boat; a row boat can be reversed almost instantaneously, but a supertanker requires more than ten miles for its course to be reversed.

Supply-Oriented Macroeconomic Policy

Stagflation is a relatively recent phenomenon. Stagflation refers to the presence of a stagnant economy accompanied by rising inflation. **Supply-side economics,** which gained popularity when President Ronald Reagan entered office in early 1981, involved attempts to shift the aggregate supply curve to the right.

As illustrated in Figure 15.7, a shift in the aggregate supply curve from AS_1 to AS_2 will increase aggregate output (and perhaps employment) from Y_1 to Y_2. Importantly, the general price level would fall from P_1 to P_2. Stagflation would be successfully countered by rising output and employment and lower general price levels.

To be successful, supply-side economics must lead to increases in technology, which increases the productivity of capital, increases the productive stock of capital in the economy, increases the supply of labor and/or labor productivity, and decreases input prices. These policies would shift the economy's production possibilities curve outward.

The Economic Recovery Tax Act (ERTA) of 1981 was the first of several pieces of tax legislation that reduced income tax rates by 25% from 1981 to 1983. These reductions were supposed to boost the incentive of labor to work more (e.g., more two-earner households). Bracket creep, or the rise in tax payments by individuals when their nominal wages or salaries rose but their real wages or salaries did not, was also eliminated with the indexation of income taxes. Tax credits for research and development were also instituted to promote the development of new technologies.

Many mainstream economists see considerable merit in the more moderate supply-side concepts.[9] The economy recorded an unprecedented period of economic growth during the majority of the 1980s. However, the large deficits that followed negated the favorable aspects of supply-side economics. Subsequent changes in the tax laws in 1986 and greater controls on government spending in the 1990s eventually led to talk of annual budget surpluses in 1999–2000.

Supply-side economics increased an inequity of income distribution. Some critics used the term *trickle down* when referring to the notion that benefits received by investors and laborers would benefit the poor by reducing inflation. This raises the age-old trade-off between efficiency and equity faced by policymakers.

---

[9]Other more radical supply-side economic concepts, such as the Laffer curve, have been less popular. The Laffer curve, named after its developer, Arthur Laffer, was based on the premise that an optimal tax rate existed, which yielded maximum income tax revenue to the U.S. Treasury. Efforts to increase income tax rates beyond this rate would lower tax receipts because of additional disincentives to work. Laffer argued this was the case and that tax cuts like those passed in 1981 would increase tax revenues and reduce annual budget deficits. Most economists correctly argued that the tax rates were lower than the rate that would maximize tax revenue to the Treasury and that tax cuts would therefore increase budget deficits.

SUMMARY

The purpose of this chapter was to examine the economic consequences of fluctuations in business activity and the potential macroeconomic policy responses to these fluctuations. The major points made in this chapter may be summarized as follows:

1. Expansionary monetary and fiscal policies can be employed to combat unemployment, and contractionary monetary and fiscal policies can be used to combat inflation.

2. In recent years, supply-side economists have proposed that demand management policies cannot attack the problems of stagflation. These economists have proposed tax cuts to shift the supply curve to the right to lower unemployment and inflation.

3. There are four phases to a business cycle: (1) expansion, (2) peak, (3) recession, and (4) trough. Farmers generally fare well when the economy enters a recovery period and inflation is low. When anti-inflationary policies are instituted, however, and interest rates rise, farmers fare rather poorly.

4. Businesses and policymakers often study published indicators of economic activity. Leading indicators, for example, are expected to indicate business fluctuations *before* they occur. Computer forecasting models are also often used for this purpose.

5. Two consequences of fluctuating business activity are unemployment of labor and inflation. Unemployment rises during periods of recession, and inflation rises in periods of expansion.

6. The unemployment rate reflects the percentage of the total civilian labor force that is unemployed. The equivalent concept for capital is the manufacturing capacity utilization rate.

7. Inflation is defined as a sustained rise in the general price level. The inflation rate for consumers is measured by the percentage change in the consumer price index (CPI). The inflation rate for producers is measured by the percentage change in the producer price index (PPI).

8. When the aggregate demand for goods and services is rising and the economy is either approaching or beyond full employment, demand-pull inflation will occur. Cost-push inflation, which is also referred to as market-power inflation, occurs when unions or imperfectly competitive firms impose an increase on the prices they charge for their goods or services.

9. The term *stagflation* refers to the existence of increasing inflation during a period when the economy is stagnant or experiencing little or no economic growth.

10. The short-run Phillips curve for an economy shows the trade-off between the rate of inflation and the rate of unemployment. Structural changes in the economy will shift this curve over time.

DEFINITION OF KEY TERMS

Business cycle: reflects the pattern of movements in the economy's real output, interest rates, or unemployment rate (also referred to as business fluctuations).

Civilian labor force: total noninstitutional population (e.g., excludes prisoners) minus total number of people not in the labor force (e.g., homemakers, students) and members of the armed services.

Coincident indicators: indicators of changes in economic activity that reflect current activity.

Consumer price index (CPI): weighted average of the prices consumers pay for goods and services.

Cost-push inflation: rise in general price level resulting from businesses and unions raising their prices and wage requests (also referred to as market power inflation).

Cyclical unemployment: unemployment of labor associated with adverse trends in the business cycle.

Demand-oriented macroeconomic policies: active policy actions that are designed to shift the economy's aggregate demand curve.

Demand-pull inflation: rise in the general price level that occurs when aggregrate demand for goods and services is rising and the economy is approaching full employment.

Expansionary period: phase of business cycle during which the nation's output is expanding.

Frictional unemployment: occurs when persons change jobs and are currently unemployed.

Full employment: high degree of employment of nation's resources (an unemployment rate of 5% to 6% for labor and a capacity utilization rate for capital of 86% to 88% is generally thought to constitute full employment).

Inflation: sustained rise in the general price level.

Lagging indicators: indicators of changes in economic activity about one or two quarters after they occur.

Laissez-faire macroeconomic policy: a decision to let the economy work its way out of its current problems; avoid taking active policy actions.

Leading indicators: indicators of changes in economic activity about one or two quarters before they occur.

Manufacturing capacity utilization rate: actual output of a nation's manufacturing firms divided by their potential output.

Nominal interest rate: market rate of interest unadjusted for the current rate of inflation.

Premature inflation: when demand-pull inflation occurs before full employment.

Producer price index (PPI): weighted average of the prices producers pay for goods and services.

Recessionary period: phase of business cycle during which the nation's output is declining.

Seasonal unemployment: unemployment of labor associated with changes in business conditions that are seasonal in nature.

Stagflation: existence of increasing inflation during a period when the economy is stagnant, or experiencing little or no economic growth.

Structural unemployment: unemployment of labor associated with structural changes in the economy.

Supply-oriented macroeconomic policies: active policy actions that are designed to shift the economy's aggregate supply curve.

Supply-side economics: promotion of macroeconomic policies that increase productivity and thereby shift the current aggregate supply curve to the right, which is believed to promote higher output at lower price levels.

Unemployment of labor: total labor force minus the number of employed persons in the economy.

Unemployment rate: number of unemployed persons divided by the size of the total civilian labor force.

REFERENCES

Barro RJ: *Macroeconomics*, 5th ed., 1997, MIT Press.

Board of Governors of the Federal Reserve System: *Chart book*, Washington, D.C., 1999, U.S. Government Printing Office.

Bureau of Labor Statistics: *Consumer Price Index*, Washington, D.C., various issues, U.S. Government Printing Office.

Economic report of the president, Washington, D.C., various issues, U.S. Government Printing Office.

EXERCISES

1. The nation, over a period of time, experiences cycles in business activity.

   a. In the space provided above, please draw and label the four phases of a business cycle.

   b. What are the two major economic consequences of business fluctuations in the nation's economy?

   c. In the space below, illustrate the relationship between these two consequences. Label all parts of this graph. What is the name given to this graph?

2. If the total labor force is comprised of 250 million people, there are 6 million people in the armed services, and there are 225 million civilians employed in the economy, calculate the nation's unemployment rate.

3. If the consumer price index was 166.6 in 1998 and 163.0 in 1999, calculate the annual rate of inflation that occurred in 1999.

16

MONEY, MONEY MARKETS, AND MONETARY POLICY

Everything, then, must be assessed in money;
for this enables men always to exchange their
services, and so makes society possible.

Aristotle
(384–322 B.C.)

TOPICS OF DISCUSSION

It is difficult to imagine a large, highly integrated economy like that of the United States without the existence of money. Money is perhaps the most important invention created by human beings. Bartering is both time-consuming and costly. Can you imagine a manufacturer of combines trying to exchange his product for the goods and services needed to manufacture additional combines? Workers certainly would not want to be paid off in combines, nor would the utility company that is supplying electrical power to the firm.

Our willingness to accept money made of paper or specific metals allows households and businesses in various regions of the country to specialize in the goods or services they provide despite a local noncoincidence of wanted goods. Given the critical role played by money in modern economies, it is important that we understand the factors that influence its demand and supply.

The purpose of this chapter is to discuss the characteristics of money, including the functions it performs; the Federal Reserve System, which is commonly called the Fed, and its role in our monetary economy; and the determinants of equilibrium in the money market. We will also discuss the monetary policy instruments the Federal Reserve uses to help achieve specific macroeconomic policy objectives.

CHARACTERISTICS OF MONEY

If you stopped people on a street corner and asked them to define the term *money*, they would probably define it as the coins in their pocket and the paper bills in their wallet. The typical definition of money, however, includes more than just currency. At a minimum, it also includes checking account balances, which is referred to as a demand deposit.

Functions of Money

Money can be defined by the functions it performs. We have already discussed its role as a **medium of exchange.** Money allows us to specialize in endeavors in which we have a comparative advantage by facilitating payments to others for the goods and services they have provided, and payments from others for our labor.

Money also serves as a **unit of accounting;** that is, it provides a basis with which we can compare the relative value of goods and services in the economy. For example, we have discussed the level of gross domestic product and national income in terms of billions of dollars. Physical measures of aggregate statistics such as these would be confronted with the age-old problem of "adding apples and oranges." Money also serves as the unit of accounting for assessing the profitability of businesses and household budgets.

Money is also an asset and, as such, has a **store of value.** You could store all your wealth in condominiums, for example. One problem with this is that condominiums are not liquid assets; that is, they cannot always be quickly converted to cash to pay for an unexpected bill without cost (realtor fees) or a loss in market value. Money is "cash," and can readily be used to meet an unexpected bill. It is partially for this reason that households and businesses store at least part of their wealth in money.

Money versus Near Monies

If we define money by the functions it performs, currency and demand deposits definitely would be considered as money. This constitutes the narrowest definition of money, a definition labeled M1 by the Federal Reserve System. Until the 1980s, only commercial banks could offer demand deposits. Today there are a variety of checking-type accounts offered by many financial institutions. The 1980 Depository Institutions Deregulations and Monetary Control Act expanded the definition of M1 balances to include these additional checkable accounts and demand deposits at mutual savings banks.

Other forms of assets may be considered **near monies** because of their relatively high liquidity. Time deposits, which must be held for a specific time before being converted to money, fall into this category. Savings accounts are particularly liquid. Although a 30-day notice of intent to withdraw funds from a savings account is required, such notice in practice is rarely required. Shares in money market mutual funds, which represent interests in holdings of government and corporate bonds, can also be classified as near money. Some of these funds allow individuals to write checks on the share they own in the fund.

The Federal Reserve System includes these near monies together with those assets previously captured in M1 in their M2 definition of money. Certain specialized overnight assets are also included in M2.[1] Certificates of deposit (CDs) at all depository institutions issued in denominations of $100,000 or more are the major factors distinguishing the Federal Reserve System's M3 definition of money.[2]

Backing of Money

What makes a piece of paper with green ink and a few numbers on it acceptable to us as payment for our labor prices? Monetary economies such as that of the United States rest on what is termed a **fiduciary monetary system;** that is, the value of money rests on the public's belief that a piece of paper with green ink on it can be exchanged for goods and services. Take a dollar bill from your wallet or purse. Is there a promise to exchange the dollar for a specific quantity of gold? No.

There are three principal reasons why money in a fiduciary monetary system has a positive value. First, money is acceptable by others when purchasing goods

---

[1]Overnight assets include overnight repurchase agreements issued by commercial banks and overnight Eurodollar deposits held by U.S. nonbank residents at Caribbean branches of U.S. banks.
[2]Term repurchase agreements issued by commercial banks and savings and loan associations are also included in M3. The Federal Reserve System also publishes a definition of liquid assets, which includes everything in M3 and financial instruments such as bankers acceptances, commercial paper, and liquid Treasury obligations.

and services. Second, demand deposits and currency have been designated as legal tender by the federal government. Most paper money will contain the phrase "this note is legal tender for all debts public and private" in the upper left-hand corner; thus, it must be accepted for payment of debts. Third, money also is a predictable value in nominal terms. A dollar is worth a dollar. Although the purchasing power of a dollar will decline in periods of inflation, it will always be accepted in exchange for goods and services.

FEDERAL RESERVE SYSTEM

Congress passed the Federal Reserve Act, and it was signed into law by President Woodrow Wilson on December 23, 1914. The Federal Reserve System began operation in the fall of 1914 as this country's central bank. Its principal domestic goals were to encourage economic growth and combat both inflationary and recessionary tendencies in the domestic economy. The Federal Reserve accomplishes these goals by regulating the quantity and cost of credit in the economy. The purpose of this section is to discuss the organization of the Federal Reserve System, the function it performs, its relationship to the banking system in this country, and the policy instruments that it has at its disposal.

Organization of the Federal Reserve System

The Federal Reserve System is organized around a Board of Governors located at the system's headquarters in Washington, D.C. The Board of Governors consists of seven full-time members appointed by the president and approved by the Senate. The Board of Governors exerts general supervision over the 12 district Federal Reserve Banks, which are located in selected major cities in the United States. The location of these district banks and their 25 branches is shown in Figure 16.1.

The Board determines reserve requirements, reviews and determines the discount rate, determines margin requirements on stock market credit, and in the past determined the ceilings on the yields paid by banks to their depositors (these ceilings have now been phased out). Each governor on the board serves one 14-year term and cannot be associated in any other way with banking during his or her term. In addition, no two governors may come from the same Federal Reserve district. The board has a chairman and a vice-chairman. Both are appointed by the President of the United States to serve 14-year terms. These officers may be reappointed over the course of their 14-year term on the board.

The **Federal Open Market Committee (FOMC),** which is composed of the seven members of the Board of Governors and five representatives of the district banks, meets periodically to determine the desired future growth of the money supply and other important issues. This committee issues directives to the manager of the System Open Market Account, which is located in New York and is responsible for seeing that the appropriate buying or selling actions are taken in the government securities market.

FIGURE 16.1   District bank boundaries in the Federal Reserve System.

Each of the 12 district Federal Reserve Banks is a federally chartered corporation. Its stockholders are the member banks in that district. These member banks pledge up to 6% of their own capital and surplus in reserve stock. The district banks are not profit-maximizing businesses, however. Instead, their objective is to service and discipline member commercial banks in their districts. Each bank is managed by a nine-member Board of Directors serving in staggered terms. The role of these directors is to appoint the president of the district bank (with approval of the Board of Governors) and to serve in an advisory capacity to the bank on economic and financial conditions in the district.

Of the approximately 15,000 commercial banks in this country, almost 5,600 are members of the Federal Reserve System. **National banks,** which receive their charter from the Comptroller of the Currency, are required to be members. There are about 4,600 national banks in this country. The remaining 1,000 member banks are **state banks.** They receive their charter from their respective state governments but have elected to be members of the Federal Reserve System for one reason or another. Usually, this decision is based on weighing the cost of membership (e.g., the requirement of non-income-producing reserve requirements) and the returns (e.g., the ability to borrow from the Fed-

Organizational Structure of the Federal Reserve

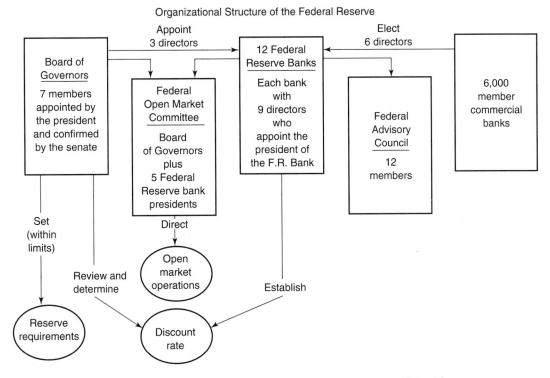

FIGURE 16.2   The Federal Reserve System consists of the Board of Governors, the Federal Open Market Committee, the 12 district Federal Reserve Banks, and approximately 6,000 member banks located throughout the country.

eral Reserve's discount window and the use of its check-clearing facilities and teletype wires to transfer funds).[3] Figure 16.2 illustrates the organizational structure of the Federal Reserve.

Functions of the Federal Reserve System

The Federal Reserve System's functions are to

- supply the economy with paper currency,
- supervise its member banks,
- provide check collection and clearing services,
- maintain the reserve balances of depository institutions,

[3]The Depository Institutions Deregulation and Monetary Control Act of 1980 virtually eliminated any differences between member and nonmember banks in the eyes of the Federal Reserve System. This act extended reserve requirements to *all* depository institutions (commercial banks, savings and loan associations, credit unions, etc.), but also made its services available to all these institutions.

- lend to depository institutions,
- act as the federal government's banker and fiscal agent, and
- regulate the money supply.

The first of the Fed's functions is to supply the economy with paper currency called Federal Reserve notes through the district banks. Every dollar bill has the words *Federal Reserve Note* printed on it. Each district Federal Reserve bank must have enough paper currency to accommodate the demands for money in its district. Although this paper currency is printed in Washington by the Bureau of Printing and Engraving, it bears the code of the originating district bank.

The Federal Reserve System also supervises its member banks, as does the Comptroller of the Currency and the Federal Deposit Insurance Corporation (FDIC). Among the things that the examiners look at are the types of loans made; the backing, or collateral, for these loans; and who borrowed the funds.

A third function performed by the Federal Reserve System is to provide check collection and clearing services. All member banks and depository institutions can send deposited checks to their district Federal Reserve Bank, which competes with private clearinghouses for their business. Suppose that Ralph Rancher, who owns a farm in the Kansas City Fed's district, travels to San Francisco to attend a convention. While in San Francisco, Mr. Rancher writes a check to Wally's Wharf restaurant for $50. Assume that Wally deposits these funds in his checking account at his commercial bank in San Francisco. Wally's bank would then deposit the check in its reserve account at the San Francisco district Federal Reserve Bank. The San Francisco Fed would then send this check to the district Federal Reserve Bank in Kansas City, which would deduct $50 from the reserve account of Ralph's bank and then send this check to Ralph's bank. The final step sees Ralph's bank deducting $50 from his checking account and sending the canceled check to Ralph.

Another function performed by the Federal Reserve System is to maintain the reserve balances of depository institutions in each district bank as required by law. Depository institutions are required to keep a certain fraction of their deposits on reserve.

The Federal Reserve System also lends to depository institutions. The interest rate charged on these loans is called the **discount rate.** Banks with a heavy seasonal demand for loans such as rural commercial banks, which make a significant volume of loans to farm businesses, may qualify for special borrowing privileges.

The Federal Reserve System also acts as the federal government's banker and fiscal agent. The U.S. Treasury has a checking account with the Fed. In addition, the Fed helps the federal government collect tax revenues from businesses and aids in the purchase and sale of government bonds.

Finally, the Federal Reserve System regulates the money supply to promote economic growth and price stability. This function has received national attention in recent years of high interest rates.

Monetary Policy Instruments

The Federal Reserve System has three major monetary policy instruments it can employ to regulate the growth of the nation's money supply. These instruments are: (1) changes in reserve requirements, (2) changes in the discount rate, and (3) changes in the direction or magnitude of its **open market operations.**

Reserve Requirements.   The Federal Reserve System alters the supply of money by changing the amount of reserves in the banking system. One way to accomplish this is to change the required reserves at depository institutions, which are required to maintain a specific fraction of their customers' deposits as reserves. Total reserves can be divided into three categories.

1. **Legal reserves** for member banks consist of deposits held at the institution's district Federal Reserve bank plus vault cash.[4]
2. **Required reserves** represent the minimum weekly average legal reserves that a depository institution must hold. This requirement is expressed as a ratio. The minimum and maximum reserve requirement ratios for depository institutions are changed infrequently.
3. **Excess reserves** represent the difference between total reserves and required reserves. The level of excess reserves determines the extent to which depository institutions can make loans.

Thus, manipulation of the reserve requirement ratio will either expand or contract the level of excess reserves at depository institutions and affect their ability to make new loans. We will see shortly how this translates into expansion or contraction of the money supply.

Discount Rate.   As indicated earlier, the discount rate represents the interest rate the Fed charges for lending reserves to depository institutions. If a depository institution wants to increase its loans but does not have any excess reserves at the moment, it can borrow reserves from the Fed at its "discount window."[5] The Fed does not have to lend all the reserves these institutions need every time they want them, however.

As an alternative, banks can borrow funds on a short-term basis to meet reserve requirements. The **federal funds market** is an interbank market trafficking in reserves. Banks with excess reserves can lend to banks that are short on a 24-hour basis at what is known as the federal funds rate.

---

[4]Nonmember banks and other depository institutions may treat as reserves their deposits with a correspondent depository institution holding required reserves, with the Federal Home Loan Bank (savings and loan associations), or with the National Credit Union Administration central liquidity facility (credit unions), as long as these reserves are passed on to a Federal Reserve bank.

[5]These reserves are referred to as *borrowed reserves.* Excess reserves minus these borrowed reserves are often referred to as *free reserves.*

Open Market Operations.   The third way the Federal Reserve System can alter the volume of reserves at depository institutions is by the sale or purchase of government securities. The directive by the Federal Open Market Committee (FOMC) to sell government securities results in a decrease in reserves at depository institutions, because deposits are withdrawn from these institutions to pay for the securities, which lowers the level of reserves in the banking system and hinders its ability to make loans.

When the FOMC issues the directive to buy government securities, an increase in reserves in the banking system occurs. This is the most frequently used monetary policy instrument. This buying and selling of government securities influences fluctuations in the federal funds rate, or the rate of interest at which banks lend money to one another on a very short-term (overnight) basis.

Other Instruments.   The Fed also uses several other policy instruments to regulate the expansion of credit in the nation's economy. The Fed also determines the margin requirements (i.e., down payment required) for loans made by stockbrokers to customers desiring to purchase stock, and the maximum rates banks can pay depositors on specific types of deposits.

CHANGING THE MONEY SUPPLY

A change in reserves in the banking system because of the use of any one of the three monetary policy instruments will affect the ability of depository institutions to make loans. How does this affect the money supply? To address this question, we must first discuss the process of money creation. Then we can discuss how changes in the three major monetary policy instruments shift the supply curve for money.

Creation of Deposits

An important relationship between the level of reserves in the banking system and the money supply exists. In the discussion to follow, we will focus on the M1 definition of the money supply: coins and currency in circulation plus checkable deposits at depository institutions. Part of this money supply consists of coins and currency, which are physical units that can be carried around in a purse or wallet. The deposits in M1, on the other hand, are merely entries in an account at depository institutions.

New deposits can be created, and the money supply expanded, by increasing the level of excess reserves in the banking system. If the level of excess reserves at depository institutions is zero, there can be no further expansion of the money supply. No new loans can be made at that point. To understand this, examine the balance sheet for a hypothetical bank, Bank Ag, in Table 16.1.

TABLE 16.1   Bank Ag's Balance Sheet before $1 Million Deposit

Assets		Liabilities	
Reserves		Deposits	$10,000,000
Required	$2,000,000		
Excess	0		
Total	$2,000,000		
Loans	8,000,000		
Total	$10,000,000	Total	$10,000,000

Table 16.1 shows how deposits can be expanded when excess reserves are greater than zero. Assume that the bank's assets consist entirely of reserves and loans, and its liabilities consist entirely of deposits. If the reserve requirement ratio is 0.20, this bank must hold at least $2 million of its $10 million of deposits in reserves. If Bank Ag already had outstanding loans of $8 million, as shown in Table 16.1, the bank would be fully "loaned up." The bank cannot increase its loan volume further because it has no excess reserves.

Assume that the depositors at this bank sell $1 million in government securities to the Federal Reserve and deposit the checks that they received from the Federal Reserve System in Bank Ag (Table 16.2). As a result of these deposits, Bank Ag's total deposits would increase to $11 million, and its required reserves would have to increase to $2.2 million (i.e., its initial $2 million required reserves plus 20% of its new $1 million deposit). Assuming that it has not had time to make any new loans, the bank would have excess reserves of $800,000.

Banks desire to keep their excess reserves as low as possible for profit-maximizing reasons. Excess reserves represent idle balances that are earning no interest income.[6] Assume that Bank Ag lends its entire excess reserves of $800,000. The bank would then be fully "loaned up" again and would remain that way until new activities altered its balance sheet position.

Now assume that the proceeds of the loans made by Bank Ag become deposits in another bank. Assume the $800,000 lent by Bank Ag shows up as deposits in Bank B.[7] Given a required ratio of 20%, Bank B must now hold $160,000 in additional reserves (i.e., 0.20×$800,000) and can now make loans of $640,000 (i.e., $800,000− $160,000).

Table 16.3 shows how the initial deposit of $1 million in Bank Ag has been used by the banking system to expand the deposits that appear in the M1 definition of the money supply. If we look at the totals at the bottom of this table, we see that the deposits of the entire banking system have increased by a multiple of the initial $1 million change in reserves at these depository institutions brought about

[6]For reasons of safety and liquidity, banks normally hold some idle balances.
[7]This process could occur easily without all the loan proceeds showing up in different banks.

TABLE 16.2   Bank Ag's Balance Sheet after $1 Million Deposit

Assets		Liabilities	
Reserves		Deposits	$11,000,000
Required	$2,200,000		
Excess	800,000		
Total	$3,000,000		
Loans	8,000,000		
Total	$11,000,000	Total	$11,000,000

TABLE 16.3   Change in Deposits in the Banking System after $1 Million Deposit

Bank	Change in Deposits	Change in Loans	Change in Reserves
Ag	$1,000,000	$800,000	$200,000
B	800,000	640,000	160,000
C	640,000	512,000	128,000
D	512,000	409,600	102,400
E	409,600	327,680	81,920
F	327,680	262,144	65,536
G	262,144	209,715	52,429
H	209,715	167,772	41,943
I	167,772	134,218	33,554
Subtotal	$4,328,911	$3,463,129	$865,782
Other banks	671,089	536,871	134,218
TOTAL	**$5,000,000**	**$4,000,000**	**$1,000,000**

by the depositors of Bank Ag selling $1 million in government securities to the Federal Reserve.

If you divide the change in deposits ($5 million) by the change in reserves ($1 million) at the bottom of Table 16.3, you will see that the change in the money supply was five times greater than the change in reserves. From this example, we can make a generalization about the extent to which the money supply will increase when the banking system's reserves are increased.

If we assume that banks will minimize their holdings of excess reserves and that all the proceeds from loans are deposited in the banking system (none are hidden in tin cans in the backyard), we may assert that:

$$MM = 1.0 \div RR \qquad (16.1)$$

in which *MM* represents the money multiplier and *RR* represents the fractional reserve requirement ratio. This equation suggests that if the **money multiplier** is equal to the reciprocal of the 20% required reserve ratio, the money multiplier would be 5 (i.e., $1.0 \div 0.20$).

We may use the definition of the money multiplier in Equation 16.1 to calculate the level of the nation's money supply as:

$$M_S = (1.0 \div RR) \times TR$$
$$= MM \times TR \qquad (16.2)$$

in which M_S represents the nominal money supply and TR represents the level of total reserves in the economy. The change in the money supply is given by:

$$\Delta M_S = MM \times \Delta TR \qquad (16.3)$$

Equation 16.3 suggests that, if the reserve requirement ratio is 0.20 and there is an increase of $1 million in total reserves in the economy, the nation's money supply would change by:

$$\Delta M_S = MM \times \Delta TR$$
$$= 5.0 \times \$1 \text{ million}$$
$$= \$5 \text{ million} \qquad (16.4)$$

or a multiple of five.[8] This is identical to the total change in the money supply reported at the bottom of Table 16.3.

In real-world situations, the money multiplier normally ranges between 2.0 and 3.0. A $1 million increase in reserves will increase the money supply by $2 million to $3 million. Several factors cause the money multiplier to fall below the reciprocal of the required reserve ratio. These factors include **currency drains** (sometimes called leakages), which refers to the desire of individuals to hold currency rather than deposit funds in a bank or other depository institution. When this occurs, funds remain outside the banking system.

If the depositors at Bank Ag received $1 million from the sale of securities to the FOMC and chose to hide the funds, the level of deposits at Bank Ag would not change and the money supply would not expand by $4 million. Thus, the greater the incidence of currency drains, the smaller the multiplier will be. A bank's desire to hold idle excess reserves for reasons of safety and liquidity will also lower the value of the money multiplier. The greater the level of excess reserves, the lower the money multiplier will be.

Monetary Policy and the Money Supply

Now that we understand how a change in reserves affects the money supply, let's return to the Fed's monetary policy instruments and examine their effect on the

---

[8]The reverse of this conclusion is also true; that is, a $1 million decrease in reserves in the banking system will reduce the money supply by $5 million, if the money multiplier is 5.0.

Expansionary and Contractionary Shifts in the Money Supply

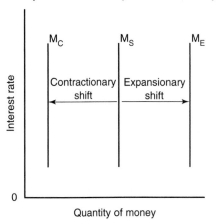

FIGURE 16.3 Expansionary monetary policy actions (i.e., the Federal Reserve buys government bonds, lowers the discount rate, or decreases the required reserve ratio) will lead to increases in the money supply, which is illustrated here by the outward shift in the money supply curve. Contractionary monetary policy actions (i.e., the Federal Reserve sells government bonds, raises the discount rate, or increases the required reserve ratio) will lead to decreases in the money supply, which will cause an inward shift in the money supply curve.

TABLE 16.4   Expansionary Effects on the Nation's Money Supply

Expansionary Actions	Effects of Action
Fed purchases securities in the open market	Increases total reserves
Fed lowers the discount rate	Increases total reserves
Fed reduces the required reserve ratio	Increases the money multiplier

money supply curve. The money supply curve M_S pictured in Figure 16.3 reflects a policy that is invariant, or unaffected by, the level of interest rates; that is, a change in interest rates is assumed to have no effect on the money supply.

Expansionary Applications.   A change in open-market operations, a change in reserve requirements, and a change in the discount rate can all affect the money supply. Table 16.4 lists the expansionary effects that monetary policy instruments have on the nation's money supply.

When the Fed purchases government securities in the open market, it increases the level of total reserves at depository institutions, which will cause the money supply curve in Figure 16.3 to shift to the right. The new money supply would be equal to the money multiplier times the new higher level of total reserves.

Lowering the discount rate will make it cheaper for depository institutions to borrow from the Fed and should increase the level of total reserves in the banking system, which will also cause the money supply to shift outward to the right. The new money supply would again be found by multiplying the existing money multiplier by the new level of total reserves.

A reduction of reserve requirements forces depository institutions to hold a smaller fraction of their deposits in reserves, which increases the money multiplier (see Equation 16.1). A reduction of reserve requirements is also an expansionary

TABLE 16.5   Contractionary Effects on the Nation's Money Supply

Contractionary Actions	Effects of Action
Fed sells securities in the open market	Reduces total reserves
Fed increases the discount rate	Reduces total reserves
Fed increases the required reserve ratio	Decreases the money multiplier

application of monetary policy and would cause the money supply curve to shift from M_S outward toward M_E. The new money supply can be found by multiplying the existing level of total reserves times the new higher money multiplier.

Contractionary Applications.   In contrast to the expansionary applications of the Fed's major monetary policy instruments, these same instruments can be used to contract the money supply. Table 16.5 lists the contractionary effects that monetary policy instruments have on the nation's money supply.

When the Fed sells securities in the open market to private investors, it reduces the total reserves in the banking system, which leads to a multiple decrease in the money supply. Therefore, the money supply curve in Figure 16.3 will shift back to the left, from M_S toward M_C.

An increase in the discount rate by the Federal Reserve makes it more costly for banks to borrow reserves through the Fed's discount window. An increase in the discount rate also reduces total reserves and leads to a multiple decrease in the money supply.

An increase in the reserve requirement ratio reduces the money multiplier (see Equation 16.1), which causes existing total reserves to expand by a smaller multiplier than before. If the Fed increases its reserve requirements at a time when all depository institutions are fully "loaned up" (i.e., each has zero excess reserves), these institutions will either have to call in loans or decrease their investments to be in compliance.

MONEY MARKET EQUILIBRIUM

Why do individuals prefer to hold part of their wealth in money rather than place all their wealth in income-earning assets? What happens to the interest rate if the demand for money increases or decreases? To answer these questions, we must understand the demand for money.

Demand for Money

Economists often identify three reasons why we hold money. The first represents our **transactions demand for money.** Households and businesses hold a certain amount of money because it is a widely accepted medium of exchange. Because their receipts of money income do not match the timing of their expenditures, households and businesses maintain some holdings of money to help finance these expenditures. The cost of not having transaction balances is the rate of interest households and businesses would have to pay when borrowing funds to finance these expenditures.

The second reason we hold money represents our **precautionary demand for money.** Some households may wish to hold money when the prices of other assets are falling because the nominal value of money is fixed. One hundred dollars in cash will always be worth $100, but $100 invested in common stock could be worth less than that tomorrow. In a recessionary economy when asset prices are falling, the **speculative demand for money** should increase.

Thus, the demand for money reflects the liquidity preferences of households and businesses. The lower the rate of interest, the lower the opportunity costs of holding cash will be, and the larger the quantity of money demanded should be. The larger the level of disposable personal income, the greater the amount of cash balances held in the economy.

Assume that the general nature of the aggregate demand for money for Lower Slobovia's economy can be expressed in equation form as:

$$M_D = c - d(R) + e(NI) \qquad (16.5)$$

in which

$$d = \Delta M_D \div \Delta R \qquad (16.6)$$
$$e = \Delta M_D \div \Delta NI \qquad (16.7)$$

in which c represents the autonomous quantity of money demanded, d reflects the interest sensitivity of the demand for money and the slope of the demand curve, R represents the market rate of interest, e reflects the income sensitivity of the demand for money, and NI represents the level of national income, which also equals the nation's gross domestic product.

Figure 16.4, *A* illustrates the inverse relationship between the demand for money and the rate of interest. When the returns on assets in general and interest rates in particular fall, households and businesses will attempt to substitute cash for such financial assets as stocks and bonds.

The Keynesian notion that the demand for money is inversely related to interest rates and income is somewhat at odds with other theories of the demand for money. The Cambridge demand for money function, which was developed by classical economists at Cambridge, England, suggests that the demand for money is equal to some constant fraction of the nominal level of income. This is the simplest of all monetarist positions. Modern monetarists such as Milton Friedman argue that the demand for money is a function of alternative rates of return, nominal income, and the expected rate of inflation.

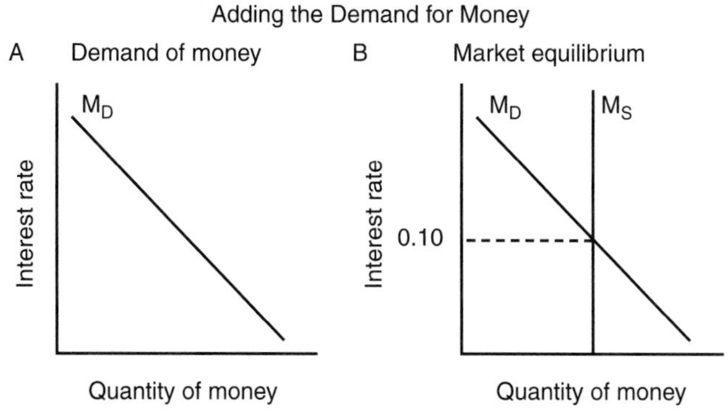

Adding the Demand for Money

FIGURE 16.4   Before we can determine the market rate of interest in the economy, we must know the demand for money. *A,* The demand for money function is downward sloping; that is, the higher (lower) the rate of interest, the lower (higher) the quantity of money demanded. This interest rate reflects the opportunity cost of holding money rather than an income-earning asset. The higher the expected income forgone by holding money, the lower the quantity of money demanded will be. *B,* The market rate of interest is determined by the intersection of the money demand and supply curves. Thus, an increase in the money supply will lower interest rates in the short run, and a reduction in the money supply will raise interest rates.

Equilibrium Conditions

An equilibrium will occur in the nation's money market when the supply curve given by Equation 16.2 intersects the demand curve given by Equation 16.4, or when:

$$M_S = M_D \qquad\qquad (16.8)$$

Figure 16.4, *B* brings together the demand and supply curves for money. The intersection of the M_D curve and the M_S curve determines the equilibrium rate of interest, which would be 10%.

EFFECTS OF MONETARY POLICY ON THE ECONOMY

We now know that the money supply affects the level of interest rates. A shift to the right of the money supply curve in Figure 16.4, *B* would lower the interest rate below 10%, and a shift to the left would raise the market interest rate.

Transmission of Policy

Figure 16.5 summarizes the mechanism through which changes in monetary policy are transmitted to the nation's aggregate demand. The change in policy first

Effects of Change in Monetary Policy on GDP

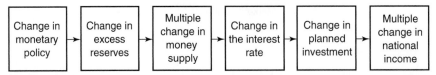

FIGURE 16.5 This figure illustrates the chain of events that takes place when a change in monetary policy is made. A change in monetary policy will directly affect the reserves that depository institutions such as banks are required to hold. This, in turn, will affect the banks' excess reserves and therefore their ability to make new loans, which ultimately affects the size of the money supply. An increase (decrease) in the money supply will put downward (upward) pressure on interest rates. A decline (rise) in interest rates will encourage (discourage) further investment expenditures by businesses in the economy.

shows up in the money market, where it affects interest rates. This, in turn, affects the equilibrium in the product market through investment and, through the income multiplier, the equilibrium level of aggregate demand and national income. Because the demand for money is affected by changes in national income, we must account for the simultaneous interaction between the money and product markets when determining the general equilibrium level of income and interest rates.

The discussion of the aggregate demand curve for Lower Slobovia in Figure 14.9 assumed a given market rate of interest. Changes in the market rate of interest resulting from the demand and supply forces operating in the nation's money market render the assumption of a fixed interest rate obsolete. Expansionary monetary policy actions that lower the rate of interest will spur investment expenditures, expand national income, and shift the economy's aggregate demand curve to the right. The opposite chain of events will occur if contractionary monetary policy actions are followed.

Combating Recessionary Gaps

The monetary policies of the federal government can be used to reduce unemployment. To combat a sluggish economy and high unemployment, policymakers use expansionary monetary policies that lead to an increase in the money supply, lower interest rates, and increased credit being available to consumers and producers. Monetary policies that reduce unemployment include:

- Federal Reserve purchasing government securities in the open market, which increases the amount of reserves in the banking system,
- lowering the Federal Reserve's discount rate, which also increases the amount of reserves in the banking system, and
- lowering the Federal Reserve fractional reserve requirements, which increases the money multiplier

An increase in total reserves in the banking system and the money multiplier will lead to an increase in the money supply over a period of time. The increased

Elimination of Recessionary and Inflationary Gaps

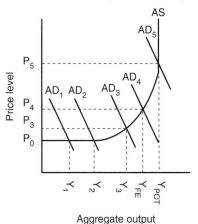

FIGURE 16.6   Monetary and fiscal policies can be used to expand or contract aggregate demand to promote economic growth and price stability.

credit availability and lower interest rates will expand consumption and investment expenditures, which will increase the need for more workers to help produce these additional goods and services.

To illustrate how expansionary demand-oriented macroeconomic policies eliminate a recessionary gap, let's examine the product market equilibrium in Figure 16.6. As consumption, investment, and government expenditures increase in response to the expansionary monetary policies, the aggregate demand curve will shift outward to the right. If aggregate demand shifts from AD_1 to AD_2, aggregate output will increase from Y_1 to Y_2.

Additional workers will also be needed to produce this additional output. Because there was plenty of slack in the economy, these policies were not inflationary (i.e., prices remained at P_0 because the AS curve was perfectly elastic over this range of the AS curve).[9] Finally, because of this increase in aggregate demand, the recessionary gap would be reduced from $Y_{FE} - Y_1$ to $Y_{FE} - Y_2$.

If aggregate demand is stimulated to the point at which the economy moves out to AD_3, aggregate output will increase to Y_3. The general price level would begin to rise, reaching the level P_3. The rate of inflation resulting from this action would be the percent change from P_0 to P_3, and the rate of growth in real GDP would be the percent change from Y_2 to Y_3. The recessionary gap would be further reduced to $Y_{FE} - Y_3$, but not totally eliminated.

Only if aggregate demand were expanded to AD_4 would aggregate output achieve full employment Y_{FE}, and the recessionary gap be completely eliminated. The general price level would reach P_4, which is somewhat inflationary, but does not swamp the growth in aggregate output from Y_3 to Y_{FE}. The increase in aggregate demand to AD_4 therefore expands the economy to full employment, or the

[9]See Chapter 14 for a further discussion of the complete AS curve.

point at which it achieves the lowest unemployment rate and highest capacity utilization rate without triggering a significant increase in inflation.

If aggregate demand was expanded to AD_5, the economy would reach its capacity to supply goods and services in the current period (Y_{POT}), and while additional output was forthcoming, it would be swamped by the increase in the general price level from P_4 to P_5. This inflation would lower the real wealth of consumers and thus self-correct aggregate demand back towards AD_4. The Federal Reserve would also likely pursue contractionary monetary policies to bring the AD curve back towards AD_4.

Combating Inflationary Gaps

Monetary policies can also be designed to reduce inflation. Contractionary monetary policies that lead to a decrease in the money supply, raise interest rates, and decrease credit available to consumers and producers combat inflation by lowering consumption and investment expenditures in the economy. Monetary policies that reduce inflation include:

- Federal Reserve selling government securities in the open market, which decreases the amount of reserves in the banking system,
- raising the Federal Reserve's discount rate, which also decreases the amount of reserves in the banking system, and
- raising the Federal Reserve fractional reserve requirements, which decreases the money multiplier.

This reduction in demand will put a downward pressure on prices in general. Unfortunately, these policies will also increase unemployment. This was precisely the tack taken by the Federal Reserve in 1980 when the rate of inflation reached double-digit levels. When inflation rates fell in the early 1980s, the annual unemployment rate began to rise, eventually reaching 10%. This inverse relationship and apparent sensitivity of employment to contractionary monetary policies raises the question of what would happen to unemployment if policymakers attempted to drive inflation rates to zero, as some members of the Federal Reserve Board of Governors would like to do. Growth in productivity in the mid-1990s has led to the situations where *both* low inflation and low unemployment were achieved.

To illustrate, let's look again at Figure 16.6. Assume that aggregate demand in the economy is represented by the AD_5 curve. If contractionary monetary policies brought about a shift in this demand curve from AD_5 to AD_4, the aggregate demand curve would intersect the aggregate supply AS curve at full employment output Y_{FE}. Prices would fall substantially from P_5 to P_4 without causing a reduction in aggregate output. If policymakers unwittingly continued to dampen demand to the point where the demand curve shifted back to AD_3, prices would fall further to P_3. But aggregate output would fall below Y_{FE} to Y_3, thereby reducing the employment of the nation's resources to below full employment levels and creating a recessionary gap.

Combating Stagflation

The demand-oriented monetary policies used to combat inflation or unemployment present the policymaker with a problem when the economy is suffering from stagflation or rising inflation in a stagnating economy. The policymaker must choose between fighting inflation or fighting unemployment. Or the policymaker must instead turn to one or more supply-oriented macroeconomic policies that shift the supply curve to the right, like that shown in Figure 16.6.

Microeconomic Perspectives

Interest rates have an impact upon investments in new equipment and buildings by businesses, and investments in new residences by households. Contractionary monetary policies that drive up interest rates will depress investment expenditures by businesses and households. Conversely, expansionary monetary policies that lower interest rates will stimulate investment expenditures in the economy. We can demonstrate how changes affect an individual business or household by examining the impact that specific interest rate levels have on the size of the loan payment and total interest expense for business and household loans.

Suppose a farmer is considering the purchase of a new combine and related equipment and seeks to finance $150,000 over a 10-year period in 10 equal annual installments. Table 16.6 shows how varying interest rates affect the value of the annual loan payments and total interest paid over the entire loan period. The data in Table 16.6 suggest that the annual total loan payment (principal and interest payments) increases significantly as interest rates (here expressed in nominal terms) rise significantly. These payments must be met annually from the revenue earned by the business and, hopefully, out of the additional revenue generated by this new asset. If we think of the combine and related equipment as the only capital used by a business providing custom harvesting services, the idea is clearer. Ignoring the cost of fuel, labor, and other related expenses, the business's revenue would have to rise in accordance with the annual interest and total loan payments if it is to continue to meet its commitments to its banker and its commitments to others providing inputs without borrowing additional funds from others or using up the business's existing equity.

Suppose the business expected to earn a revenue of $30,000 above all other expenses. Although the average annual interest payment at a 20% interest rate

TABLE 16.6   Effects of a Varying Interest Rate on a 10-Year Loan

Rate	Annual Total Payment	Annual Interest Payment	Total Interest Payment
8%	$22,354.69	$7,354.69	$73,546.90
14%	28,757.67	13,757.67	137,576.68
20%	35,782.44	20,782.44	207,824.40

TABLE 16.7   Effects of a Varying Interest Rate on a 20-Year Mortgage

Rate	Monthly Total Payment	Monthly Interest Payment	Total Interest Payment
8% rate	$848.78	$432.08	$103,707.46
12% rate	1,115.73	699.06	167,773.46

could "service its debt," or meet its interest payment, it could not make its total loan payment of $35,782.44.

Suppose a household is considering entering into an agreement to purchase a new residence now under construction. This household needs to borrow $100,000 to finance the purchase of this new home. Assume the household can obtain a 20-year loan calling for monthly payments. Table 16.7 shows how varying interest rates affect the size of the monthly mortgage payments and total interest paid over the entire loan period. If this household could incorporate no more than a $900 monthly mortgage payment into its budget, it could work the monthly mortgage payment into its budget if the interest rate were 8%, but not if the interest rate were 12%. In the latter instance, the monthly mortgage payments would be $266.95 higher than under the lower interest rate. In both instances, higher interest rates may lead to lower levels of investment expenditures—or at least postponed investment expenditures by businesses and households.

Potential Problems in Policy Implementation

Four problems encountered when using monetary policy to achieve specific aims in the economy are:

1. knowing the value of the money multiplier,
2. knowing the exact size of the inflationary or recessionary GDP gap to be eliminated,
3. knowing whether fiscal policy will also be used to eliminate such gaps, and
4. knowing how long it will take for these policies to be fully reflected in the product markets.

There are several factors that cause a delay in the full effectiveness of needed policy changes. The first is a recognition lag, or the time between when the policy is needed and when this need is recognized by policymakers. The second is an implementation lag, or the time between when the need for the policy change is recognized and when it is actually implemented. Finally, there is an impact lag, or the time between when the policy is implemented and when it becomes fully effective.

Congress and the president are often opposed to the Federal Reserve's position on the economy. In the early 1980s, when Congress and the president were pushing an expansionary fiscal policy, the Federal Reserve was following contractionary monetary policies to eliminate inflation. The result was extraordinarily high interest rates (the prime commercial bank rate at one point reached 21%) and a marked recession during the 1981–1982 period. When the Federal Reserve switched to a more expansionary monetary policy stance, however, the economy took off, registering an unprecedented number of quarters of economic expansion (i.e., real GDP growth) during a peace-time period.

SUMMARY

The purpose of this chapter was to acquaint you with the functions of money in a monetary economy, the role of the United States' central bank (the Federal Reserve System) in regulating the supply of money to achieve specific national economic objectives, how interest rates are influenced by expansionary versus contractionary monetary policies, and the effect of interest rates on national income. The major points made in this chapter may be summarized as follows:

1. Money serves as a medium of exchange, a unit of accounting, and a store of value. These functions help us define money. Money has value because it is widely acceptable in exchange for goods and services and because its nominal value is fixed.

2. The Federal Reserve System is comprised of the Board of Governors, the 12 district Federal Reserve Banks, and its member banks. An omnibus banking act in 1980 gave the Fed regulatory powers over the reserves held by all depository institutions in this country, including savings and loan associations and credit unions.

3. The Fed regulates the supply of money to the private sector, holds the reserves required of depository institutions, provides a system of check collection and clearing, supplies fiduciary currency (Federal Reserve notes) to the economy, acts as the banker and fiscal agent for the Treasury, and supervises the operations of its member banks.

4. Money is created by the banking system through the multiple expansion of deposits. The full effect of an increase in reserves is found by multiplying these new reserves by the money multiplier, which is equal to the reciprocal of the reserve requirement ratio. The Federal Reserve can influence the money supply by changing the reserve requirement ratios for deposits. It can also use the discount rate and its open-market transactions for government securities to influence the level of reserves and the money supply.

5. A decrease in the discount rate, a lowering of reserve requirements, or the purchase of government securities by the Fed all represent an expansionary monetary policy because they will expand the money supply. Conversely, an increase in the discount rate, a raising of reserve requirements, or the sale of government securities by the Fed all represent a contractionary monetary policy.

6. Monetary policy affects aggregate demand in the economy during the current period through the market rate of interest. The higher the rate of interest, the lower aggregate demand will be, and vice versa. The effectiveness of monetary policy will be reduced by the presence of currency drains and bank holdings of excess reserves.

DEFINITION OF KEY TERMS

Currency drain: currency (paper bills and coins) held by the public, whether it is in their wallets or stored in a nonbank location (e.g., hidden in a mattress).

Discount rate: rate the Federal Reserve charges when it lends to member commercial banks (all national banks plus those state banks that choose to be members of the Federal Reserve System).

Excess reserves: difference between legal reserves and required reserves.

Federal funds market: an interbank market from which banks borrow the excess reserves of other banks to meet their own reserve requirements.

Federal Open Market Committee (FOMC): consists of members of the Board of Governors of the Federal Reserve System plus selected district Federal Reserve Bank presidents who meet periodically to assess the appropriate action the Fed should take (buy or sell) in the open or private secondary bond market for government securities.

Fiduciary monetary system: value of currency issued by government based on the public's faith that currency can be exchanged for goods and services.

Legal reserves: deposits at district Federal Reserve banks plus vault cash.

Medium of exchange: benefit where money allows businesses and individuals to specialize in their endeavors and to purchase the goods and services they need with money they receive for their efforts.

Money multiplier: the reciprocal of the fractional required reserve ratio if there are no currency drains and no excess reserves.

National bank: bank that is chartered by the federal government.

Near monies: assets that are almost money or can be converted to money quickly with little or no loss in value.

Open market operations: the buying and selling of government securities by the FOMC's fiscal agent at the New York District Federal Reserve Bank.

Precautionary demand for money: demand to hold money to pay for unexpected expenditures that may arise.

Required reserves: minimum amount of deposits that banks and other depository institutions must hold in reserve.

Speculative demand for money: demand to hold money as an asset in expectation that other asset prices will fall.

State bank: bank that is chartered by a state government.

Store of value: value of holding money as an asset.

Transaction demand for money: demand to hold money to pay for expected expenditures.

Unit of accounting: money provides a basis with which the relative value of goods and services can be assessed and with which the profitability and financial position of businesses can be assessed.

REFERENCES

Mankiw NG: *Principles of macroeconomics,* New York, 1997, Dryden Press.

Board of Governors of the Federal Reserve System: *Federal Reserve bulletin,* Washington, D.C., various issues, U.S. Government Printing Office.

EXERCISES

1. Please fill in the blanks in the following table with the appropriate response; "lower" if the variable will decline in value as a result of the policy action or "raise" if the variable will increase in value as a result of the policy action.

Policy Effects on	Expansionary *Monetary Policy*	Contractionary *Monetary Policy*
Farm input prices	_____	_____
Farm crop prices	_____	_____
Farmland prices	_____	_____
Net farm income	_____	_____

2. Suppose the Federal Reserve buys $4,000,000 in government securities from someone who is a depositor at the First National Bank of North Zulch. Let's assume the person deposits this money in this bank. Further assume that the current reserve requirement ratio is 20%.

   a. Please indicate below what will initially happen to this bank's balance sheet as a result of this transaction.

   Change in reserves   _____

   Change in loans   _____

   Change in deposits   _____

b. Please indicate below what will eventually happen to the *nation's banking system* as a result of this transaction.

Change in reserves   _____

Change in loans   _____

Change in deposits   _____

3. Review the graph below and answer questions a, b, and c.

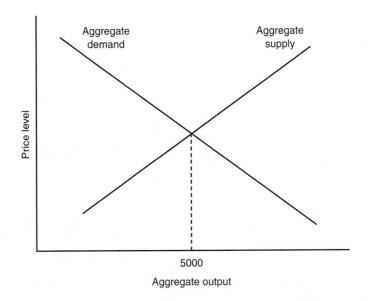

a. If the full employment level of output is $4,000, is the economy depicted in an inflationary or recessionary gap? What *specific* monetary policies would you recommend, if any, to correct this?

b. If the full employment level of output is $6,000, is the economy depicted above in an inflationary or recessionary gap? What *specific* monetary policies would you recommend, if any, to correct this?

c. If the full employment of output is $5,000, is the economy depicted above in an inflationary or recessionary gap? What *specific* monetary policies would you recommend, if any, to correct this?

17

FISCAL POLICY AND BUDGET DEFICITS

And it came to pass in those days, that there went out a decree from Caesar Augustus, that all the world should be taxed.

Bible (Luke 2:1)

Read my lips: no new taxes.

Vice President George H. Bush
1988 presidential nomination acceptance speech

TOPICS OF DISCUSSION

The second form of macroeconomic policy available to governments is the power to spend and to tax—policies generally referred to as fiscal policy. The Balanced Budget and Emergency Deficit Control Act, better known as the Gramm-Rudman-Hollings Act (or simply Gramm-Rudman), is an example of activist fiscal policy. The Gramm-Rudman Act was passed in the late 1980s to address the growing annual federal budget deficits and subsequent growth in the national debt. Subsequent legislation in the 1990s required demonstration of budget neutrality by new spending proposals.

The purpose of this chapter is to discuss the nature of the federal budget, the financing of annual budget deficits, and the growth of the national debt.

THE FEDERAL BUDGET

The annual **federal budget** is one of the largest publications printed each year by the Government Printing Office. The annual federal budget sets forth the president's spending and revenue plans for the coming year. Although some politicians summarily say that the budget is "dead on arrival," the budget often represents a starting point for further negotiations. Often this is due to the overly optimistic assumptions underlying the revenue projections. During the 1990s, however, surprises such as the Persian Gulf War have meant additional budgetary burdens that were not planned for.

Federal Expenditures

Total spending by the federal government, or **federal expenditures,** has grown dramatically over the post–World War II period. Nominal government spending grew twelvefold from 1950 to 1999. Much of this growth can be attributed to inflation. Real government spending in 1999 represented only a threefold increase over 1950 levels. Per capita government spending in 1999 was approximately $6,008 in nominal terms, but only $1,828 in real terms. Real government spending per capita has remained remarkably stable.

The largest single expenditure item in the federal government's 1999 budget was social security benefit payments (22.9%), followed by national defense (16.0%) and net interest payments on the national debt (13.5%). Medicare and income security programs amounted to 11.2% and 14.0% of the 1999 budget, respectively. Agricultural programs accounted for less than 1.0% of total federal expenditures in 1999, down more than one-half from the 2.6% of the federal budget spent on agriculture in 1987. Total government spending in the United States in 1999 amounted to approximately 18.4% of the nation's GDP.

Although many would argue that government spending in this country is too high and should be cut, it is substantially lower than in other developed countries. Total government spending is 60% of GDP in Sweden, 55% of GDP in the Netherlands, 45% of GDP in the United Kingdom, and 43% of GDP in Germany.

The federal government expenditure totals in Table 17.1 conform with the National Income and Product Account (NIPA) accounting rules followed when

TABLE 17.1   The Federal Budget for 1999

Item	Billion Dollars	Percent of Total
Receipts		
Personal income taxes	879.5	48.1
Social Security taxes	611.8	33.5
Corporate income taxes	184.7	10.1
Excise taxes	70.4	3.9
Customs duties and fees	18.3	1.0
Estate and gift taxes	27.8	1.5
Miscellaneous receipts	34.9	1.9
Total Expenditures	1,827.4	100.0
Social Security payments	390.0	22.9
Income security payments	237.7	14.0
National defense	274.9	16.0
Interest on public debt	229.7	13.5
Education, training, health	56.4	3.3
Medicare	190.4	11.2
Agriculture	23.0	1.4
Natural resources and the environment	23.9	1.4
Commerce and housing	2.6	0.2
Transportation	42.5	2.5
All other (net)	231.9	13.6
Total	1,703.0	100.0

Source: *2000 Economic Report of the President.*

measuring the nation's GDP in Table 14.1. The complete set of receipt and expenditure transactions of the federal government represents one sector in the set of NIPAs maintained by several government agencies.[1]

Why does Table 14.1 include only purchases of goods and services, but Table 17.1 includes these expenditures and transfer payments to domestic and foreign recipients, grants to state and local governments, subsidies and net interest payments (payments made on the national debt less interest received from government lending)? The answer lies in the fact that, when all sectors are combined into a single entity as is done in Table 14.1, the transfer payments and other financial transactions not involving the purchase of goods and services cancel or "net out" (i.e., payments by one sector represent receipts by another sector and hence cancel each other when the two sectors are combined).

---

[1]The U.S. Department of Commerce maintains the NIPAs, which measure the nation's gross national product (GNP) and national income. The Federal Reserve System maintains a national flow of funds account, which accounts for the flow of funds between agriculture, government, and other sectors of the economy. The nation's central bank also measures a national wealth or balance sheet account, which captures the assets and liabilities of agriculture and other sectors in the economy.

Federal Receipts

Believe it or not, no federal income tax existed before 1913. **Federal receipts** from taxes on liquor and tobacco, and tariffs on imported goods were enough to offset government expenditures. Before 1913, government spending was less than 10% of the nation's GDP. Today, government spending by federal, state, and local governments is more than 18% of GDP.

Figure 17.1, *A* shows that the growth in federal revenue was approximately equal to the growth in federal spending until the mid-1970s when the economy experienced the most severe recession since the Great Depression of the 1930s. By the late 1970s, the growth in federal revenue and expenditures converged again until the 1980s, when federal revenues fell because of the 1981–82 recession and the Economic Recovery Tax Act of 1981, which slashed personal income tax rates.

The vast majority of federal government revenue comes from tax payments made by households. In 1999, 48.1% of federal receipts were in the form of personal tax payments. Another 33.5% were in the form of social security contributions.[2] Corporate income taxes were 10.1%, and indirect business taxes, which are taxes that businesses treat as expenses (i.e., licensing fees, business property taxes, federal excise taxes, customs duties, and sales taxes), accounted for 8.3%.

Budget Deficit

Most people have heard the terms *budget deficit* and *surplus,* yet few can define what they are. In general terms, the federal budget deficit is the annual shortfall of federal receipts relative to federal expenditures.[3]

The **federal budget surplus** for calendar year 1999 was $124.4 billion, as opposed to the deficits recorded during much of the 1990s. The value of this surplus, which represents the unified approach to measuring budget receipts and expenses, can be calculated using the information in Table 17.1.[4] Federal receipts of $1,827.4 billion in Table 17.1 exceed federal government expenditures of $1,703.0 billion by $124.4 billion. The federal government funds its operations on a fiscal year basis, which covers the 12-month period beginning on October 1st. Table 17.1 reports the expenditures and receipts recorded by the federal government for the October 1, 1998 to September 30, 1999 period.

Federal budget deficits (Figure 17.2) are financed by the issuance of government bonds by the U.S. Treasury and are sold in the government bond market. These

---

[2]Individuals contribute a specific percent of their wage and salary income through payroll taxes, with their employer matching this amount. In 1990, for example, individuals paid 7.51% of their first $56,000 of wage and salary income. Self-employed persons face a similar tax.

[3]The opposite of a federal budget deficit is a federal budget surplus. Because the United States has experienced only one federal budget surplus since 1960 (until 1999), it is commonplace to focus on deficits rather than surpluses.

[4]There are at least four different perspectives on how to define the federal budget deficit. Until recent years these differences did not lead to substantially different deficit numbers. Thus, it is possible to hear one policymaker say the budget deficit is $250 billion and another to say the budget deficit is $300 billion, with both telling the "truth."

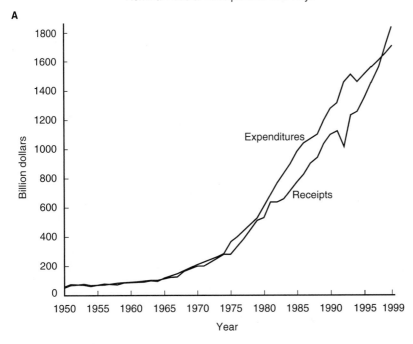

Nominal Federal Receipts and Who Pays

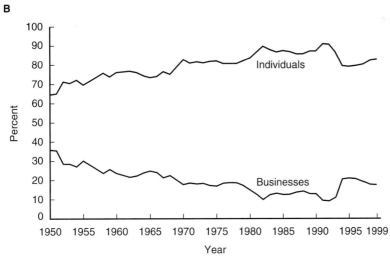

FIGURE 17.1 *A,* A relatively slower growth in nominal federal receipts occurred over the last four decades. *B,* Households have increasingly carried the burden of financing federal spending.

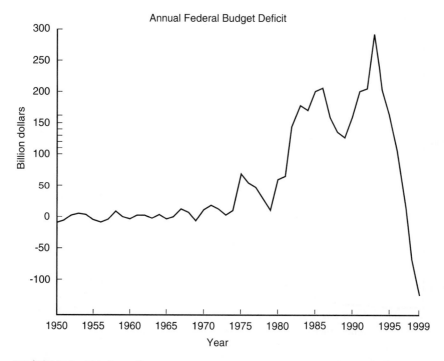

FIGURE 17.2   This figure illustrates the dramatic growth in the annual federal budget deficit, or the amount that federal government expenditures exceed federal receipts. The deficit reported here is that based on the National Income and Product Account (NIPA) approach and a calendar year average. Hovering around zero during the 1950s and 1960s, the deficit rose sharply in the mid-1970s when the economy suffered a severe recession. The 1980s saw the annual budget deficits rise to unprecedented highs when the economy experienced a recession in 1981–82 and when tax cuts were enacted without equal cuts in government spending. Reductions in annual deficits in the mid-1990s and surplus in the late 1990s reflected growth in the economy and budget-neutral government spending.

securities are purchased by individuals, businesses, commercial banks, foreigners, and the Federal Reserve System. The use of the proceeds from the sale of these securities to finance government expenditures represents an increase to the **national debt.**

THE NATIONAL DEBT

The national debt is (approximately) the sum of the outstanding federal government securities upon which interest and principal payments must be made. The federal budget deficit is the annual addition to the national debt. If the federal budget deficit is $100 billion, the national debt will grow by $100 billion at year's end.

The national debt makes headlines. It first rose to a significant level during World War II, reaching about $300 billion. It remained at that level for more than two decades. The national debt rose somewhat during the Vietnam era. Before the 1980s, growth in the national debt was tied largely to either wars or recessions. The growth in national debt in the 1980s, however, reflects the impact of cuts in per-

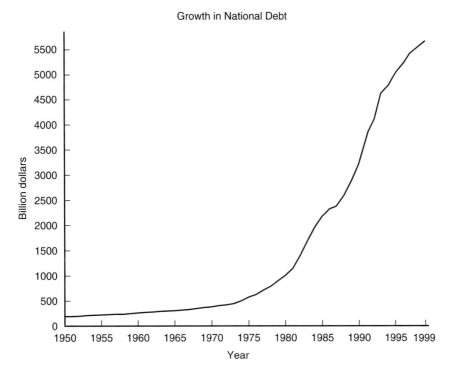

Growth in National Debt

FIGURE 17.3   This figure illustrates the dramatic growth in national debt in the 1980s relative to previous decades. This growth has caused some to question the nation's financial soundness. Are we going broke? Is today's generation overburdening future generations? And who owns this debt?

sonal income taxes that included lower marginal tax rates, indexing taxes for inflation, and other changes that contributed heavily to the growth in national debt. Corporate tax rates were also cut.

From 1776 to 1980, a period of 204 years, the United States accumulated its first trillion dollars in national debt. It recorded its second trillion dollar indebtedness by 1984 and its third trillion dollar indebtedness by the end of the decade. This growth alarmed some economists. Should we be concerned? Is the United States going bankrupt?

There are several approaches to evaluating the size of the national debt. We can study the relationship between national debt and GDP, and interest payments and GDP, and judge the ownership of national debt to determine if we are burdening future generations.

National Debt and GDP

The growth in national debt illustrated in Figure 17.3 suggests a federal financing strategy that is out of control. Or does it? Looking at this figure alone, one could certainly find cause for concern. Per capita debt figures tell the same story. In 1950,

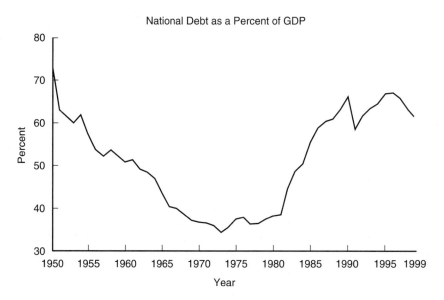

FIGURE 17.4 National debt has grown dramatically since the early 1980s, but this should not necessarily be alarming if the nation's income and output are growing also. This figure shows that national debt as a percent of GDP was actually higher in the early 1950s than it was in the late 1980s. However, the downward trend during the 1950–73 period was reversed when the economy was in the throws of the 1973–75 recession, and rose sharply in the 1980s when personal and corporate income tax rates were slashed, without making similar reductions in government spending. The flattening and eventual decline in this ratio in the 1990s reflects the budget-neutrality rules practiced with federal government spending.

the ratio of the national debt to the U.S. population would suggest that every man, woman, and child theoretically owed approximately $1,500 to holders of U.S. government bonds. By 1999, this figure had risen to about $20,535 per capita! Whether the public debt is too large can be judged by the same rules one uses to judge private debt—namely, by the size of an individual's existing debt and the individual's earning potential. We can evaluate the size of the national debt by examining the ratio of national debt to gross domestic product.

Figure 17.4 shows that, although the trend in the ratio of national debt to GDP until recently has been going in the wrong direction, the levels of these annual percentages are not out of line with levels experienced over the post–World War II period.

Interest Payments and GDP

A second approach to evaluating the size of the national debt is to look at the relationship between the value of interest payments on the national debt relative to the value of the nation's GDP. Interest payments increased dramatically from 1975, when they totaled $23 billion. By comparison, interest payments on the national debt totaled $229.7 billion in 1999. The share of the annual federal budget devoted to interest payments to bondholders has risen from single-digit magnitudes in the 1950s, 1960s, and 1970s to double-digit magnitudes in the 1980s. By 1999, interest payments represented about 13.5% of the annual federal budget.

We can assess the burden this places on the economy by looking at the ratio of interest payments to our GDP. The ratios of interest payments to GDP in 1975 and 1989 were 1.4% and 3.2%, respectively, which means that in 1989 the federal government had to collect more than double the tax dollars of 1975 to pay interest on its debt. By 1999, this percentage had fallen to 2.5%, a welcomed trend. Interest payments on the national debt have been a growing claimant on this nation's output. They remain a small percentage of GDP, however.

Large deficits, which feed a growing national debt, become increasingly difficult to manage. Higher deficits lead to higher future interest payments, which makes it harder to control future deficits. Interest payments become a greater percentage of GDP. Declines in interest rates, coupled with a growing economy and slowed growth in government spending relative to receipts, have led to the first surpluses since the 1960s.

Ownership of National Debt

About one-fourth of the national debt is held by government agencies and the Federal Reserve. The remaining three-fourths is held by state and local governments, individuals, commercial banks, life insurance companies, and others. Only 13% of the national debt is held by foreigners. This particular percentage is important for one reason: interest payments made to domestic individuals, firms, and state and local governments are payments to ourselves. They are not a drain on the nation's resources, but rather a transfer of income from one group of our society (taxpayers) to another (bondholders and those who have ownership claims on firms that hold bonds).

The fact that only 13% of the national debt is held by foreigners, and only that portion of interest payments leaves the country, suggests the drain on the national resources is small at present.

Burdening Future Generations?

Finally, there is the matter of the charge that the current generation is spending/borrowing at a rate that represents a burden on future generations. There are several perspectives one can take on this issue. Do you expect General Motors or IBM to retire all the stock and debt they accumulated when expanding their operations? The answer is, of course, no. These firms transcend generations, and they need never retire all capital claims on the business. The real issue is whether these firms can service the claims on the business (dividends to stockholders, interest payments to bondholders, and scheduled interest payments to lenders) from the proceeds of their ongoing operations.

The same thing can be said for the federal government. We should not expect to retire the national debt, but instead keep a healthy relationship between it and the nation's GDP, which means covering the growing interest costs on the national debt with increased tax revenues. With respect to the principal outstanding, the federal government pays off bondholders when their bonds mature, but it does so by refinancing this debt through selling new securities. This is exactly what large firms such as General Motors do.

One way in which today's growing national debt can represent a burden to future generations is through something called crowding out. Crowding out occurs when an increase in government spending either replaces private spending or pushes up interest rates and crowds out some private investment that would have taken place at lower interest rates.

The first type of crowding out is **direct crowding out,** in which the government activity supplants private business activity on a dollar-for-dollar basis. Some examples are government expenditures on rail service, health care, and parks that could be operated by the private sector.[5]

The second type of crowding out is **indirect crowding out** and is of particular interest to us in the present context. When the U.S. Treasury sells new government bonds to finance federal budget deficits, it increases the demand for funds in the nation's money markets and drives up interest rates.[6] These higher rates may mean that some businesses may suddenly find that scheduled investment projects based on a cost of capital of 8% may no longer be feasible from an economic standpoint if the cost of capital jumps to 9.5%. Figure 17.5 summarizes the chain of events leading to crowding out.

Crowding out, therefore, results in a smaller capital stock of manufacturing assets passed on to future generations and fewer future jobs associated with the operation of these investment goods.[7] This also may translate into lower future productivity growth than otherwise would have been the case if these investments were targeted for modernization of manufacturing operations that capture new technologies.

---

[5]The sale of AMTRAK by the government and the private operation of much of the U.S. postal service represents an effort to avoid direct crowding out.

[6]U.S. government securities are viewed as one of the safest investment outlets in the world. The U.S. Treasury typically has no difficulty in finding buyers for its new security issues, although it may have to offer a higher interest rate to buyers to meet its needs. The issuance of new government securities by the U.S. Treasury does not change the nation's money supply. When the buyers of the bonds write out a check to the U.S. Treasury, the demand deposits at the banks at which these checks are written are reduced. The effect at the national level is neutralized, however, when the government spends these funds and the proceeds from the transaction wind up back in the commercial banking system. The Federal Reserve can purchase additional securities (it is prohibited from purchasing new issues but purchases of securities in the open market have the same effect) to hold down interest rates and avoid serious indirect crowding out in the economy. This, of course, does represent an increase in the money supply, which might later prove to be inflationary.

[7]We can also think of the effects of crowding out on the economy by recalling the production possibilities curve discussed in Chapter 8. Instead of boxes of canned fruit and boxes of canned vegetables, we can talk instead of broad sets of goods and services (e.g., guns and butter). The greater the nation's productive capital stock, the more guns and more butter we can produce, and the more the production possibilities curve will lay to the right of the economy's existing curve. Crowding out negates this expansion.

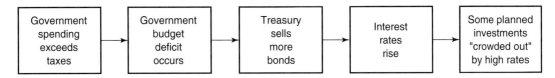

FIGURE 17.5 Chain of events leading to crowding out.

FISCAL POLICY OPTIONS

Fiscal policy refers to the act of changing government spending and/or taxes to achieve specific policy goals, such as promoting economic growth and stability. To achieve these goals, Congress and the president have certain fiscal policy instruments at their disposal. These fiscal policy instruments can be divided into two distinct categories:

1. **Automatic fiscal policy instruments** are policies that take effect without an explicit action by the president or Congress.
2. **Discretionary fiscal policy instruments** are policies that require the legislative and executive branches of government to take explicit policy actions that would not occur otherwise (e.g., pass a law).

In the discussion to follow, we will assume that the relevant goal for fiscal policy is to eliminate inflationary and recessionary gaps.

Automatic Policy Instruments

Several types of automatic fiscal policy instruments, which are constantly at work in the U.S. economy, act as built-in stabilizers, automatically watering down the punch bowl just when the party suddenly picks up steam (i.e., economy suddenly enters into a boom period) and automatically spiking the punch when the party is moving too slowly (i.e., economic activity suddenly starts to slow down).

The progressive income tax system in the United States serves the function of a built-in stabilizer. When taxable income goes up in a boom period, the marginal tax rate also increases. Consider the tax liability of the six different families, each consisting of four people and each having $13,000 in income tax deductions, in Table 17.2. The family of four with an adjusted gross income of $10,000 would owe no federal income tax. The family of four with an adjusted gross income of $30,000 has a federal tax liability of $2,550. And the family of four with an adjusted gross income of $150,000 would have a tax liability of $37,071. Each of these families fell into different marginal tax brackets because of their income level. This represents a progressive tax system because, when their taxable income rose, the rate at which their income was taxed also rose. This automatic progressive tax rate structure causes aggregate demand in the economy to be less than it would have been if all families of four were taxed at the same flat rate of 20%.

TABLE 17.2   Tax Liability for a Progressive Income Tax System*

Adjusted Gross Income Over	But Not Over	Tax Base		Marginal Tax Rate	Of Amount Over
0	43,050	0		15.0%	0
43,050	104,050	6,457	+	28.0%	43,050
104,050	158,550	23,538	+	31.0%	104,050
158,550	283,150	40,432	+	36.0%	158,550
283,150	....	85,288	+	39.6%	283,150

*The tax paid by the family having a taxable income of $77,000 is equal to $6,457 plus 28% of the taxable income exceeding $43,050. The tax paid by the family with $137,000 in taxable income is equal to $23,538 plus 31% over $104,050.
Source: Internal Revenue Service.

If the economy suddenly slowed down and wages and salaries fell, taxable income would fall and thus be taxed at a lower marginal rate than before. As a result, aggregate demand would not fall by as much as it would have if all families of four were taxed at the same flat rate. In short, our progressive income tax system acts as a built-in stabilizer, buffering the economy from any sudden changes in economic activity.

The program of unemployment compensation in this country also serves automatically to stabilize aggregate demand. When economic activity slows down, workers who have been laid off automatically become eligible to receive unemployment compensation, which keeps their disposable income from dropping to zero until they find other work. When economic activity is strong, however, there is less unemployment, which means that fewer unemployment payments would be necessary.

Both built-in stabilizers tend to dampen the effects of sudden changes in disposable income and, hence, shifts in the equilibrium level of national income. Because automatic fiscal policy instruments may remove only part of a recessionary or inflationary gap in the economy, their use is normally complemented by other policy actions.

Discretionary Policy Instruments

The passage of a law cutting taxes or an increase in government spending authorized by Congress are examples of discretionary fiscal policy instruments or actions. Both involve specific actions for these fiscal policies to take effect.

Discretionary Tax Policies.   The Economic Recovery Tax Act (ERTA) of 1981 is an example of a discretionary fiscal policy. This act, which put into place the taxation side of President Ronald Reagan's activist fiscal policies, introduced major changes in the U.S. fiscal policy. The act sought to reduce individual taxes by providing across-the-board rate reductions and indexation of tax brackets, reducing the maximum tax rate on taxable income, and introducing individual retirement accounts (IRAs) for virtually all workers, in which contributions were deductible from federal income taxes.

Specific incentives for savers, including the "all savers" certificate, were also a part of this act. Major changes were also made in the estate and gift tax rules to lessen the tax burden on smaller estates, such as farm-related estates. The act also contained provisions for businesses such as the Accelerated Cost Recovery System, which allows businesses to recover the cost of eligible assets more quickly than was previously allowed. This act was accompanied by cuts in government spending for specific programs but increased spending for others.

The aim of the Reagan administration was to lessen the tax burden on savings and investment and to promote economic recovery. The hope was that a revitalized economy would eventually result in an increased individual and corporate tax base that could offset or exceed the projected tax revenue losses from the lower tax rate.[8]

The ERTA of 1981 provided the largest overall tax reduction in U.S. history. The size of the ensuing federal budget deficits gave rise to a number of significant tax reforms. The Tax Equity and Fiscal Responsibility Act of 1982 turned out to be the largest revenue-generating bill in this country's history. The Tax Reform Act of 1984 was the most comprehensive and complex revision of the federal tax system that had ever been attempted. The Tax Reform Act of 1986 attempted to create a fairer and simpler tax system that would not inhibit economic growth. This act reduced the number of tax brackets and also broadened the tax base, or income subject to taxation, by tightening items such as medical and miscellaneous deductions and eliminating loopholes. The act also eliminated specific tax shelters that were encouraging investments in areas that were economical only because of tax incentives. This act also restricted the deductibility of IRA contributions introduced in the ERTA of 1981 to those earning less than $50,000 annually.

Discretionary Spending Policies.   The annual formulation of the federal budget by the president and the spending programs authorized by Congress represent discretionary spending policy actions. Perhaps the most noteworthy discretionary spending policy action, the Balanced Budget and Emergency Deficit Control Act first passed in 1986 and later amended in 1987, has an automatic trigger involved. Otherwise known as the Gramm-Rudman-Hollings Act (or simply Gramm-Rudman),

---

[8]The notion that a reduction in tax rates will actually lead to an increase in government tax revenues has been promoted by Arthur Laffer, a chief proponent of supply-side economics. The Laffer curve captures the relationship between tax rates and tax revenues and reflects the notion that a specific tax rate will maximize tax revenues. It rests on the assumption that a cut in tax rates will lead to an increase in tax collection resulting from an increase in work effort, saving, and investment, and a decrease in tax avoidance. See the Advanced Topics section at the end of Chapter 15 for a discussion of the Laffer curve.

TABLE 17.3   Effects That Expansionary Actions Have on U.S. Product Markets

Expansionary Actions	Effects of Action
Lower income tax rates	Increases disposable income, which shifts the AD curve to the right
Raise government spending	A component of total spending (G), this shifts the AD curve to the right

this act was an explicit response to the record levels of deficit spending taking place during the 1980s. The Gramm-Rudman act set forth annual maximum allowable federal budget deficits, with the ultimate goal of balancing the budget by 1993 (a zero deficit). While this target date was missed, the combination of zero-based budgeting, reductions in subsidies, lower interest rates and growth in the economy, budget surpluses and how to spend them became the topic of discussion in 1998. The automatic trigger referred to above was the act's **sequester** feature for cutting spending whenever deficits were projected to exceed the target by more than $10 billion. These spending cuts were to be across the board with the exception of interest on the debt, social security benefits, and certain other entitlement programs. Congress typically enacts an across-the-board sequester by discretionary actions designed to either cut spending or raise (enhance) revenue.

Issues likely to be debated in the coming years include the integrity and uses of the Social Security Trust Fund and specific spending acts targeted to sectors of the economy such as agriculture.

Fiscal Policy and Aggregate Demand

The application of fiscal policy has a direct effect on the nation's aggregate demand and, hence, on the equilibrium level of GDP and interest rates. This is different from monetary policy, which targeted the rate of interest. These policies can be applied either in an expansionary context or a contractionary context.

Expansionary Applications.   **Expansionary fiscal policy actions** involve the fun things that politicians like to do, which is to increase government spending or cut taxes. Table 17.3 summarizes the effects these actions have upon the nation's product markets. Both of these expansionary actions will shift the aggregate demand curve in Figure 13.10 to the right.

Contractionary Applications.   **Contractionary fiscal policy actions** are far less popular with politicians, particularly in an election year. Instead of increasing spending back in their districts or cutting taxes, they are asked to cut spending and/or raise taxes.

TABLE 17.4   Effects That Contractionary Actions Have on U.S. Product Markets

Contractionary Actions	Effects of Actions
Raise income tax rates	Decreases disposable income, which shifts the AD curve to the left
Cut government spending	A component of total spending (G), this shifts the AD curve to the left

Table 17.4 summarizes the effects of contractionary fiscal policy actions on the nation's product markets. Both of these contractionary actions will shift the aggregate demand curve in Figure 13.10 to the left.

Combating Recessionary Gaps

The fiscal policies of the federal government can be designed to reduce unemployment. Policymakers can adopt expansionary fiscal policies to stimulate aggregate demand in the economy and promote the employment of workers currently seeking jobs. These policies would include reducing income taxes by lowering marginal income tax rates, increasing tax depreciation allowances, adopting other tax credits such as investment tax credits (which stimulates investment) and increasing government expenditures on goods and services.

Reducing income taxes increases the personal disposable income of consumers and retained earnings of businesses, which should lead to an expansion of consumption and investment expenditures. In Table 17.1, we saw that these two components of aggregate demand constitute more than 80% of the nation's GDP.

Increasing government expenditures directly stimulates aggregate demand. Government purchases of goods and services represent approximately 20% of the nation's GDP. To illustrate how these expansionary demand-oriented policies eliminate a recessionary gap, let's examine what happens to product market equilibrium depicted in Figure 17.6.

If aggregate demand shifts from AD_1 to AD_2, aggregate output will increase from Y_1 to Y_2. Additional workers will also be needed to produce this additional output. Because there was plenty of slack in the economy, these policies were not inflationary (i.e., prices remained at P_0 because the AS curve was perfectly elastic over this range of the AS curve).[9] Because of this increase in aggregate demand, the recessionary gap would be reduced from $Y_{FE} - Y_1$ to $Y_{FE} - Y_2$.

If aggregate demand is stimulated to the point at which the economy moves out to AD_3, aggregate output will increase to Y_3. The general price level would begin to rise, reaching the level P_1. The rate of inflation resulting from this action would be the percent change from P_0 to P_1, and the rate of growth in real GDP

---

[9]See Chapter 14 for a further discussion of the complete aggregate supply (AS) curve.

Elimination of Recessionary and Inflationary Gaps

FIGURE 17.6   Fiscal policies can be used to expand or contract aggregate demand to promote economic growth and price stability.

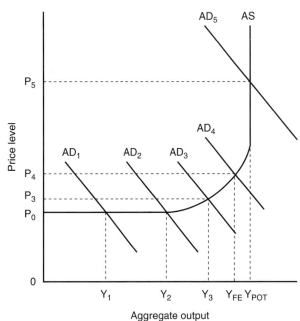

would be the percent change from Y_2 to Y_3. The recessionary gap would be further reduced to $Y_{FE} - Y_3$, but not totally eliminated.

Only if aggregate demand were expanded out to AD_4 would aggregate output achieve full employment Y_{FE}, and the recessionary gap be completely eliminated. The general price level would reach P, which is somewhat inflationary, but does not swamp the growth in aggregate output from Y_3 to Y_{FE}. The increase in aggregate demand to AD_4 therefore expands the economy to full employment, or to the point where it achieves the lowest unemployment rate and the highest capacity utilization rate without triggering a significant increase in inflation. When consumption, investment, and government expenditures increase in response to the expansionary demand-oriented policies, the aggregate demand curve will shift outward to the right. This leads to greater levels of output in the economy, expands the number of persons employed in the economy, and lowers the unemployment rate.

If aggregate demand were expanded out to AD_5, the economy would reach its capacity to supply goods and services in the current period (Y_{POT}), and while additional output was forthcoming, it would be swamped by the increase in the general price level from P_4 to P_5. This inflation would lower the real wealth of consumers and self-correct aggregate demand back toward AD_4. Congress and the president would also likely pursue contractionary fiscal policies to bring the *AD* curve back toward AD_4.

This cartoon unfortunately is all too true. The politician here is promising huge subsidies for tobacco growers, billions in price supports for dairy farmers, and federal grants for his state's water project. And, by the way, he is also a proud sponsor of a balanced budget amendment. The first three panels in this cartoon involve increases in federal government spending. The only way the federal budget can be balanced (i.e., a zero annual budget deficit) is to raise federal government receipts (or that nasty "T" word, *taxes*).

Combating Inflationary Gaps

Fiscal policies can also be designed to reduce inflation. Contractionary fiscal policies dampen aggregate demand in the economy and thus reduce existing inflationary pressures. These policies would include increasing income taxes by raising marginal income tax rates, by decreasing tax depreciation allowances, or by reducing or eliminating other tax credits like investment tax credits (which stimulate investment) and decreasing government expenditures on goods and services.

Increasing taxes will reduce personal disposable income of consumers and retained earnings of businesses and retard the expansion of consumption and investment expenditures. Decreasing government spending on goods and services will also have a dampening effect on aggregate demand.

To illustrate, let's look again at Figure 17.6. Assume that aggregate demand in the economy is represented by the AD_5 curve. If contractionary fiscal policies brought about a shift in this demand curve from AD_5 to AD_4, the aggregate demand curve would intersect the aggregate supply curve at full employment output Y_{FE}. Prices would fall substantially from P_5 to P_4 without causing a reduction in aggregate output.

If policymakers continued to dampen demand to the point at which the demand curve shifted back to AD_3, prices would fall further to P_3. But aggregate output would fall below Y_{FE} to Y_3, thereby reducing the employment of the nation's resources below full employment levels and creating a recessionary gap.

Combating Stagflation

The demand-oriented fiscal policies used to combat inflation or unemployment present the policymaker with a problem when the economy is suffering from stagflation, or rising inflation in a stagnating economy. The policymaker must choose between demand-oriented macroeconomic policies, which either fight inflation or fight unemployment—but not both. Policymakers may instead turn to one or more supply-oriented macroeconomic policies that shift the supply curve to the right, such as is shown in Figure 17.6, which theoretically combats stagflation by lowering *both* inflation and unemployment.

Difficulties in Policy Implementation

Monetary and fiscal policies do work to reduce inflation or unemployment, but neither work well at reducing stagflation. Supply-side policies that increase efficiency and aggregate output have both benefits and costs associated with them. To become fully effective, fiscal policies need time. Also, to be effective, expansionary fiscal policies should not be practiced at a time when the Federal Reserve is pursuing contractionary monetary policies, or vice versa, because at best it can be offsetting and at worst create undesirable consequences.

There are several factors that cause a delay in the full effectiveness of needed policy changes. The first is a recognition lag, or the time between when the policy was needed and when this need was recognized by policymakers. The second is an implementation lag, or the time between when the need for the policy change was recognized and when it was actually implemented. Finally, there is an impact lag, or the time between when the policy was implemented and when it becomes fully effective.

There is no delay in the effectiveness of automatic fiscal policy instruments. When real income rises, tax receipts of the U.S. Treasury increase. The delay in effectiveness of monetary policy is also relatively short, particularly in the case of the Federal Reserve's FOMC open market operations. A change in reserve requirements also has a quick effect on credit availability at depository institutions. The length of the lag for discretionary fiscal policy is often frustratingly long. New tax legislation requires the joint consensus of both Congress and the president.

Congress and the president sometimes pursue different policies than the Federal Reserve. In the early-1980s, when Congress and the president were pushing an expansionary fiscal policy, the Federal Reserve was following contractionary monetary policies to eliminate inflation. The result was extraordinarily high interest rates (the prime commercial bank rate at one point reached 21%) and a marked recession from 1981 to 1982. When the Federal Reserve switched to a more expansionary monetary policy stance, however, the economy took off, registering an unprecedented number of quarters of economic expansion (i.e., real GDP growth) during a peace-time period.

WHAT LIES AHEAD?

This chapter now puts into place all the necessary policies the macroeconomic policymakers employ to address fundamental weaknesses in the economy when they occur. Chapter 14 discussed the nation's product markets and the determination of gross domestic product. Chapter 15 discussed the consequences of business fluctuations, including the different concepts of inflation and unemployment found in government literature, and the broad macroeconomic policy options policymakers can consider. Chapter 16 discussed the functions of money and the role of the Federal Reserve System in managing the nation's money supply in a manner that promotes economic growth and price stability. And this chapter discussed the formulation and implementation of fiscal policy to achieve a similar set of objectives. The final chapter in this part of the book illustrates what this all means for the nation's farmers and ranchers and for the food and fiber industry as a whole.

SUMMARY

The purpose of this chapter was to discuss the federal budget, the budget deficit, and the national debt. This chapter also examined the fiscal policy options available to Congress and the president when they strive to achieve specific macroeconomic objectives. The major points made in this chapter may be summarized as follows:

1. The federal budget reflects the federal government's revenue expectations and expenditures for a specific calendar or fiscal year. Real government spending has grown steadily over the post–World War II period.

2. The difference between federal spending and expenditures is the federal budget deficit. There are several definitions of the budget deficit, including the deficit based on national income and product account concepts, and the unified budget deficit.

3. The national debt and the annual federal budget deficit are interrelated. The national debt at the end of the year is approximately equal to the national debt at the beginning of the year plus the federal budget deficit.

4. Two measures of the burden of the national debt are: (1) the ratio of national debt to the nation's GDP, and (2) the ratio of interest payments on the national debt to the nation's GDP.

5. Foreign investors own only 13% of the national debt. State and local governments, banks, life insurance companies, and individuals hold much of the balance of U.S. government securities.

6. The national debt does not necessarily represent a burden on future generations. Just as there is no reason to believe that General Motors should retire its indebtedness, there is no need for the national government to retire its debt. It is important that the relationship between the national debt and national income and output remain healthy.

7. Crowding out because of rising interest rates associated with larger federal budget deficits can retard investment expenditures and slow the growth of the economy's productive capital stock.

8. Automatic fiscal policy instruments are those policies that take effect without explicit action by policymakers. Two examples are the progressive income tax system and unemployment compensation. Discretionary fiscal policy instruments are those policies that require explicit actions on the part of Congress and the president. Two examples of discretionary fiscal policy are the 1986 Tax Reform Act and the Emergency Deficit Control Act.

9. Expansionary fiscal policy actions involve lowering income tax rates or raising government payments, which would shift the economy's aggregate demand curve to the right.

10. Contractionary fiscal policy actions involve raising income tax rates or lowering government payments, which would shift the economy's aggregate demand curve to the left.

11. A balanced budget requires that federal government receipts equal federal government expenditures. The federal budget deficit in this instance would be equal to zero.

DEFINITION OF KEY TERMS

Automatic fiscal policy instruments: those taxation and spending policies that take effect without an explicit action by the president and Congress. Examples are the progressive income tax system and unemployment compensation program.

Contractionary fiscal policy actions: actions taken to contract aggregate demand; examples include raising income tax rates or lowering government spending.

Direct crowding out: when government activity supplants private business activity on a dollar-for-dollar basis.

Discretionary fiscal policy instruments: those taxation and spending policies that require an explicit action by Congress and the president before they are effective.

Examples include the Tax Reform Act of 1986 and the Balanced Budget and Emergency Deficit Control Act.

Expansionary fiscal policy actions: actions taken to expand aggregate demand; examples include lowering income tax rates or raising government spending.

Federal budget: statement of the federal government's receipts and expenditures for the year.

Federal budget surplus: amount by which federal government receipts during the year exceed federal government expenditures.

Federal expenditures: sum of social security and income security payments, national defense expenditures, medicare expenditures, interest on the national debt, and the costs of other federal government programs during the year.

Federal receipts: sum of personal and corporate income taxes, social security taxes and contributions, and other sources of revenue received by the federal government.

Indirect crowding out: when the U.S. Treasury sells new government bonds to finance federal budget deficits, which leads to higher interest rates in money markets and reduces scheduled private investment expenditures.

National debt: sum of the outstanding federal government securities and other claims on the federal government.

Sequester: the setting aside of funds previously authorized to fund federal government expenditures required under the Gramm-Rudman Act, if projected deficits exceed targets.

REFERENCES

Economic report of the president, Washington, D.C., various issues, U.S. Government Printing Office.

U.S. Department of Commerce: *Business conditions digest,* Washington, D.C., various years, U.S. Government Printing Office.

EXERCISES

1. Please place a T or F in the blank appearing next to each statement.

_____ Only a small level
of inflation occurs
as the economy
moves from AD_1
to AD_2.

_____ A tax cut repre-
sents one approach
to reducing prices
below P_2.

_____ This graph depicts
the effects of
stagflation.

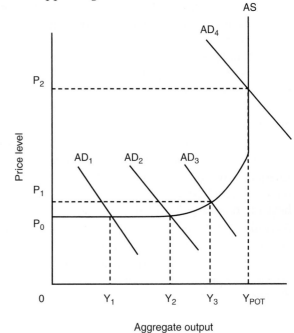

2. Review this graph and answer questions a, b, and c.

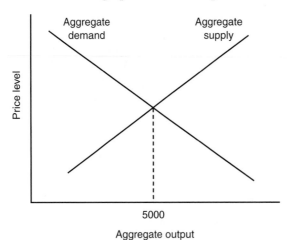

a. If the full employment level of output is $6,000, is the economy in an infla-
tionary or recessionary gap? What *specific* fiscal policies could be used to
correct this?

b. If the full employment level of output is $4,000, is the economy in an inflationary or recessionary gap? What *specific* fiscal policies could be used to correct this?

c. If the full employment level of output is $5,000, is the economy in an inflationary or recessionary gap? What *specific* fiscal policies could be used to correct this?

3. Given the following figure, explain what type of inflation is taking place and the specific fiscal and monetary policies that may be used to combat this type of inflation if, during a period of years, the economy moves from AD_1 to AD_4.

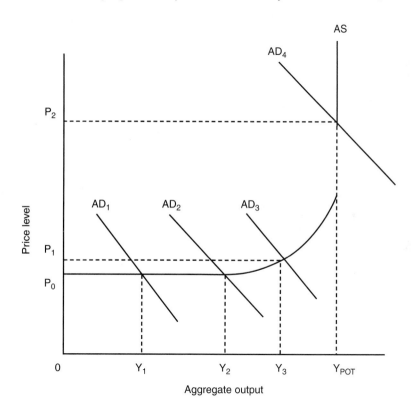

18

MACROECONOMIC POLICY AND AGRICULTURE

To describe the failures of government policy as failures of economics is the very opposite of the truth.

Allan H. Meltzer

TOPICS OF DISCUSSION

This chapter focuses on what happens to agriculture and the food and fiber industry when policymakers respond to unwanted macroeconomic trends. Macroeconomic policy has a large impact on the nation's farmers, particularly those who produce crops such as wheat that rely on a significant export demand to complement domestic demand. The chapter concludes with a summary of the lessons learned in the 1970s and 1980s, and the issues facing agriculture and the food and fiber industry over the balance of the 1990s and beyond.

IMPACTS OF POLICY ACTIONS ON THE GENERAL ECONOMY

Several trends in the general economy that all economists, including agricultural economists, watch closely are trends in the market rate of interest, the rate of growth in both real GDP and personal disposable income, the civilian unemployment rate, the rate of inflation, and the foreign exchange rate. Each of these variables gives economists some insight into what is taking place in the domestic economy and helps them determine if changes in the monetary and fiscal policies are needed to achieve desired macroeconomic objectives with respect to real economic growth, employment, price stability, and international competitiveness.

Market Interest Rates

The direction and level of market rates of interest, such as the prime interest rate or the rate commercial banks charge their best customers for new loans, is directly influenced by the growth of the money supply. As indicated in Chapter 16, **contractionary monetary policy** actions will shift the money supply curve to the left, and **expansionary monetary policy** actions will shift this curve to the right. When combined with the demand for money curve, alternative money supplies in the short run will shift interest rates either higher or lower, respectively.

Fiscal policy, as reflected by the need for the U.S. Treasury to finance larger or smaller budget deficits, can also cause interest rates to change. **Expansionary fiscal policies,** which lead to higher budget deficits, for example, will absorb more of the available funds in money markets and cause interest rates to rise. **Contractionary fiscal policies,** which reduce the government's need for funds, can cause interest rates to fall in the short run. The effects of these alternative macroeconomic policy actions on market rates of interest are summarized in the first row of Table 18.1.

Real GDP and Disposable Income

The level of economic activity and national income are also influenced by the direction taken by macroeconomic policymakers. For example, if a recessionary gap exists in the economy, such as that occurring during the 1991 to 1992 recession, monetary policymakers can adopt expansionary monetary policies, such as reducing the discount rate, in an effort to expand the money supply, lower interest rates,

TABLE 18.1   Short-Run Effects of Policy Actions on Macroeconomy

Policy Effects On	Expansionary		Contractionary	
	Monetary Policy (1)	Fiscal Policy (2)	Monetary Policy (3)	Fiscal Policy (4)
Market interest rates*	Lower	Raise	Raise	Lower
Real GDP	Raise	Raise	Lower	Lower
Real disposable income	Raise	Raise	Lower	Lower
Unemployment rate†	Lower	Lower	Raise	Raise
Inflation rate‡	Raise	Raise	Lower	Lower
Foreign exchange rate§	Lower	Raise	Raise	Lower

*The market rates of interest here refer to short-term interest rates such as the three-month Treasury bill rate or the prime interest rate commercial banks charge their best customers.
†The unemployment rate followed for the economy as a whole is the civilian unemployment rate as computed in Equation 15.1.
‡Some of the measures of inflation that economists follow are the rate of inflation for consumers or the percent change in the consumer price index (see Equation 15.4), the rate of inflation for businesses or the percent change in the producer price index, or the rate of inflation for the entire economy as reflected by the percent change in the implicit GDP price deflator.
§The foreign exchange rate reflects the price paid when converting U.S. dollars into another country's currency, such as the Japanese yen.

and enhance spending by producers and consumers. This expansion of economic activity will mean more jobs in the economy and a higher level of national income and product (see Figure 16.6). Thus, real GDP and disposable personal income will rise in the short run.

The opposite chain of events would occur if an inflationary gap existed and the Federal Reserve were following contractionary monetary policies.

Expansionary fiscal policies, such as those adopted in the 1980s, can provide an economic stimulus to the levels of spending by both consumers and producers. A tax cut, for example, increases disposable income of consumers and retained earnings of producers. With more after-tax dollars to spend, consumers and producers will spend more. An increase in the level of spending by government itself will also expand the level of economic activity in the economy. These are the directions fiscal policy would take if Congress and the president wanted to take an active role in eliminating a recessionary gap in the economy. The opposite would occur if contractionary fiscal policy (tax increase or government spending cut) were used to combat an inflationary gap in the economy. The effects of these alternative macroeconomic policy actions on real GDP and personal disposable income are summarized in the second and third row of Table 18.1.

Unemployment Rate

Expansionary monetary and fiscal policies that expand the economy's aggregate demand will increase employment and lower the civilian unemployment rate (see

Figures 16.6 and 17.6, respectively). The demand for more goods and services will necessitate a larger workforce to supply these goods and services. The ultimate size of this increase will depend, in part, on what happens to imports.

Contractionary monetary and fiscal policies would have the opposite effect on the unemployment rate, causing it to rise as the demand for goods and services in the economy falls. The effects of these alternative macroeconomic policy actions on the civilian unemployment rate are summarized in the fourth row of Table 18.1.

Inflation Rate

Perhaps one of the most closely watched statistics reported by various federal government agencies throughout the year is the rate of inflation. Depending upon where the economy is on its current aggregate supply curve (see Figures 16.6 and 17.6), expansionary monetary and fiscal policies, which shift the aggregate demand curve to the right, will result in either no inflation (AD_1 to AD_2), modest inflation (AD_2 to AD_3), or substantial inflation (AD_4 to AD_5).

Contractionary monetary and fiscal policies would have the opposite effect on the rate of inflation, causing a decrease in inflation as the demand for goods and services in the economy falls. The effects of these alternative macroeconomic policy actions on the rate of inflation are summarized in the fifth row of Table 18.1.

Foreign Exchange Rate

The final major macroeconomic variable watched closely by all economists is the foreign exchange rate, also referred to as the exchange rate. This rate reflects how many U.S. dollars it takes to equal a unit of foreign currency. For example, if the exchange rate between the U.S. dollar and the British pound is 1.50, it takes $1.50 to acquire one British pound. An increase in this exchange rate from 1.50 to 1.75 reflects what is referred to as a weaker dollar, because it now takes more U.S. dollars to acquire a British pound. A weaker dollar tends to decrease the level of imported goods into the United States because they are now more expensive even if the price of the good itself did not change. It would, *ceteris paribus,* tend to increase foreign investment in property in the United States simply because foreign currency in this example would have more purchasing power in U.S. markets. It would also tend to increase exports of U.S. products to foreign markets for the very same reason.

A decrease in the exchange rate from 1.50 to 1.25 reflects a stronger dollar. Now fewer U.S. dollars are needed to acquire one British pound. This economic event would tend, *ceteris paribus,* to increase imports of foreign-produced goods into the United States, decrease foreign investment in U.S. property, and decrease exports.

The level of foreign exchange rates is influenced by relative economic trends in the U.S. and foreign economies, which in turn affects the demand for these currencies in foreign currency markets. These relative trends include what is happening to market rates of interest, inflation, and **economic growth** as measured by GDP both in the United States and abroad. A strong dollar, for example, is generally associated with relatively high U.S. real interest rates (nominal interest rate minus the rate of inflation) and relatively strong macroeconomic performance abroad. A weak dollar typically is associated with relatively lower real U.S. interest rates and a relatively weak macroeconomic performance in the United States.

Contractionary monetary policies and expansionary fiscal policies, both of which result in higher interest rates, will also lead to higher foreign exchange rates or a stronger dollar. Expansionary monetary policies and contractionary fiscal policies, which tend to result in lower interest rates, will lower foreign exchange rates or create a weaker dollar. The effects of these alternative macroeconomic policy actions on foreign exchange rates are summarized in the sixth row of Table 18.1.

IMPACTS OF POLICY ACTIONS ON AGRICULTURE

In addition to the macroeconomic variables followed closely by economists, a number of agriculturally related variables that agricultural economists, farm lending institutions, and farm policymakers monitor are the level of farm product prices, farm input prices, farm interest rates, net farm income, and farm land prices. When conditions change in both the U.S. economy and abroad, macroeconomic variables will be affected. In many instances, the same can be said for these farm sector variables. Each of these variables can have a major impact on the economic performance of farmers and ranchers and agribusiness firms. The effects of monetary and fiscal policies on each of these variables are summarized in Table 18.2.

Farm Product Prices

Monetary and fiscal policy actions taken by the federal government can affect the prices farmers receive for their crops and livestock in several ways. Part of this impact on agriculture can be found in the effects that macroeconomic policies have on the domestic and export demand for U.S.-produced crops. The application of expansionary monetary policies, for example, will increase export demand by lowering the strength of the dollar and hence the exchange rates between the dollar and foreign currencies.[1]

The net effects of these monetary policy actions on farm crop prices are straightforward. Expansionary monetary policies will lower interest rates and the

---

[1]The value of the dollar in foreign currency markets varies directly with U.S. interest rates. Higher interest rates on U.S. government securities will increase the demand for dollars in foreign currency markets. To buy U.S. wheat, a foreign firm must pay for this wheat in U.S. dollars. The firm, which we will assume is located in France, must convert its francs into U.S. dollars. Thus, if interest rates on U.S. government securities are rising, it will take more francs to buy the necessary number of dollars than it would have if the demand for U.S. dollars in foreign currency markets were weak. The exchange rate in this case is simply the price of U.S. dollars divided by the price of the French franc. The higher the exchange rate, the more francs this French firm will need to complete this international transaction, and the more depressed export demand will be.

TABLE 18.2   Short-Run Effects of Policy Actions on Agriculture

Policy Effects On	Expansionary		Contractionary	
	Monetary Policy (1)	Fiscal Policy (2)	Monetary Policy (3)	Fiscal Policy (4)
Farm crop prices				
Domestic demand	Raise	Raise	Lower	Lower
Export demand	Raise	Lower	Lower	Raise
Net impact	Raise	Lower*	Lower	Raise*
Farm livestock prices	Lower	Raise*	Raise	Lower*
Farm input prices	Raise	Raise	Lower	Lower
Farm interest rates	Lower	Raise	Raise	Lower
Net farm income	Raise	Lower†	Lower	Raise†
Farmland prices	Raise	Lower‡	Lower	Raise‡

*Direction of change is not unambiguous, but instead conditional upon the relative market share for export demand and relative price elasticity of export demand. The direction of change here assumes a relatively high market share for exports and/or relatively high price elasticity of export demand (see Equation 13.1).
†Assumes the impact on crop prices and interest payments offsets the impact on livestock prices when calculating net farm income for both crop and livestock producers. Deficiency payments are held constant here.
‡Assumes the impact on crop prices and interest rates offsets the impact on livestock prices when calculating farmland prices.

value of the dollar and thus boost export demand and expand the domestic demand for all goods and services, including food and fiber products. Both events shift the market demand curve for crops in Figure 18.1 to the right and raise U.S. crop prices from P_1 to P_2. Higher crop prices will raise feed costs for livestock producers and trigger sell-off of herds, which will depress livestock prices in the short run. Both of these farm level effects are summarized in column 1 of Table 18.2.

Conversely, contractionary monetary policies will raise interest rates and foreign exchange rates, which decrease export demand and contract the domestic demand for all goods and services. These events will shift the demand curve for farm crops in Figure 18.1 to the left and lower U.S. crop prices from P_2 to P_1. Higher crop prices will mean cheaper feed costs to livestock producers and lead to herd expansions, which will raise livestock prices in the short run. Both of these effects are summarized in column 3 of Table 18.2.

It is more difficult to determine the direction of change in selected agricultural variables when policymakers change fiscal policy. As Equation 13.1 suggested, the net effects of fiscal policy actions on the prices of exportable crops will depend on what happens to the relative market share associated with export demand and the relative magnitude of the price elasticity of export demand. For those crops with a relatively high export market share, however, we can conclude that expansionary fiscal policies, which increase federal budget deficits, will lower U.S. crop prices because higher interest rates increase the value of the dollar in foreign currency markets, lower export

Impact of Policy Changes on Crop Prices

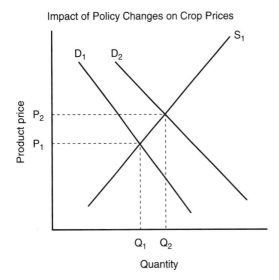

FIGURE 18.1 Expansionary monetary policies will shift the market demand curve for wheat to the right, as will contractionary fiscal policies. Both actions lower interest rates and the value of the dollar, which expands export demand. Fiscal policies will have the opposite effects because they lead to higher interest rates and a stronger valued dollar.

demand for U.S. crops, and shift the market demand curve for U.S. crops in Figure 18.1 to the left.[2] Lower crop prices (P_1 instead of P_2) will lower feed costs for livestock and lead to herd expansions, which will raise livestock prices in the short run. Both of these farm market effects are summarized in column 2 of Table 18.2.

Conversely, contractionary fiscal policies, which reduce budget deficits, lower interest rates, and lower the value of the dollar, will enhance the demand for exportable crops such as wheat and corn and shift the demand curve for these crops in Figure 18.1 to the right. Higher crop prices (P_2 instead of P_1) will raise feed costs for livestock producers and trigger sell-off of herds, which will depress livestock prices in the short run. Both of these farm level effects are summarized in column 4 of Table 18.2.

Farm Input Prices

Another way in which farmers are affected by macroeconomic policy actions is through changes in the prices they pay farm input suppliers for goods and services. Expansionary monetary policies will increase the demand for inputs used by both farm and nonfarm businesses in the economy. As suggested in column 1 of Table 18.2, this will bid up the prices of these goods and services. In the short run, input manufacturers will increase the quantity of manufactured inputs supplied but will be limited by their current manufacturing capacity. Thus, the demand for

---

[2]The discussion here focuses on fiscal policies broadly defined. It ignores farm commodity policies that support specific farm commodity prices and income. This assumption will be relaxed when we discuss the effects of macroeconomic policies on net farm income.

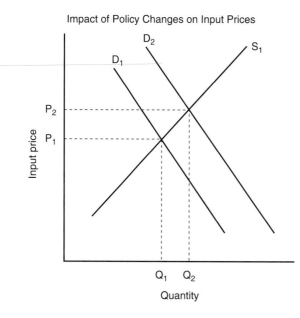

Impact of Policy Changes on Input Prices

FIGURE 18.2 Expansionary monetary and fiscal policies will shift the demand curve for inputs to the right and increase input prices. Conversely, contractionary monetary and fiscal policies will shift the market demand curve for inputs to the left, causing input prices to decline.

inputs in Figure 18.2 will shift to the right, causing input prices to rise from P_1 to P_2 when capacity utilization rates for input supplies rise. Expansionary fiscal policies will also bid up input prices, particularly if they are designed to stimulate investment (see column 2 in Table 18.2).

The overall effects will differ somewhat from that brought about by expansionary monetary policies because interest rates are rising rather than falling and muting input demand and prices to some extent.

We would expect the opposite effects if contractionary macroeconomic policies are followed. For example, contractionary monetary policies that lead to higher interest rates and more restrictive credit availability will cause the market demand curve for inputs to shift to the left and prices to fall from P_2 to P_1. The investment schedule in Figure 13.7 showed that investment expenditures are inversely related to the rate of interest. This will lead to lower input prices for both farm and nonfarm businesses as suggested in column 3 of Table 18.2. Similarly, contractionary fiscal policies can reduce investment incentives by lowering depreciation allowances, eliminating investment tax credits, raising marginal tax rates, or lowering transfer payments to businesses. These effects will cause the market demand curve for inputs to shift to the left and input prices to fall from P_2 to P_1, as illustrated in Figure 18.2.

Farm Interest Rates

Farm interest rates will follow general interest rate movements we would expect to see in the general economy because agriculture is a small player in the nation's money markets. Expansionary monetary policies that expand the money supply will lower interest rates, as suggested in column 1 of Table 18.2.[3] Contractionary monetary policies that contract the money supply will have the opposite effect: interest rates will rise (see column 3).

Interest rates will rise under expansionary fiscal policies because the U.S. Treasury must offer additional government securities on the open market to finance higher federal budget deficits (column 2 of Table 18.2).[4] If contractionary fiscal policies are adopted, however, interest rates should fall when federal budget deficits decline (see column 4).

The direction and magnitude of changes in farm interest rates are extremely important because of the capital-intensive nature of agricultural operations. There is more capital invested per worker in agriculture than there is for the economy as a whole.[5] A one percentage point increase in interest rates will, therefore, increase production expenses more in agriculture than in other production sectors in the economy. In addition, many major crops such as wheat are sensitive to interest rates because they affect the relative value of the dollar in foreign currency markets. A rise in interest rates therefore will lower export demand for wheat, which in turn lowers cash receipts by wheat producers.

Another sector in the U.S. economy that is both capital intensive and sensitive to foreign exchange movements is the steel manufacturing industry. This sector is also very capital intensive and was battered during the 1980s by imported steel and steel-based products.

Net Farm Income

Perhaps no statistic is watched more closely at the U.S. Department's National Outlook Conference in Washington, D.C. each November than the department's projection of net farm income for the coming year. Based on the expected direction of macroeconomic policy and the effects it will have on farm product prices, farm input prices, and farm interest rates, the U.S. Department of Agriculture projects what we can expect to see happen to the net farm income of U.S. crop and livestock producers.

Net farm income can be defined as:

> Cash receipts from farm marketings
> + Other farm income
> + Government payments
> = Gross farm revenue

---

[3]See the discussion of money market equilibrium in Chapter 16, including the effects that expansionary and contractionary monetary policies have on the money supply curve in Figure 16.4 and what this means for equilibrium interest rates in Figure 16.5.

[4]Consult the discussion regarding crowding out of investment expenditures in Chapter 17.

[5]See Chapter 5 in the *Economic Report of the President* for 1988, which substantiates this claim.

Feed costs and other variable production expenses
+ Interest payments
+ Other fixed cash expenses
= Total cash expenses
+ Noncash expenses
= Total production expenses

Finally, net farm income is equal to:

$$\text{net farm income} = \text{gross farm revenue} - \text{total production expenses} \quad (18.1)$$

With this accounting definition in mind, let's examine the impact that alternative macroeconomic policies would have upon net farm income.

Expansionary monetary policies that raise crop prices, raise farm input prices, and lower interest rates will likely lead to higher revenue from crop marketings and relatively stable total production expenses. The definition of net farm income given by Equation 18.1 suggests therefore that net income of crop producers will rise (column 1 of Table 18.2). The extent to which this is true will depend upon the relative export market share for crops and the relative price elasticity of export demand. The higher these two values, the more net income of crop producers will rise. Net income of livestock producers would fall in the short run under these policies; when feed prices are rising, livestock prices are falling (livestock prices do not benefit as much as crop prices do from a weaker dollar in foreign currency markets).

Contractionary monetary policies will have the opposite effect, as suggested in column 3 of Table 18.2. Higher interest rates will increase production costs and lower crop revenue, thus squeezing net income of crop producers. Livestock producers, on the other hand, should benefit from lower feed costs, lower farm input prices, and higher livestock prices.

Fiscal policies affect net farm income in somewhat different ways. For example, expansionary fiscal policies that lead to higher interest rates and a stronger dollar will lower crop prices and cash receipts from crop marketings. Lower farm input prices help reduce the squeeze on net crop income, but higher interest rates will raise interest payments, a major expense category. The net effect is therefore likely to be a lower net income for crop producers. Livestock producers will benefit from both lower feed costs and input prices in general, and should expect to earn a higher net income.

What is unusual when talking about fiscal policy is whether or not these general expansionary fiscal policies include an expansion, a contraction, or no change in the income support payments to crop producers. If government payments to crop farmers are raised to offset lower net income from market sources, then the net income of crop producers could remain unchanged.[6]

---

[6]Keep in mind that not all crops are covered by federal commodity programs. Major exportable crops such as wheat and corn are covered, but other crops such as fruits and vegetables are not. Refer to Chapter 11 for an expanded discussion of federal farm commodity programs.

Contractionary fiscal policies will have the opposite effects, as suggested by column 4 in Table 18.2. Lower interest rates leading to a weaker dollar should expand export demand for U.S. crops and increase crop prices. Higher cash receipts from crop marketings, coupled with lower input prices and interest rates, should lead to a significant increase in net income for crop producers. If these contractionary fiscal policies involve cutting government payments to crop farmers, however, this will have a negative effect on their net income. Livestock producers will be confronted by higher feed costs and falling livestock prices in the short run, which will squeeze net income of livestock producers.

Farmland Prices

There are several complex approaches to estimating the market price for an acre of farmland that capture factors such as ordinary and capital gains tax rates, year-to-year variations in the cost of debt and equity capital, and other factors that influence buyer and seller decisions. A simple approach to calculating the capitalized per acre value of farmland is:

$$\text{farmland price} = \text{expected per acre net income} \div \text{discount rate} \qquad (18.2)$$

in which the discount rate is the rate of return on the owner's next best alternative. Let's assume the next best alternative is to lend to others at the current interest rate. Thus, if the expected net income per acre is $150 and current interest rates are 10%, the price of farmland would be $1,500 per acre (i.e., $1,500 = $150 ÷ 0.10).

Equation 18.2 suggests that farmland prices will rise if expected net farm income rises and/or interest rates fall. Conversely, farmland prices will fall if expected net farm income falls and/or interest rates rise.

Expansionary monetary policies raise net income of crop producers and lower farm interest rates. According to Equation 18.2, both trends will lead to higher farmland prices in crop producing areas. Land prices in livestock producing areas would not fare so well. Net income of livestock producers will fall in the short run, but interest rates are lower. Thus, according to Equation 18.2, the impact of these policies on land prices in livestock-producing areas will depend upon the relative changes in the numerator and the denominator. Contractionary monetary policies will have the exact opposite effects.

It was suggested earlier that expansionary fiscal policies in general would, by expanding budget deficits and raising the value of the dollar, lower net income of crop producers and raise farm interest rates. Equation 18.2 implies that farmland prices in cropping areas would fall because both the numerator and denominator are moving in an unfavorable direction. Expansionary fiscal policies aimed specifically at farmers would tend to nullify this trend. Although livestock net income under these conditions should rise, higher farm interest rates will mute farmland prices in livestock-producing areas. Contractionary fiscal policies, in general, will have the exact opposite effects.

A BRIEF HISTORY LESSON

The decades of the 1970s, 1980s, and 1990s in agriculture can be described by a passage from Charles Dickens' *A Tale of Two Cities:* "It was the best of times, it was the worst of times." The 1970s represented an era of rising real net farm incomes. Furthermore, the combination of rising real net farm incomes and declining real farm interest rates made farm owners "paper" millionaires.

The 1980s saw a reversal of fortunes for many farmers when interest rates rose, real net farm incomes fell, and farmland prices tumbled. The purpose of this section is to provide today's student with a historical perspective of these two important decades.

The 1990s saw a major shift in farm policy in an effort to return the sector to the free market. A dramatic decline in crop commodity prices as a result of the Asian Crisis required a substantial infusion of government payments late in the decade. By 1999, government payments accounted for almost one-half of net farm income.

The 1970s: A Decade of Expansion

The 1970s truly represented a decade of expansion for agriculture. The sector accepted the challenge to produce for world markets. Farm sizes expanded, machinery and equipment were modernized, and farmers borrowed heavily. Let's look at the macroeconomic policies that existed during the 1970s, the federal farm program to which farmers were responding, and other externalities that affected agriculture and the nation's food and fiber industry.

Macroeconomic Policies.   The decade of the 1970s can be characterized by cheap credit. Seven years during this decade recorded negative real short-term interest rates, which means that the rate of inflation was greater than the nominal rate of interest on such financial instruments as short-term government bonds (see Figure 18.3, *A*). The annual real interest rate on short and intermediate farm loans made by the Farm Credit System fell six times during the 1970s.

In Chapter 16, we learned that declining interest rates are characteristic of expansionary monetary and contractionary fiscal policies. The value of the dollar as measured by the Federal Reserve's real **trade-weighted exchange rate** fell sharply during the 1970s (see Figure 18.3, *B*). The annual federal budget deficit showed more variability during the 1970s than it did in the previous decade. The only substantial increases in annual budget deficits corresponded to efforts to stimulate the economy out of the 1974–75 recession (see Figure 18.3, *C*).

Federal Farm Policies.   The farm legislation that was in force during the 1970s was targeted toward markets. The strong export demand for wheat and other crops in 1973 enabled real direct government payments to farmers to drop to zero.

Trends in Selected Macro Variables in the 1970s

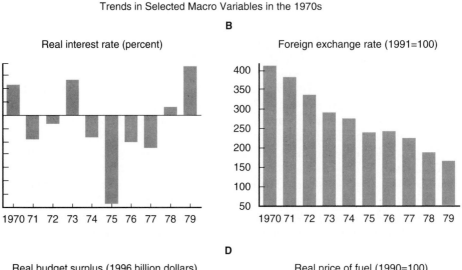

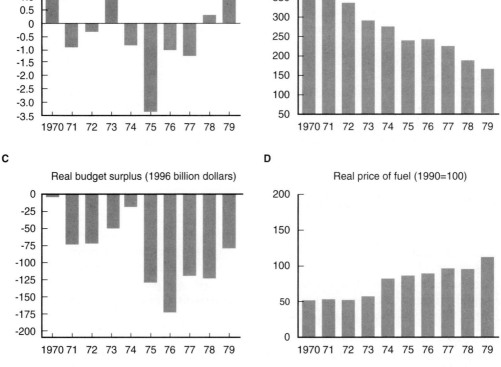

FIGURE 18.3   This figure depicts the trends in four selected macroeconomic variables during the 1970s. These variables are: (1) real short-term interest rate, (2) real trade-weighted exchange rate between the dollar and the currencies of our 10 major trading partners, (3) the real annual federal budget deficits, and (4) the real price of fuel.

Although these payments rose over the remainder of the decade, they ended the decade at less than $1 billion in real terms (see Figure 18.4, C).

Other Externalities.   Perhaps the biggest "externality" confronting agriculture and the rest of the economy during the 1970s was the rise in the price of fuel. Figure 18.3, *D* illustrates the sharp increase in the real price of fuel paid by farmers in the 1970s initiated by OPEC or Organization of Petroleum Exporting Countries, which withheld crude oil from international markets. By the end of the decade, the real price of fuel doubled from the pre-OPEC days in the early 1970s.

Trends in Selected Agriculture Variables in the 1970s

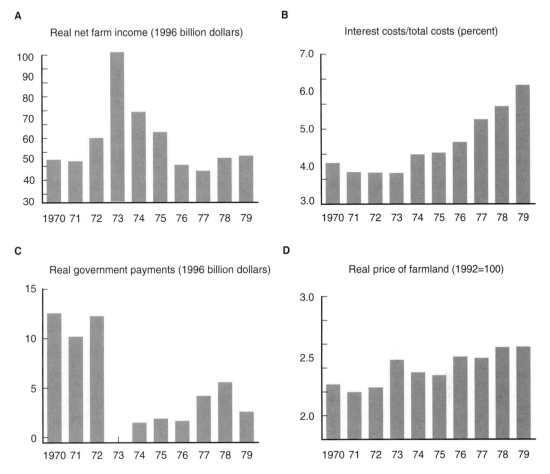

FIGURE 18.4   This figure depicts the trends in four selected agriculture variables during the 1970s. These variables are: (1) real net farm income, (2) ratio of farm interest payments to total farm production expenses, (3) real government payments to farmers, and (4) real price of farmland.

Trends in Agriculture.   Real net farm income, a broad measure of economic performance in agriculture, showed considerable strength over much of the decade. The declines noted in real net farm income in the late 1970s in Figure 18.4, *A* were heavily influenced by the rise in the farm price of fuel (Figure 18.3, *D*) and the rise in interest payments when farmers began to service the new farm debt they undertook during the decade (Figure 18.4, *B*). Another factor was the rising inflation in the general economy during the late 1970s, which caused farm input prices in general to grow at a time when crop and livestock prices were declining.

Finally, real farmland prices ended the decade higher than they were at the start (Figure 18.4, *D*). Equation 18.2 suggested that farmland prices were positively related to expected net farm income and negatively related to farm interest rates. Much of the period experienced growth in real net farm income and negative real farm interest rates. Therefore, it is not surprising that farmland prices rose as they did. The combination of real net farm income and rising farmland prices left many farmers in a strong net worth position at the end of the decade.

Implications for Food and Fiber Industry.   The real price of food rose during the decade, reflecting to a certain extent the rising price of agricultural products. The level of real food expenditures also rose during the 1970s, increasing from approximately $120 billion in 1970 to about $140 billion by 1979. Thus the real value of the output of food products produced by the nation's food and fiber industry rose by almost 17% during the decade. However, since the percentage of disposable income spent for food declined from 19.4% in 1970 to 16.6% in 1979, spending on other goods and services by consumers must be rising at a faster pace. Finally, real cash outlays for variable production inputs and real investment expenditures for farm machinery and motor vehicles both rose dramatically during the 1970s when farmers adopted new technologies and expanded their operations to meet an expected growing world demand.

The 1980s: A Decade of Contraction

The 1980s started on an ominous note for agriculture. The Federal Reserve got serious about fighting inflation. For a sector that is highly capital-intensive, such as agriculture, contractionary monetary policies that lead to higher interest rates and a stronger dollar in foreign currency markets are relatively harsh.

Macroeconomic Policies.   The Federal Reserve began in late 1979 to introduce contractionary monetary policies in response to the high double-digit rates of inflation that existed in the economy. The Monetary Decontrol Act of 1980 also gave the Fed broader control over depository institutions in the economy. The combination of broadened powers and contractionary policy actions designed to reduce the nation's real money supply drove real interest rates to post–World War II highs.

President Ronald Reagan and Congress also put into place a series of tax acts in the early 1980s that reduced taxes paid by businesses and individuals (see Chapter 15) more than government spending was reduced. These expansionary fiscal policies caused annual federal budget deficits to rise sharply (see Figure 18.5, *C*), which put further upward pressure on interest rates in the economy. The **nominal interest rate** reached 21% at one point during 1983.

Rising **real interest rates** in the United States versus the rest of the world attracted capital from foreign investors, which helped finance the United States'

Trends in Selected Macro Variables in the 1980s

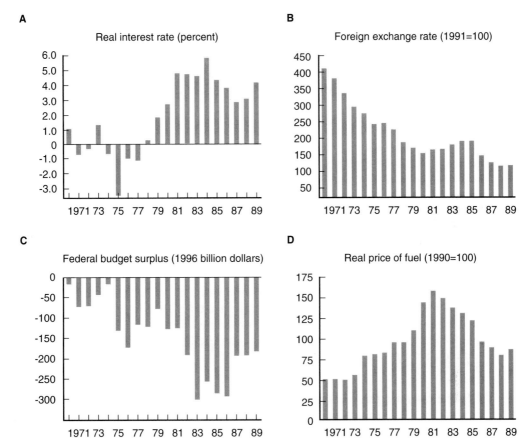

FIGURE 18.5 This figure depicts the trends in four selected macroeconomic variables during the 1980s. These variables are: (1) real short-term interest rate, (2) real trade-weighted exchange rate between the dollar and the currencies of our 10-major trading partners, (3) the real annual federal budget deficits, and (4) the real price of fuel. The values previously reported for the 1970s are repeated here for comparison purposes.

growing annual federal budget deficits and private investment projects in the United States. Real interest rates in the economy rose markedly also (see Figure 18.5, *A*, which repeats the 1970 experience for comparison purposes). The stronger dollar in foreign currency markets (see Figure 18.5, *B*), also increased our purchasing power in international markets, and resulted in higher, relatively cheap imports that helped reduce domestic inflation. However, it also depressed the export demand for U.S. produced products that had to be paid for with more expensive U.S. dollars. The growth in imports and decline in exports led to a growing international trade deficit, which reached unprecedented levels in the late 1980s.

As inflation rates came down to low, single-digit levels in the mid-1980s (see Figure 15.4), the Federal Reserve began to adopt more expansionary monetary policies. The combination of *both* expansionary monetary and fiscal policies led to a prolonged period of economic growth in the U.S. economy that continued until the fourth quarter of 1990 (see Figure 14.4). Did the phenomenal period of real economic growth in the general economy help agriculture? Before answering this question, let's review the federal income and price support policies in place during the 1980s.

Federal Farm Policies.   The features of the 1980 and 1985 Farm Bills continued the practice of attempting to control the supply of surplus commodities such as wheat, corn and other feed grains, cotton, and other commodities by requiring producers to set aside a portion of the land historically planted to these crops in exchange for deficiency payments that made up the difference between the announced target price for the year and the current market price (see Chapter 13 for a discussion of production control policies, set-aside requirements, and target prices).

The federal government also enacted a payment-in-kind (PIK) program for specific crops, such as corn and wheat, in 1983 and 1984 under which farmers were paid in the form of surplus commodities rather than cash when participating in government income and price support programs. Real direct government payments to farmers increased dramatically in the late 1980s (see Figure 18.6, C) when commodity prices fell.

Other Externalities.   Other nations began to intensify support of their agricultural sectors, both for self-sufficiency reasons in their food supply, and to gain foreign currency (including dollars) to purchase needed manufactured goods from the United States and other technologically advanced nations. Countries such as Argentina and Brazil greatly expanded their production of wheat and soybeans, respectively, further adding to world surpluses and depressing world prices for these commodities. China went from a rice-importing country to a rice-exporting country in a matter of a few years. The United States responded by further idling production resources previously dedicated to these commodities.

Trends in Agriculture.   The level of real net farm income fell to Great Depression levels in 1983 after surpluses of specific crops grew in the early 1980s (see Figure 18.6, A). Farm revenue was suppressed by declining export demand for U.S. crops due to a rising value of the U.S. dollar. Rising interest rates also increased farm production expenses. Interest payments on farm debt outstanding rose to 19% of cash production expenses in the 1980s, as opposed to 7% in the 1970s (see Figure 18.6, B). These trends put downward pressure on real net farm incomes and farmland prices (see Figure 18.6, D).

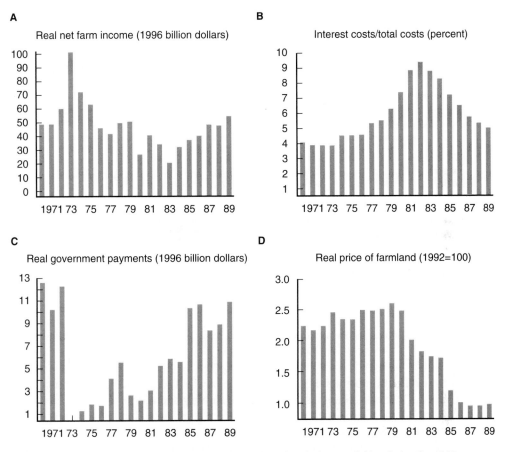

FIGURE 18.6   This figure depicts the trends in four selected agriculture variables during the 1980s. These variables are: (1) real net farm income, (2) ratio of farm interest payments to total farm production expenses, (3) real government payments to farmers, and (4) real price of farmland. The values previously reported for the 1970s are repeated here for comparison purposes.

The paper millionaires created by rising farmland prices in the 1970s saw these unrealized gains eroded in the 1980s. Many farmers who bought high-priced farmland late in the 1970s at low interest rates on variable rate loans became insolvent (i.e., the value of their liabilities exceeded the value of their assets) in the 1980s, when interest rates rose and both net farm income and farmland prices fell. These farmers are no longer in business (Figure 18.7).

Implications for Food and Fiber Industry.   The combination of production control policies that constrained the supply of specific crops below free market levels, declining real net farm incomes, and rising real interest rates led to lower purchases

Des Moines
Sunday Register

■ Des Moines, Iowa ■ March 25, 1984   ■ $1.00 single copy from dealer or vendor
■ 25¢ by motor route; 58¢ by carrier

SECTION A

THE WEATHER — Increasing cloud-
iness. Highs 40 to 48s; lows in 20s to
40s. Sunrise 6:30 a.m.; sunset 6:33
p.m. Details 12.

Economic squeeze may force thousands of farmers off land

Big debts, high interest, falling prices create chaos

FIGURE 18.7   The headlines from the March 25, 1984, issue of the *Des Moines Register* (Iowa) newspaper summarize the economic conditions facing many farmers and rural communities during the 1980s.

of farm production inputs. Figure 18.8, *A* illustrates the general expansion of farm machinery (tractors, combines, and other machinery and equipment) and motor ve- hicles (for business use) during the 1970s, when farmers modernized and expanded their farming operations.

The squeeze on real net farm income and rising interest rates in the 1980s and problems of insolvency (debts exceeding assets) for many farmers caused the an- nual real investment expenditures for farm machinery and motor vehicles to fall markedly. Real capital outlays for farm machinery and motor vehicles expressed in 1996 dollars fell by more than one-half, from $22.3 billion in 1979 to about $9.6 billion in 1989. The upturn in real net farm income in the late 1980s began to turn this trend around again, however.

These expenditures, of course, represent the revenues received by John Deere and other farm machinery and motor vehicle manufacturers and the level of sales activities by merchants and dealers in rural communities throughout the country.

Figure 18.8, *B* shows a similar trend in the purchases of nondurable farm pro- duction inputs and services during the year. There was a marked expansion of variable production input use during the 1970s, when the sector modernized and expanded, and an equally marked contraction during much of the 1980s, at which time farm profits were squeezed when the cost of borrowing to purchase these in- puts rose.

These expenses of farmers also represent the volume of sales activities of rural businesses and farm input manufacturers. Figure 18.8, *A* and *B* clearly

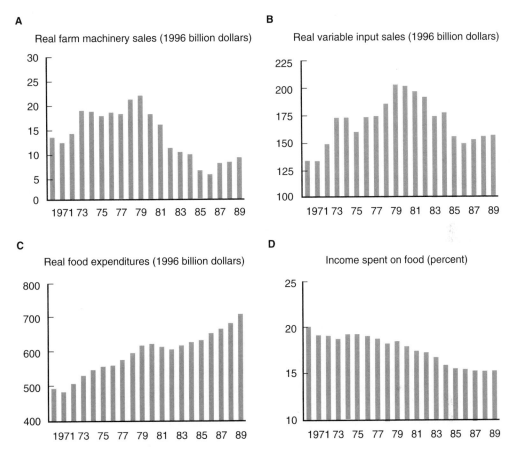

FIGURE 18.8 This figure depicts the trends in four selected variables in the nation's food and fiber industry from 1970 to 1989. These variables are: (1) real expenditures for farm machinery and motor vehicles, (2) real purchases of nondurable farm inputs and services from merchants and dealers in rural communities, (3) real expenditures by consumers for food and related items in the domestic economy, and (4) the percent of total disposable income Americans devote to food and related items.

make the point that farm input manufacturers and local merchants and dealers are very much dependent on a healthy and expanding agriculture. The adverse trends in the 1980s caused a contraction of the number of implement dealerships and similar firms supplying inputs to agriculture, and increased rural unemployment.

Rural communities that had three or four full-line farm implement dealers at the beginning of the 1980s may have only a local parts store today. Farmers now must drive to a larger community to see a full line of farm machinery and motor vehicles.

An increasing number of farmers becoming insolvent also caused financial stress for many of the financial institutions that were actively lending to agriculture.[7] The declining deposit growth and increasing number of bad loans at rural commercial banks had ripple effects throughout rural communities.

Figure 18.8, *C* illustrates the general uptrend in real consumer expenditures on food and related items in the U.S. economy. The only major periods of softness in this trend occurred during the 1981–82 recession in the general economy. This illustrates the relative stability enjoyed by food processors and manufacturers during this 20-year period in the nation's food and fiber industry.

Finally, Figure 18.8, *D* shows that U.S. consumers continue to enjoy the benefits of cheap food made possible by a productive agriculture. This graph depicts the ratio of expenditures for food and related items to total personal disposable income of consumers (income after taxes). By the end of the 1980s, U.S. consumers had to devote only 16% of their take-home pay to food and related items, compared with almost 20% in the early 1970s. In contrast, the average family in the former Soviet Union spends approximately 50% of its income on food and related items.

The 1990s: A Decade of Change

The U.S. economy entered the 1990s with unprecedented levels of both public and private debt. The level of private debt in the economy dampened consumer expenditures as households struggled to meet existing installment debt commitments before taking on additional commitments. And the level of public debt fueled by rising annual federal budget deficits gave rise to suggestions on various ways to cut government spending and/or raise taxes. These twin debts represented a drag on the growth of the economy in the early-1990s.

Status of the Macroeconomy.   High real interest rates, a strong dollar, high federal budget deficits, high trade deficits with the rest of the world, a major crisis in the savings & loan industry, commercial bank failures, and a heavily-debt-burdened consumer. Twenty years ago all of these trends would have seemed impossible. As we entered the new millennium, this seemed like ancient history to some. Yet these are exactly the conditions the economy was saddled with as it entered the decade of the 1990s!

The Federal Reserve throughout the 1990s was forced to walk a tightrope. It had to pursue a monetary policy that kept interest rates low enough to encourage economic growth in the domestic economy and export demand for U.S. products, and yet high enough to keep inflationary expectations low. The Federal

---

[7]Approximately 20% of the nation's farmers in the mid-1980s were considered to be in serious financial trouble. The increase in credit problems, loan delinquencies, foreclosures, and bankruptcies reached significant levels, and had rippling effects for commercial banks making agricultural loans and the Farm Credit System. Some commercial banks in rural areas became insolvent, and the Farm Credit System required financial assistance from Congress in the late 1980s.

Reserve wants to see Congress and the president adopt fiscal policies that reduce federal budget deficits, which would lower domestic interest rates and promote export demand. The president wants to see lower interest rates, positive economic growth and low unemployment. And the Congress wants to reduce federal budget deficits without cutting their favorite spending projects or increasing taxes, which of course is extremely difficult to do as suggested by the four-panel cartoon in the previous chapter! Because of the high annual budget deficits observed over the first half of the 1990s, most of the burden for stimulating the economy rested more on monetary policy rather than fiscal policy. Congress could no longer justify the forms of expansionary fiscal policy that were practiced in the 1980s.

By the end of the 1990s, however, the economy was achieving a steady rate of growth with low interest rates and virtually no inflation. In direct contrast to much of the post–World War II period, the federal government was faced with the issue of how to spend budget surpluses rather than how to reduce budget deficits. Interest rates fell during the middle of the decade as the Federal Reserve eased monetary policy coming out of the 1991 recession, reaching almost zero in real terms (nominal rate minus the rate of inflation) in 1993. Late in the decade, the Federal Reserve moved to add liquidity to the banking system during the Asian Crisis, which saw the financial institutions in many Asian countries experience great difficulty and their economies fall into recession. At the peak of the crisis, approximately 40% of the world's economies were in a recession.

Growth in the economy, fueled by productivity from technological advances, resulted in a dramatic increase in federal government revenues, causing the federal budget deficit by the end of the decade to actually become a surplus (see Figure 18.9). Other favorable trends during the 1990s relative to the previous two decades are the stable real price of fuel, low real wage rates, and low inflation rates facing consumers and producers. The low foreign exchange rate also helped offset other factors depressing export demand over the last four years of the decade.

Status of the Agricultural Economy.

Real net farm income during the decade of the 1990s was more stable than the earlier two decades, but not without the help of a substantial increase in direct government payments to farmers. By 1999, direct government payments reached almost half of net farm income as commodity prices tumbled, reflecting a continued weak export demand and a world awash in surplus commodities. Interest payments as a percentage of total operating costs spiked to 12% in the early 1990s, eventually settling at about the 9% level toward the end of the decade. Farmland prices stabilized in the early part of the decade after falling sharply during the 1980s. The growth in government payments stabilizing net farm incomes in the late 1990s, real short-term interest rates in the 3% range, and a strong macroeconomy helped cause farmland prices nationally to increase,

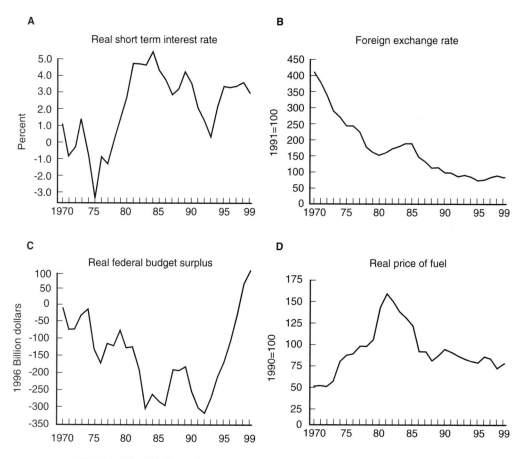

FIGURE 18.9   This figure depicts the trends in the same four selected macroeconomic variables over the 1970–1999 period.

but remain dramatically below the values experienced during the 1970s (see Figure 18.10).

The farm sector approached the end of the decade confronted by increasing uncertainty. Perhaps the major contributor to uncertainty over the future debt repayment capacity for many farmers is the FAIR Act signed into law by President Clinton in April 1996. Termed "freedom to farm" during its developmental stages, the FAIR Act replaced target prices and annual variable deficiency payments with annual fixed transition payments that *declined in value* over the 1996–2002 period. These payments were de-coupled from planting decisions, giving participating farmers the flexibility to switch to another crop without jeopardizing their eligibility. Exactly what will happen after 2002 is unclear at this point, although one probable outcome is freezing transition payments at 2002 levels. The FAIR Act also eliminated ad hoc disaster payments to farmers. It

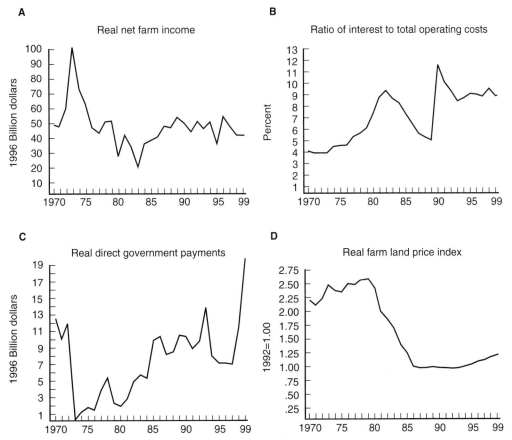

FIGURE 18.10   This figure depicts the trends in the four agricultural variables over the 1970–1999 period.

is now the responsibility of farmers to acquire private crop insurance to protect against such peril. In cases of widespread natural disasters, aid will be in the form of disaster loans rather than disaster payments. The payments made to farmers following the 1988 drought and 1993 floods, for example, will no longer be available. As some have suggested, "freedom to farm" means both freedom to succeed *and* freedom to fail.

Another contributor to uncertainty is international trade, which is influenced by economic and political events in both competitor and client nations. Most economists had been forecasting an extremely rosy outlook for agriculture based upon an expected strong export demand for U.S. commodities. The building Asian financial crisis, however, has caused many economists to revise their estimates downward over the balance of the decade. The big unknown in 1998 is the degree to which the Asian crisis spills over to Japan, the world's second largest economy, and a major importer of U.S. agricultural products.

The *bottom line* is the strong likelihood of greater price variability and more downside risk associated with net income and asset values as we move into the next decade and a new century. *Gone* are the deficiency payments that made up the differences between announced target prices and spot market prices. *Gone* are the federal supply management tools designed to stabilize farm prices and incomes. *Gone* are the ad hoc disaster payments that assisted farmers in periods of floods and droughts.

U.S. agriculture today is thus more reliant than ever before on *a growing world demand* for food and fiber products to keep commodity prices at levels that do not result in another financial crisis like that faced by farmers in the 1980s.

The spin-off effects of events in agriculture on related sectors in the food and fiber industry are depicted in Figure 18.11. Expenditures for farm machinery and equipment during the 1990s were stable but low in real terms relative to expenditures in the 1970s. Expenditures in 1999 of $9.35 billion were less than half of the $22.3 billion expenditure level observed 20 years earlier. This translates into lower sales by John Deere and other farm equipment manufacturers, as do lower variable input sales by chemical, seed, and other input manufacturers reflected in lower operating costs (see Figure 18.11).

Real food expenditures have risen steadily during the 1970s, 1980s, and again in the 1990s. However, gains in real disposable personal income have been even greater, resulting in a declining percentage of income spent on food. By the end of the last decade, food expenditures represented only 14% of income as compared with 20% in the early 1970s. As indicated earlier in Chapter 2, the percentage of income spent on food in many developing countries is dramatically higher.

Federal Farm Policies.   The growing pressures to bring federal budget deficits under control led to the design of the 1990 farm bill that was directly linked to budget reconciliation measures. In fact, the Food, Agriculture, Conservation and Trade Act of 1990 (which we shall simply refer to as the 1990 farm bill) was tied to the Budget Reconciliation Act of 1990.

The cost of farm programs to the U.S. Treasury in 1990 ($6.5 billion) was approximately one-fourth of what it was in the mid-1980s. Stocks of grains were down to 30 percent of total use in 1990 as compared to almost 70 percent in the mid-1980s. Total agricultural exports were approximately $40 billion in 1990 as compared to only $26 billion in 1986. U.S. agricultural exports were a record $60.2 billion in calendar year 1996. High prices for grain and soybeans helped bulk exports reach $28 billion, the highest level since 1981. High-value product exports reached a record $32 billion.

The 1990 Farm Bill gave way to the FAIR Act in 1996. Unlike the 1990 Farm Bill which made up the difference between announced target prices and market prices with deficiency payments, the FAIR Act phased out subsidies in fixed annual rates over the 1996–2002 period (see discussion in Chapter 13). The strong economic performance recorded by farmers in 1996 made the transition from the 1990

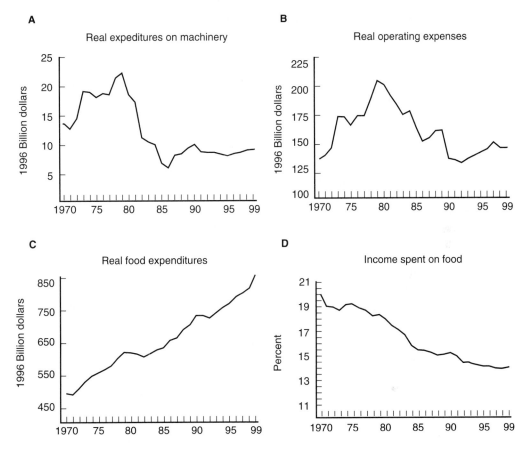

FIGURE 18.11   This figure depicts the trends in four variables affecting the entire food and fiber industry.

legislation to the FAIR Act in 1996 extremely smooth. The concept of phasing out government payments according to a fixed schedule has been severely tested in the late 1990s, as commodity prices fell markedly below previous target price levels and deficiency payments are not there to make up the difference.

Congress elected in 1998 and again in 1999 to address low farm commodity prices and weak yields in selected commodities by passing supplemental appropriations. Under the 1996 FAIR Act, farmers received $6 billion in flexibility contract payments in 1998 and $5.1 billion in 1999. The appropriations bill passed in October 1998 included an additional $5.8 billion in total assistance for farmers. Direct government payments to farmers in 1998 by the federal government totaled $12.2 billion, topped by another $22.7 billion in direct payments to farmers in 1999. By comparison, direct payments to farmers over the 1990–1997 period averaged $7 billion. Thus, as commodity prices declined, payments under the existing 1996 FAIR Act had to be dramatically increased by the federal government to avoid a substantial increase in financial stress challenging the level observed during the 1980s.

Policymakers are confronted with the issue of how best to modify existing safety net concerns as we approach 2002, the last year of the 1996 FAIR Act. The increased reliance on year-to-year appropriations to supplement existing program benefits increases the level of uncertainty already at high levels.

SUMMARY

The purpose of this chapter was to examine the manner in which macroeconomic policy actions affect agriculture, and the major problem confronting agriculture in the 1980s. The major points made in this chapter can be summarized as follows:

1. Expansionary monetary policies raise crop prices, lower livestock prices, raise farm input prices, lower farm interest rates, raise net farm income and raise farm land prices.

2. Contractionary monetary policies lower crop prices, raise livestock prices, lower farm input prices, raise farm interest rates, lower net farm income and lower farm land prices.

3. Expansionary fiscal policies would lower crop prices, raise livestock prices, raise farm input prices, raise farm interest rates, lower net farm income, and lower farm land prices. If these expansionary fiscal policies, however, involved increases in farm price and income support programs, both net farm income and farm land prices could rise rather than fall.

4. Contractionary fiscal policies would raise crop prices, lower livestock prices, lower farm input prices, lower farm interest rates, raise net farm income, and raise farm land prices. If these contractionary fiscal policies, however, involved cutbacks in farm price and income support programs, both net farm income and farm land prices could fall rather than rise.

5. The sector is indirectly affected by these policies as they affect the prices farmers receive for their products, the prices they pay for their inputs, and the interest rates they are charged on loans.

6. The 1970s represented a decade of expansion for agriculture. Farmers responded to negative real interest rates and an expanded export demand for crops by increasing their purchases of farm inputs, including farm machinery and motor vehicles.

7. The 1980s, on the other hand, represented a decade of contraction for agriculture. Farmers responded to high real interest rates and a declining export demand for crops by decreasing their purchase of farm inputs, including farm machinery and motor vehicles.

8. The 1990s have represented a period of improved farm balance sheets, policy changes, and increased uncertainty as the FAIR Act phases out subsidies for specific groups of producers.

DEFINITION OF KEY TERMS

Contractionary policies: monetary and fiscal policies that lead to reduced economic growth of the economy.

Economic growth: increase in the economy's real level of output (real gross domestic product).

Expansionary policies: monetary and fiscal policies that lead to greater economic growth of the economy.

Nominal interest rate: market rate of interest unadjusted for the current rate of inflation.

Real interest rate: nominal interest rate minus the rate of inflation.

Trade-weighted exchange rate: price of foreign currencies in terms of the U.S. dollar weighted by the relative importance of trade flows with our leading trading partners.

REFERENCES

Economic report of the president, Washington, D.C., various years, U.S. Government Printing Office.

U.S. Department of Agriculture: *Agricultural statistics,* Washington, D.C., 1982, U.S. Government Printing Office.

U.S. Department of Commerce: *Business conditions digest,* Washington, D.C., various years, U.S. Government Printing Office.

P A R T

VI

INTERNATIONAL AGRICULTURAL TRADE

19

AGRICULTURE AND
INTERNATIONAL TRADE

*If we take care of our imports, our exports will
take care of themselves.*

Anonymous

TOPICS OF DISCUSSION

This chapter examines recent trends in world agricultural trade, focusing on the growth and volatility of U.S. agricultural trade. The importance of trade to the farm and nonfarm sectors of the economy is emphasized. The changing product composition of world and U.S. agricultural trade is examined, and an overview of the changing direction of trade, the factors affecting market potential, and international competition is given. Agricultural trade performance is analyzed, and why trade performance is important to the economy is discussed.

GROWTH AND INSTABILITY IN AGRICULTURAL TRADE

U.S. agricultural exports experienced unprecedented growth during the 1970s, rising from $6.7 billion in 1970 to almost $44 billion in 1981 (Figure 19.1). Agricultural export volume more than doubled from 60 million tons to 160 million tons during this same period. This surge in farm exports occurred because of four global events that caused world food demand to increase. These events were: (1) an oil price

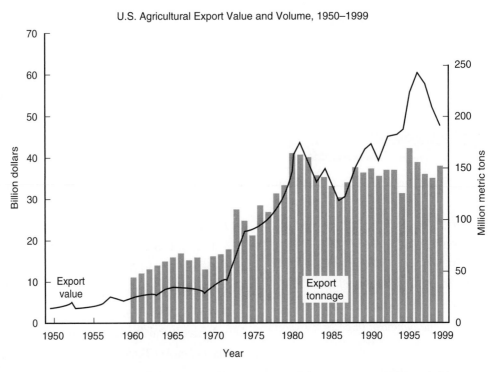

FIGURE 19.1 U.S. agricultural exports reached record levels in both value and volume during the 1970s. Steep declines in exports caused a financial crisis in U.S. agriculture during the mid-1980s. Exports have since recovered with sales reaching $40 billion, but are expected to remain volatile, creating additional uncertainty for farm and agribusiness managers and rural communities. (*Source:* U.S. Department of Agriculture, *1970*, and U.S. Department of Agriculture, *1999*.)

boom, (2) worldwide inflation, (3) easy credit, fueled by **petrodollars** and loaned to developing countries, and (4) a series of weather-related crop shortages. During the 1970s, U.S. agriculture expanded and became the world's largest exporter of agricultural products. The U.S. share of world agricultural trade increased from 10% in the 1950s to 17% in 1981. During this same period, world agricultural exports increased by 14%, and U.S. agricultural exports expanded by 22%.

Export Boom and Bust

The expansionary trend of the 1970s led to U.S. dominance in the world trade of major farm commodities. The United States became the world's largest exporter of soybeans, with 92% of world trade. In addition, the United States captured 53% of the world wheat market, 76% of the coarse grain market, and 40% of the world market for cotton. This expansion in trade resulted in higher farm prices and incomes but increased exposure to uncontrollable world market forces and greater market instability.

The export boom ended abruptly in the early 1980s, when the world economy slipped into recession. World agricultural trade growth stagnated near 3% and the U.S. share of that total trade declined to 9.8%. Sharply lower oil prices, declining purchasing power among developing countries, and greater protection of agricultural markets only added to the demise of agricultural trade worldwide. U.S. agriculture was especially hard hit by these events because of the combined forces of a strong dollar against foreign currencies, uncompetitive prices of major program crops, and the conversion of the European Union (EU) from a major net grain importer to a major net grain exporter.[1] This final event was accomplished mainly through the use of trade-distorting domestic agricultural policies and trade barriers designed to restrict EU imports of grains. The main results were reduced demand by the EU for food and fiber and increased competition for U.S. farm products. The U.S. share of world wheat trade fell below 40%, and trade shares for soybean meal and soybean oil dropped to 18% and 20%, respectively. Growth in U.S. agricultural trade fell to only 2% for all of the 1980s, compared with more than 20% during the previous decade. U.S. agricultural exports subsequently fell from the previous high in 1981 to only $26 billion in 1986.

Since the mid-1980s, farm exports rebounded to $60 billion in 1996. This turnaround in agricultural trade can be attributed to a **weaker dollar** and global economic recovery that led to stronger demand for U.S. agricultural exports. In addition, implementation of the Food Security Act of 1985 allowed more market-oriented farm policies and a more aggressive export marketing stance for U.S. agriculture. Together, these factors made U.S. agricultural exports more competitive on the world

---

[1]The European Union was established on November 1, 1993 to coordinate policy among the 15 members in the areas of economics, defense, and justice. The EU includes the 12 members of the European Economic Community (Belgium, Denmark, France, Germany, Greece, Ireland, Italy, Luxembourg, Netherlands, Portugal, Spain, and the United Kingdom) and, as of January 1, 1995, Austria, Finland, and Sweden were admitted to membership.

market and increased foreign demand. While exports fell during the late 1990s, trade is still a major force affecting the farm economy.

U.S. agricultural trade has experienced more instability during the past two decades. In contrast to the stable markets of the 1950s and 1960s, the 1970s and 1980s were a virtual rollercoaster for most U.S. farm and agribusiness industries. Figure 19.1 shows that the destabilizing effects of world market forces on U.S. agriculture began in the early 1970s and have continued to the present. Greater uncertainty for agriculture has also meant increasing instability for input supply businesses, marketing firms, and related sectors of the agricultural and nonagricultural economies. Rural communities have also shared in the boom or bust fortunes of U.S. agriculture.

Moves toward Trade Liberalization

Increased world market instability, declining commodity prices, and lower farm incomes led to record levels of government support to the farm sector during the mid-1980s. Other major players, such as the EU, also spent more to support agriculture. When distortions to agricultural trade continued to drive up farm program costs, world leaders called for major reform of the agricultural trading system. The result was the initiation of the eighth round of multilateral trade negotiations under the General Agreement on Tariffs and Trade (GATT). This negotiation was called the Uruguay Round because the talks began in Punta del Este, Uruguay, in September 1986. For the first time in the history of GATT, major reform of agricultural trade was attempted. All forms of trade distorting agricultural and trade policies were placed on the table for possible modification or elimination.

The Uruguay Round Agreements (URA), which involved 125 nations, were completed on December 15, 1993, ratified by the U.S. Congress, and signed into law on April 15, 1994. The URA was implemented on January 1, 1995, and concludes in the year 2001. The URA represents the most substantial reform of agricultural trade undertaken by GATT parties and an important first step to additional trade liberalization in subsequent rounds of negotiations. Major provisions of the URA include the following:

1. Create the World Trade Organization (WTO) to implement the URA provisions, provide a dispute settlement body, and provide a trade policy review mechanism.
2. Provide market access by converting quotas to tariff equivalents, then phase in a reduction of tariffs by an average of 36% over six years, with a minimum of 15% reduction for each tariff and minimum market access growing from 3% to 5% of domestic consumption.
3. Reduce export subsidies by 36% in expenditure and 21% in tonnage over six years.
4. Lower trade-distorting domestic agricultural support by 20% over six years (the United States and the European Union are already in compliance so no additional reductions in farm support programs are required by those countries).

5. Use accepted international standards for food safety and animal and plant health regulations, harmonize those standards in order to prevent their use as nontariff barriers to trade, and establish disease/pathogen free zones within geographic boundaries of countries affected by certain diseases and pests. Countries may maintain standards more stringent than the international standards if they are based upon sound scientific evidence.

The USDA has estimated that the URA will increase U.S. agricultural exports by $1.6 to $4.4 billion per year by the year 2000 (Office of Economics, Economic Research Service). Grains and animal products account for three-fourths of the expected increase. Export-related employment is projected to grow by 112,000 jobs and net farm income to increase by as much as $1.3 billion per year.

In addition to GATT, the United States has negotiated free trade agreements with three other countries, such as the U.S.–Israel agreement, which began in 1985. In 1989, the Canada–U.S. Free Trade Agreement (CUSTA) went into effect. CUSTA specifies the phased reduction and eventual elimination of all duties and quotas between the two countries during a 10-year period. However, because of the politically sensitive nature of agriculture, many important issues were not resolved or were excluded from CUSTA.

A new round of multilateral trade negotiations (MTNs) began in 2000 under the auspices of the World Trade Organization. This new MTN includes only agriculture and services. Under the Uruguay Round Agreements (URA) of GATT, agriculture was obligated to begin new negotiations in the year 2000 or revert to more restrictive trade measures agreed to before the end of the URA. While attempts were made in November 1999 to launch a comprehensive MTN in Seattle, Washington, involving all sectors, these efforts failed because of disagreement between the United States and the Cairns group in one camp and the European Union, Japan, and some developing countries in the other. In addition, many environmental groups staged disruptive demonstrations that precluded delegation members from attending some key sessions and further hindered progress at the meetings. The result of the "Battle in Seattle," as it became known in the popular press, was a failed attempt to open a new MTN under the WTO and a reversion to the previous agreement to negotiate only in agriculture and services.

The United States has negotiated another agreement with both Canada and Mexico, called the North American Free Trade Agreement (NAFTA). This agreement eliminates tariffs, converts quotas to tariff equivalents, and institutes an investment reform over a 15-year time frame. NAFTA allows special treatment for agriculture by providing a 10- to 15-year phase-in of tariff reductions for import-sensitive fruits and vegetables grown in the United States and Mexico. In return, Mexico would be allowed to gradually eliminate **import quotas** on corn and some other sensitive crops. The most difficult areas of negotiation appeared to be centered on environmental concerns, labor rights, animal health, plant health, and safety. NAFTA expanded U.S. agricultural exports to Mexico by more than $2.0 billion annually.

The United States is also pursuing further trade liberalization within the Western Hemisphere by negotiating trade and investment agreements with 33 countries under the Free Trade Area of the Americas.

The move toward regional free trade areas has resulted in the formation of the South American Common Market (MERCOSUR) among Argentina, Brazil, Paraguay, and Uruguay. The objective of MERCOSUR is to create a zone of free trade, free movement of labor, and capital mobility. The effects of free trade areas will be discussed in Chapter 23.

THE IMPORTANCE OF AGRICULTURAL TRADE

U.S. agriculture's increased dependence on foreign markets has made U.S. farmers and rural communities more sensitive to forces and events beyond U.S. borders. Weather in other countries; foreign, domestic, and trade policies; and political events in importing nations all have a direct and profound impact on the well-being of U.S. agriculture. Increased uncertainty associated with greater dependence on international trade means that U.S. producers and agribusinesses need to become better informed about global events affecting their operations. Improved understanding of global forces will be crucial for those managers trying to make better decisions about the impacts of U.S. macroeconomic policies, export competitor policies, or changing price trends on their businesses.

Increased Export Dependence

Crop agriculture has become **export** dependent. More than 30% of all U.S. agricultural output was exported in 1998, accounting for 25% of farm income. Table 19.1 shows that large shares of many agricultural commodities have been exported in recent years. Wheat, rice, cotton, almonds, and hops are examples of crops depending on the export market as a major source of demand. In 1997 the United States exported 42% of its wheat production and 34% of its almond production. Since 1987, more than one-half of the U.S. rice, cotton, and hops crops have been shipped to other countries.

Horticultural products and meats are less export dependent than row crops. Exports of apples, poultry, and beef represent much smaller shares of production than traditional row crops. However, the export dependence of these products is increasing. The share of beef production exported has increased from 1.0% in 1985 to 8.0% in 1997. Similar trends exist for poultry and apples. The results of increased export dependence mean that U.S. farmers are more closely linked to forces and events beyond U.S. borders.

Greater Dependence on Imports

During the 1980s, a growing share of domestic food consumption was met by greater **imports** (Table 19.2). For example, canned fish and shellfish imports grew from 20% of consumption in 1980 to 33% in 1997. Frozen citrus juice consumption

TABLE 19.1   U.S. Agricultural Exports as a Share of Production, 1985–1997

Commodity	1985	1990	1995	1997
		Percent Exported		
Wheat	42	53	49.5	41.9
Rice, milled	45	49	58.2	46.6
Corn	24	31	24.6	16.3
Grain sorghum	34	46	35.6	33.5
Soybeans	44	44	36.2	33.4
Sunflower seeds	66	40	11.6	7.4
Cotton, raw	45	63	44.3	39.9
Tobacco	37	40	15.1	27.0
Almonds	60	76	81.8	34.0
Peanuts	21	27	23.8	19.25
Dried beans	27	37	26.0	26.4
Hops	48	52	38.0	32.0
Apples, fresh	5	8	12.9	14.0
Poultry	3	6	14.5	17
Pork	1	1	2.3	6.0
Beef and veal	1	3	45.1	8.3

Source: U.S. Department of Agriculture, 1999.

TABLE 19.2   U.S. Food Imports as a Share of Domestic Consumption, 1980–97

Commodity	1980	1985	1990	1995	1997
			Percent Imported		
Red meat (excl. offals)	6.5	7.7	8.1	6.5	7.1
Beef	8.8	8.1	9.8	8.3	9.2
Pork	3.3	7.2	5.6	3.8	3.8
Lamb	9.5	9.4	10.3	18.4	24.9
Fish and shellfish	43.7	56.2	56.3	55.3	62.1
Fresh and frozen	55.7	68.3	65.8	66	74.3
Canned	20.1	33.9	36.0	30.8	33
Dairy products	1.7	2.0	1.9	1.9	1.9
Tropical oils (palm and coconut)	100.0	100.0	100.0	100	100
Bananas	100.0	99.9	99.8	100	100
Grapes	14.2	25.9	37.6	38.8	41.2
Frozen citrus juices	13.0	50.9	N/A	13.0	16.2
Fresh vegetables					
Artichokes	20.6	23.2	25.7	N/A	N/A
Vegetables for processing					
Broccoli	9.1	22.2	57.8	N/A	N/A
Cauliflower	7.8	23.8	46.6	N/A	N/A
Tomatoes	1.4	7.0	5.7	N/A	N/A
Coffee	99.9	99.9	99.9	100	100
Tea	100.0	100.0	100.0	100	100
Cocoa	100.0	100.0	100.0	100	100
Spices and herbs	86.6	89.9	88.3	96.2	86.1

Source: U.S. Department of Agriculture, August 1999.

increased from 13% to 50% during the 1980s. Imports of broccoli and cauliflower for processing expanded from very low levels to account for 53% and 47% of domestic consumption, respectively. Much of the increase in import supply represents direct competition for U.S. producers, effectively lowering domestic prices, and depressing grower returns.

In some cases, however, imports play a vitally important role in supplying the processing needs and capacity of U.S. food processing plants. While imported tomatoes and peas may compete with U.S. production during some parts of the year, without imports many U.S. food processors could be forced to reduce plant capacity, relocate processing facilities, or close the plant. These alternatives would limit the market for U.S. grown products.

Imported foods satisfy consumer demand for a diverse market bundle of goods. Imports represent a lower-cost food product. By purchasing imports at relatively lower prices, consumers are actually working to increase their disposable incomes. Imports also represent more variety of choice to the consumer, thereby increasing the competition for U.S. produced goods and effectively lowering the cost to the consumer.

Noncompetitive imports such as tropical oils, coffee, tea, and cocoa are major sources of essential food products. These imports represent additional product variety for consumers at a lower cost. In many cases, limiting imports would reduce or eliminate available supplies, resulting in substantially higher food costs and a lower standard of living.

Farm and Nonfarm Impacts of Exports

U.S. agricultural trade has important implications for both the farm and nonfarm sectors. U.S. agricultural exports generate employment, income, and purchasing power throughout the economy. Farm exports were valued at $57 billion in 1997, but these exports generated an additional $82.8 billion of economic output in the supporting activities required to produce them. Farm purchases of fuel, fertilizer, and other inputs to produce for the export market stimulated economic activity in manufacturing, wholesale and retail trade, transportation, and other services. When combined, these activities represent an additional $101.3 billion in economic activity for the farm and nonfarm sectors. Exports create jobs and additional business activity for truckers, railroads, shippers, food processors, marketing firms, and others involved in handling commodities and products for export.

USDA economists estimate that a minimum of $1.38 in additional business activity was generated for every dollar of agricultural exports. This multiplier reveals that value added exports generate an additional $1.61 per dollar of exports, compared with only $1.02 for bulk commodities. Of this $1.61, 26% of the additional business activity occurred in the farm sector, 30% in other services, and 14% in wholesale trade and transportation. In fact, $.74 from every dollar of additional business activity generated by exports occurred in nonfarm sectors of the economy.

Employment Attributed to U.S. Agricultural Exports, 1997 (thousands of jobs)

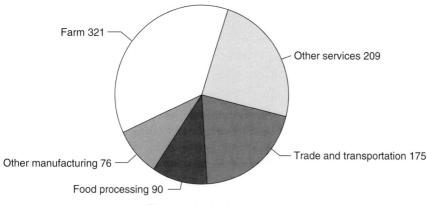

Farm 321

Other services 209

Other manufacturing 76

Food processing 90

Trade and transportation 175

Total employment 871

FIGURE 19.2   U.S. agricultural exports generate jobs in both the farm and nonfarm sectors of the economy. Nonfarm employment accounts for 60% of all jobs generated by farm exports. U.S. agricultural exports account for 769,000 jobs nationwide, employing people on the farm, in food processing, and in wholesale and retail trade. (*Source:* U.S. Department of Agriculture, October/November/December 1996.)

These figures stress not only the importance of exports to the nonfarm economy, but also the importance of value added exports in generating additional economic activity.

Agricultural exports generate additional jobs for the economy. Recent estimates indicate that 871,000 people are employed in jobs that directly or indirectly support agricultural exports. One new job is created for every $62,000 in agricultural exports. Figure 19.2 shows that most of the employment benefits from exports occur in the nonfarm sector. In 1997, 64% of the employment generated from agricultural exports benefited sectors of the economy associated with nonfarm activities such as food processing, wholesale and retail trade, transportation, other services, and other manufacturing. As with the economic activity multipliers discussed above, 562,000 jobs representing 63% of total agricultural export employment resulted from value added exports and related activities. More than 300,000 jobs, representing 11% of the total farm workforce, were involved in production of crops and livestock for export.

Indirect economic impacts of export activity are also important. With additional export earnings, farm households can purchase durable goods, equipment, building supplies, and other capital and consumer products; thus, additional purchasing power is spread throughout the total economy. These spinoff effects should always be considered when estimating economic impacts of agricultural trade. Indirect economic benefits to services, such as transportation and wholesale and retail trade, lead to increased investment, greater operating capacity, and heightened competitiveness for U.S. agriculture.

THE COMPOSITION OF AGRICULTURAL TRADE

The composition of agricultural exports and imports has changed dramatically over the last two decades. U.S. food and fiber exports have become more value added, and imports have become essential high quality food products. The role of both exports and imports is discussed in the following section. Important changes affecting the traded product mix are emphasized, and special attention is given to the growth and market potential for value added food trade.

The Role of Agricultural Exports

During the trade expansion of the 1970s, world agricultural trade doubled in volume, reaching 300 million tons. Much of this growth occurred in the trade of grains. Grain trade grew to become a key component of world agricultural trade, representing 76% of total world agricultural trade tonnage in 1980. The United States accounted for about 60% of world grain trade in 1982. By 1990 world grain trade had stagnated at less than 215 million tons, with annual growth less than 5%. The U.S. share of total world grain trade declined to 37%.

The fastest growth in world agricultural trade during the 1980s was the value added component composed of intermediate processed and consumer-ready food products.[2] Value added products trade has shown impressive gains in growth, increasing ninefold from $28 billion in the early 1960s to more than $150 billion in 1984 (Figure 19.3). The U.S. share of world trade in value added products peaked at 10.1% in 1981, then declined to a low of 8.4% in 1986. Since then, the United States has regained its competitive advantage, with its share of value added trade reaching 9.8% in 1990.

In 1970, wheat, corn, and soybeans accounted for 68% of the **value** of all U.S. agricultural exports. In **tonnage,** or based on the weight of commodities traded, bulk commodities represented more than three-fourths of total U.S. agricultural trade. Figure 19.4 shows that by 1999 this trend had reversed. Value added products represented 63% of total U.S. agricultural export value, and bulk commodities represented only 37% of the value. Bulk commodities are still important in tonnage terms, representing 80% of exports. However, because of relatively low unit prices and storage costs and no processing, bulk commodities have declined in importance when compared with value added products.

---

[2]Value added exports are products that have been processed to some degree or unprocessed goods that have high transport and storage costs and relatively high per unit values. Value added products consist of: (1) intermediate products including vegetable oils and meals, flour, live animals, and hides and skins that have been partially processed, but are not ready for final consumption; and (2) consumer-ready products that include meats, poultry, dairy products, dried, frozen, canned, or fresh fruits and vegetables, tree nuts, and beverages. Bulk exports include commodities that require no processing but have low per unit values, such as wheat, corn, and soybeans, and commodities that have higher per unit values but have high storage costs associated with value added products, such as cotton and tobacco.

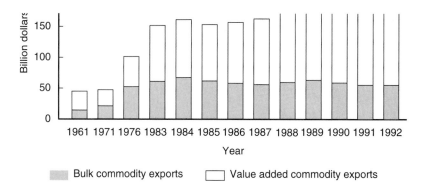

FIGURE 19.3 Value added export growth greatly exceeded trade growth in bulk commodities during the 1980s. Value added trade accounts for three-fourths of total world agricultural trade. The U.S. share of world trade in value added products has grown from 8.4% in the mid-1980s to 9.8% in 1990. (*Source:* United Nations Food and Agricultural Organization.)

The Role of Agricultural Imports

U.S. agricultural imports increased fourfold over the last two decades from $5.6 billion in 1970 to $38 billion in 1999. Agricultural imports have become an important source of low cost, high quality foods for consumers, and they are a major source of foreign competition for some U.S. producers. Import growth marked the emergence of a U.S. market for a large variety of high quality food products at reasonable prices, regardless of their source or national origin. Growth in imported food products also heightened consumer awareness of food safety issues and international environmental concerns, and focused attention on the important role other countries play in supplying critical U.S. food needs.

Imports may be classified under two broad categories: competitive and noncompetitive. Competitive imports are substitutes for domestic production and create competition for U.S. farmers, thereby lowering prices and potentially reducing grower returns. Imports that complement U.S. production are considered noncompetitive. These products do not compete with U.S. production, and they represent supplies of goods that would not otherwise be available. Imports represent goods that are available in greater variety and at lower cost to consumers than if produced domestically. Over the last three decades, competitive products have increased from 49% to 75% of total U.S. agricultural imports during the late 1980s. Since 1980, noncompetitive imports have declined about 2% annually, and competitive imports have increased 4.7% annually.

The United States imported $8.2 billion in noncompetitive products in 1996, representing 24% of total agricultural imports. Major noncompetitive imports and

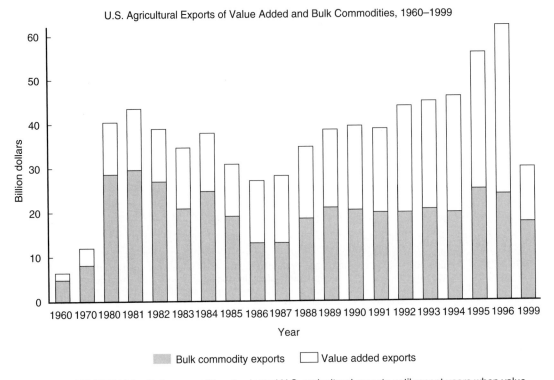

U.S. Agricultural Exports of Value Added and Bulk Commodities, 1960–1999

Bulk commodity exports   Value added exports

FIGURE 19.4   Bulk commodities dominated U.S. agricultural exports until recent years when value added sales began to increase. For 1994, U.S. value added exports represented 57% of all U.S. farm export sales. Value added exports generate an additional $1.61 for every dollar of export sales. (*Source:* U.S. Department of Agriculture, selected years.)

their values were coffee, $2.8 billion; cocoa and bananas, each $1.0 billion; rubber, $1.5 billion; and tea, spices, essential vegetable oils, and crude drugs, $1.1 billion.

In 1999, more than three-fourths of all U.S. agricultural imports were competitive products, valued at $28.0 billion. These competitive imports were live animals, mainly feeder cattle, $1.5 billion; dairy products, primarily cheese and casein, $1.3 billion; meats, beef and pork, $2.3 billion; fruits and preparations, $1.9 billion; vegetables and preparations, including sugar, $3.5 billion; and beverages, mainly wine and beer, $2.9 billion. Other major agriculturally related imports included fertilizers, $2.3 billion; fish, both fresh and shellfish, $6.7 billion; and farm machinery, mainly tractors and machinery, $2.9 billion.

Some competitive imports, such as dairy, were restricted by import quotas under provisions of U.S. law. **Section 22 of the Agricultural Adjustment Act of 1933** was designed to give the executive branch authority to limit imports of products that would interfere with the operation of domestic farm income support or price support programs. This authority has been used to restrict imports of cotton, peanuts, dairy products, and sugar. The sugar quota was modified to comply with GATT rules by allowing sugar imports duty free up to certain limits, after which a

16 cent per pound duty is required. Section 22 protection was modified under the Uruguay Round Agreements of GATT, negating its use as a trade barrier.

Meat imports have been limited by the **U.S. Meat Import Act of 1964.** This act was amended in 1979, allowing import quotas on fresh, chilled, or frozen beef, veal, mutton, and goat when imports were expected to exceed stated minimums. This quota is counter-cyclical so that more meat is imported when U.S. meat production is low and less meat is imported when U.S. meat production is high. Since 1980, the president has imposed quotas only rarely because meat imports were not projected to exceed minimum amounts. This law was modified and liberalized to allow more imports due to the Uruguay Round. In reality, **Voluntary Export Restraint (VER)** agreements, limiting exports by major meat suppliers, were normally negotiated before the imposition of the U.S. meat import quota. VERs constitute voluntary export limits that are followed by major meat suppliers. This protection has been modified to allow additional imports under the URA. In fact, Argentina, Brazil, and Uruguay have all become more important meat suppliers since the URA was implemented in 1995.

DIRECTION OF U.S. AGRICULTURAL TRADE

The direction of U.S. agricultural trade has shifted dramatically over the last two decades. Major markets for U.S. agricultural exports are changing. Figure 19.5 indicates developing nations have become larger markets, purchasing 51% of all U.S. agricultural exports in 1999. Several key factors have been instrumental in making developing countries more important. First, developed countries such as Canada, Australia, and Germany are mature economies that do not exhibit the robust growth rates of developing nations, hence increases in demand for imports are less. Second, developing countries such as Taiwan are growing more rapidly, and any increase in per capita incomes results in greater demand and, in many cases, increased imports. Third, developing nations are more likely to rapidly diversify diets, switching from plant protein to animal protein sources. This results in greater demand for feed grains and animal products, and in most cases this increased demand must be met by imports. In 1970, sales to developed countries in Western Europe, Japan, Australia, and Canada accounted for 69% of all U.S. agricultural exports and were valued at $4.4 billion. Exports to centrally planned economies such as the former Soviet Union and China represented only 3% of the total, while sales to developing countries in Asia, Latin America, and Africa accounted for 29% and were valued at $1.3 billion.

Emerging Export Market Trends

During the 1970s and 1980s, traditional trading patterns reversed. When developed countries became mature markets, their population growth stabilized, and even declined in some cases. More importantly, these markets became oversupplied with

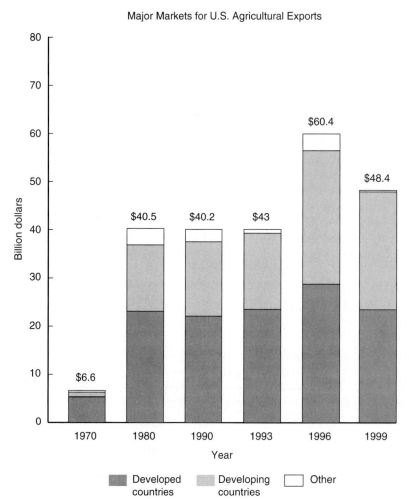

Major Markets for U.S. Agricultural Exports

FIGURE 19.5 Developing nations account for 46% of total U.S. agricultural export markets. Developing countries represent the fastest growing market for U.S. agricultural exports. Although important, developed country markets have exhibited only slight growth over the last decade. (*Source:* U.S. Department of Agriculture, FATUS.)

food products because basic food needs were met and increases in income led to little or no increase in consumption. In addition, subsidized agricultural production transformed the EU from a net importer of grains into a net exporter. This not only reduced the import demand for grains, but also increased the competition for U.S. grain exports in other countries.

The food situation in developing countries was different. Population growth was relatively high, diets were inadequate, and small increases in income led to large increases in food consumption and imports. People switched from basic staple grains to more animal protein, and consumption of both grains and meat rapidly increased. Figure 19.5 shows that developing countries emerged as a major growth market for U.S. food products, purchasing 51% of all U.S. agricultural exports in 1999.

Many developing countries face a declining capability to supply their own consumption needs in coarse grains, and **self-sufficiency** is expected to decline from 90% in 1985 to 60% by the year 2000. If this occurs, developing countries will emerge as a larger net importer of U.S. feed grains and meats. For trade expansion in developing countries to occur, however, will require continued economic development and broad based income growth, along with trade policies favorable to the expansion of food imports, while ensuring a reasonable degree of food security for the population.

Key Regional Markets

Trade patterns can also be viewed on a regional basis. Table 19.3 shows U.S. agricultural exports by region. Major shifts in the export of food and fiber products have occurred since 1970. All regional markets increased their imports of U.S. products, but Asia, Western Europe, Latin America, and Canada have become the dominant markets. The former Soviet Union (FSU) and China are important, but highly unstable because of economic and political uncertainty. Recent events in the

TABLE 19.3   U.S. Agricultural Exports by Region and Selected Countries, 1970 and 1999

	1970		1999	
Region	Value (Million $)	Percent of U.S. Exports	Value (Million $)	Percent of U.S. Exports
Western Europe	$2,369	35.2	6956	14.5%
European Economic Community*	2,087	31.1	6416	13.4%
Eastern Europe	133	2.0	171	.4%
Former Soviet Union	17	0.3	907	1.9%
Asia	2,452	36.5	20,163	42.0%
Japan	1,089	16.2	8919	18.6%
China	0	0	852	1.8%
Rep. of Korea	171	2.5	2453	5.1%
Hong Kong	55	.8	1209	2.5%
Singapore	15	.2	214	.5%
Taiwan	114	1.7	1955	4.1%
Other Asia	1,008	15	6769	14.0%
Canada	767	11.4	7073	15.0%
Africa	229	3.4	2093	4.4%
Latin America	649	9.7	10,078	21.0%
Mexico	137	2.0	5637	11.8%
Oceania	56	0.8	484	1.0%
Other	50	0.7	374	.8%
Total[†]	6,721	100.0	48,299	100.0%

*Adjusted to reflect purchases by EC-12 member countries in 1970, now referred to as the European Union (EU).
[†]Individual country percentages not included in total percentage figures.
Source: USDA, Foreign Agricultural Trade of the United States, 1999.

FSU and China suggest that this uncertainty will continue and even increase before normal trade growth is resumed.

The EU accounted for almost one-third of all U.S. agricultural export sales in 1970. By 1999, the EU represented only 15% of the market, even though exports had doubled. The implementation of the EU's Common Agricultural Policy (CAP) in 1957, with the stated goals to achieve food self-sufficiency and reduced dependency on imports, has reduced the EU's import demand for food; therefore, the EU is no longer a major market for U.S. farm exports.

The development of Asia as an export market is unprecedented in recent history. Devastated by World War II, Asia has emerged to rival and even surpass Europe as a growth market for U.S. food products. Japan is the largest single country market for U.S. agricultural exports, with purchases valued at $9.0 billion in 1999. This market now represents one-fifth of all U.S. agricultural exports, but it is expected to increase when trade is liberalized and incomes increase. Other important Asian markets are the Four Tigers: Hong Kong, Singapore, South Korea, and Taiwan. Together, these countries purchased more than $6.0 billion in U.S. agricultural products in 1999. Rising per capita incomes and a growing number of two-income families have created higher living standards and increased import demand for food. Consequently, although bulk products have represented more than one-half of all U.S. exports to these markets in the past, recent data suggest that the demand for consumer-ready food and fiber products is growing rapidly.

China also has become a major importer of U.S. agricultural products, purchasing $1.0 billion in 1999. Economic reform and liberalization of agricultural markets since 1987 have converted China into a stronger export market. With a population of more than 1.2 billion, China is expected to become the world's largest mass market. However, slow income growth in some sectors of the economy, uneven income distribution among the population, inconsistent import policies, the self-sufficiency policy in grains, and a limited infrastructure have slowed China's market potential.

Canada is the second largest importer of U.S. food and fiber products. The Canadian market expanded fivefold during the last two decades, reaching $7.0 billion in 1999. Most of this market growth is attributable to higher per capita incomes and increasing consumer demand for more diversity in food products. Although the United States and Canada entered a bilateral trade agreement in 1989, agricultural trade was not effectively liberalized, so little of the increased import demand can be attributed to a general lowering of agricultural trade barriers. Canada also is a major U.S. competitor in grain trade.

Africa represents a $2.0 billion market, with most of the growth in food imports coming from Egypt and Algeria. Together, these two countries account for 52% of all U.S. agricultural exports to Africa. Sub-Saharan Africa purchased $1.0 billion in U.S. agricultural products in the late 1990s, mostly concentrated in South Africa, former Ethiopia, and Nigeria. Market growth in this region continues to be limited by extreme poverty, disease, high infant mortality, poor transportation and storage infrastructure, political uncertainty, and civil strife. Average annual per capita incomes in the Sub-Saharan region were less than $490 in the 1990s. Many countries face declining income prospects in the near future. Until the most severe of these conditions are corrected, the Sub-Saharan will hold little potential as a major growth market for U.S. agriculture.

Latin America has emerged as a major customer for U.S. agricultural products. After effectively managing the external debt problems of the 1980s, declining incomes, and measures limiting food imports in the early 1980s, many Latin American nations are poised to become strong markets for U.S. exports. For example, until joining the GATT in 1986, Mexico was a closed economy with import duties at 100%, highly restrictive trade policies such as import licensing requirements, and stated goals of self-sufficiency in staple grain production. Since then, import duties have been lowered, licensing procedures have been liberalized, and attempts have been made to reform expensive farm programs and allow for imports of meats, feed grains and wheat, and other consumer-ready food products. Because of these measures and attempts to control inflation and stimulate employment, Mexico has become a $5.0 billion market for U.S. agricultural exports. Most of this trade growth has occurred since 1985.

Major Import Suppliers

Major suppliers of food and fiber products to the United States have changed significantly since 1970. Figure 19.6 indicates that developed countries are actually supplying a growing portion of food and fiber needs for the U.S. market. In 1970, developed countries exported $1.8 billion worth of agricultural products to the United States, with a market share of 33%. Developing countries supplied 66% of the U.S. market, with sales valued at $3.6 billion. Although imports expanded threefold to $17.3 billion in 1980, import supplier trends remained stable, with developed countries supplying 31% of the U.S. market. By 1999, however, the developed country share was 49% and the developing country share had declined to 51%.

Several major developed regions of the world have been responsible for supplying the growing import demand in the United States. Table 19.4 shows that the European Community, Canada, Australia, and New Zealand supplied 48% of the U.S. food import market in 1999, compared with only 28% in 1970. Their combined sales were $18 billion. Of total competitive imports shipped to the United States, developed countries supplied one-half.

More than 80% of all EU food and fiber exports to the United States are supplied by eight countries: the Netherlands, the United Kingdom, Belgium, Luxembourg, France, Germany, Spain, and Italy. Over 42% of the products shipped from these countries were beverages—mainly wine and beer. The EEC also supplies pork, dairy products, olives and other prepared vegetables, confectioneries, and tobacco.

Canada is a major supplier of pork, and Australia and New Zealand are major suppliers of beef and mutton. From 1970 to 1999 Canadian exports to the United States expanded to $8.0 billion. Much of this import growth occurred during the 1980s. Since the Canada/U.S. Trade Agreement went into effect in 1989, Canada has sold additional amounts of live cattle, wheat, and corn to the United States.

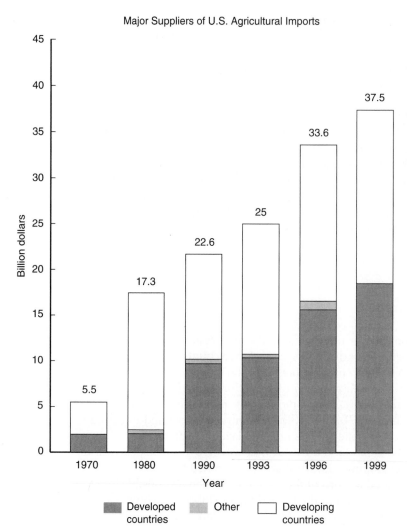

Major Suppliers of U.S. Agricultural Imports

FIGURE 19.6 Developed countries account for an increasing supply of imports into the U.S. market. Over the last decade, U.S. imports from developed countries doubled to $10 billion, accounting for 45% of all U.S. farm imports. Developing nations supply relatively less to the United States than in previous years, but represent an important and potentially large source of imported food products. (*Source:* U.S. Department of Agriculture, 1970 and 1999.)

Developing countries in Asia and Latin America are supplying an increasing amount of food and fiber to the U.S. market. Indonesia, the Philippines, Thailand, and Malaysia supplied 56% of all Asian exports to the United States in 1990. The most important products supplied to the United States were prepared and preserved vegetables, sold mainly in the ethnic Asian markets in major metropolitan areas. Three countries, Mexico, Brazil, and Colombia, supplied 60% of total Latin American exports to the United States. Mexico's exports to the United States have expanded sixfold since 1970, and it is now the number one supplier of feeder/stocker cattle, coffee, and fresh vegetables and the number two supplier of fresh fruit and orange juice to the United States. Brazil is the largest supplier of frozen concentrated orange juice and the number two supplier of coffee. Most de-

TABLE 19.4   U.S. Agricultural Imports by Region and Selected Countries, 1970 and 1999

	1970		1999	
Region	Value (Million $)	Percent of U.S. Imports	Value (Million $)	Percent of U.S. Imports
Western Europe	$864	15.8	8180	21.6
European Economic Community*	698	12.7	8002	21.1
Eastern Europe	83	1.5	227	.6
Former Soviet Union	0.4	—	63	.2
Asia	871	15.9	5845	15.4
Indonesia	129	2.4	1061	2.8
Philippines	293	5.3	481	1.3
Thailand	30	0.5	690	1.8
Malaysia	147	2.7	440	1.2
China	0	—	767	2.0
Canada	275	5.0	8,000	21.0
Africa	581	10.6	853	2.3
Latin America	2,214	40.4	12,395	33.0
Mexico	435	7.9	4883	13.0
Colombia	159	2.9	1190	3.1
Brazil	526	9.6	1482	3.9
Other	1,094	19.9	4840	12.8
Oceania	592	10.8	2301	6.0
Australia	395	7.2	1278	3.4
New Zealand	182	3.3	950	2.5
Other	0	0	0	—
Total[†]	5,480	100.0	37,865	100.0

*Adjusted to reflect sales by EEC-12 member countries in 1970, now referred to as the European Union (EU).
[†]Individual country percentages not included in total percentage figures.

veloping country suppliers of foods to the United States tend to specialize in those products that require hand labor to produce and harvest, thereby capitalizing on their natural advantage, low cost labor.

U.S. AGRICULTURAL TRADE PERFORMANCE

The importance of international trade to agriculture and other sectors of the economy has been well documented in previous sections of this chapter. What may not be apparent, though, is the importance of trade to the overall economy and recent trends in trade performance.

U.S. merchandise trade as a share of gross domestic product has been rising steadily since the mid-1960s. In 1999, trade accounted for 20% of U.S. gross domestic product (GDP). These figures mirror the importance trade had to the economy from 1880 to 1914, when trade represented 16% of GDP. In subsequent years, trade restrictions were used to limit imports, thereby protecting domestic jobs from

TABLE 19.5 U.S. Trade Balance, Agriculture and Total, 1950–1999

Year	U.S. Exports				U.S. Imports				U.S. Trade Balance		
	Ag.	Nonag.	Total	Ag-Share of Total %	Ag.	Nonag.	Total	Ag-Share of Total %	Ag.	Nonag.	Total
	Million Dollars				Million Dollars				Million Dollars		
1950	2,986	7,118	10,104	30	3,177	3,849	7,026	45	−191	3,269	3,078
1951	3,411	9,187	12,598	27	5,147	5,607	10,754	48	−1,736	3,580	1,844
1952	4,053	11,518	15,571	26	4,699	5,753	10,452	45	−646	5,765	5,119
1953	2,819	12,307	15,126	19	4,303	6,584	10,887	40	−1,484	5,723	4,239
1954	2,936	12,307	15,243	19	4,176	6,324	10,500	40	−1,240	5,983	4,743
1955	3,144	11,785	14,929	21	3,781	6,649	10,430	36	−637	5,136	4,499
1956	3,496	13,411	16,907	21	4,086	8,010	12,096	34	−590	5,401	4,811
1957	4,728	15,964	20,692	23	3,800	8,879	12,679	30	928	7,085	8,013
1958	4,003	14,744	18,747	21	3,929	8,858	12,787	31	74	5,886	5,960
1959	3,719	13,638	17,357	21	4,004	9,853	13,857	29	−285	3,785	3,500
1960	4,628	14,482	19,110	24	4,010	11,486	15,496	26	618	2,996	3,614
1961	4,946	15,561	20,507	24	3,645	10,516	14,161	26	1,301	5,045	6,346
1962	5,142	16,305	21,447	24	3,762	11,925	15,687	24	1,380	4,380	5,760
1963	5,078	16,560	21,638	23	3,907	12,491	16,398	24	1,171	4,069	5,240
1964	6,068	18,650	24,718	25	4,096	13,680	17,776	23	1,972	4,970	6,141
1965	6,097	20,234	26,331	23	3,986	15,745	19,731	20	2,111	4,489	6,600
1966	6,676	22,210	28,886	23	4,454	18,841	23,295	19	2,222	3,369	5,591
1967	6,771	24,048	30,819	22	4,453	21,974	26,427	17	2,318	2,074	4,392
1968	6,331	26,426	32,757	19	4,993	26,755	31,688	16	1,398	−329	1,069
1969	5,751	29,637	35,388	16	4,831	30,000	35,131	14	920	−663	257
1970	6,958	34,337	41,295	17	5,686	33,379	39,065	15	1,272	958	2,230
1971	7,955	35,928	43,883	18	6,128	38,744	44,872	14	1,827	−2,816	−989
1972	8,242	36,633	44,875	18	5,936	45,926	51,862	14	2,306	−9,293	−6,987
1973	14,984	47,759	62,743	24	7,737	57,521	65,258	12	7,247	−9,672	−2,515
1974	21,559	69,423	90,982	24	10,031	81,995	92,026	11	11,528	−12,572	−1,044

TABLE 19.5 U.S. Trade Balance, Agriculture and Total, 1950–1999 (continued)

Year	U.S. Exports Ag.	Nonag.	Total	Ag-Share of Total %	U.S. Imports Ag.	Nonag.	Total	Ag-Share of Total %	U.S. Trade Balance Ag.	Nonag.	Total
	Million Dollars				Million Dollars				Million Dollars		
1975	21,817	83,178	104,995	21	9,435	89,251	98,686	10	12,382	−6,073	−6,309
1976	22,742	89,047	111,789	20	10,492	103,743	114,235	9	12,250	−14,696	−2,446
1977	23,974	95,144	119,118	20	13,357	129,061	142,418	9	10,617	−33,917	−23,300
1978	27,289	104,270	131,559	21	13,886	152,095	165,981	8	13,403	−47,825	−34,422
1979	31,979	135,639	167,618	19	16,186	177,424	193,610	8	15,793	−41,785	−25,992
1980	40,481	169,753	210,234	19	17,276	219,305	236,581	7	23,205	−49,552	−26,347
1981	43,788	185,415	229,203	19	17,218	237,469	254,687	7	26,570	−52,054	−25,484
1982	39,097	176,308	215,405	18	15,485	233,349	248,834	6	23,612	−57,041	−33,429
1983	34,769	159,373	194,142	18	16,373	230,527	246,900	7	18,396	−71,154	−52,758
1984	38,027	170,014	208,041	18	18,916	297,736	316,652	6	19,111	−127,722	−108,611
1985	31,201	179,236	210,437	15	19,740	313,722	333,462	6	11,461	−134,486	−123,025
1986	26,312	179,291	205,603	13	20,884	342,846	363,730	6	5,428	−163,555	−158,127
1987	27,876	202,911	230,787	12	20,650	367,374	388,024	5	7,226	−164,463	−157,237
1988	37,080	270,934	308,014	12	20,955	416,185	482,140	5	16,125	−145,251	−129,126
1989	40,029	307,498	347,527	12	21,879	446,483	468,362	5	18,150	−138,985	−120,835
1990	39,598	335,635	375,193	11	22,404	468,071	490,975	5	16,654	−132,436	−115,782
1991	39,365	361,421	400,786	10	22,864	460,164	483,028	5	16,529	−98,771	−82,242
1992	43,132	389,137	432,269	10	24,790	500,470	525,260	5	18,351	−111,342	−92,991
1993	42,911	396,329	439,240	10	25,165	549,698	574,863	4	17,756	−153,379	−135,623
1994	46,244	436,402	482,646	10	27,074	630,242	657,286	4	19,183	−193,823	−174,640
1995	56,348	491,813	548,161	10	30,336	709,256	789,592	4	26,024	−217,455	−191,431
1996	60,445	524,207	584,652	10	33,655	761,634	795,289	4	26,802	−237,427	−210,637
1997	57,245	587,784	645,029	9	36,300	826,126	862,426	4	20,945	−242,424	−221,479
1998	51,829	584,626	636,455	8	37,073	868,637	905,710	4	14,756	−288,374	−273,616
1999	48,299	593,890	642,189	8	37,865	979,571	1,017,436	4	10,434	−385,681	−375,247

Source: U.S. Department of Agriculture.

473

foreign competition. Recent efforts to open the U.S. economy and increased U.S. dependence on trade make it critical to understand the performance of trade and what it means for agriculture.

The Balance of Trade

The **balance of trade** is commonly used to measure overall trade performance. The merchandise trade balance measures the difference between total revenues earned from exports and total expenditures on imports.[3] If exports exceed imports, the nation is said to have a **trade surplus.** If more is imported than exported, the nation has a **trade deficit.** A trade deficit may be portrayed as a bad thing, but the merchandise trade account is only one component of a nation's total balance of trade. Placing emphasis on monthly or even quarterly reports of merchandise trade statistics may mislead policymakers, resulting in misdirected macroeconomic policy, overcorrection, and futile attempts to reduce short-term imbalances in trade. This has become especially important since the early 1980s as international capital flows and trade in services have become more important components in the overall balance of payments. Balance of payments and its importance will be discussed in more detail later in this book.

During the 1950s and 1960s, the United States experienced moderate surpluses in the merchandise trade balance, averaging about $5 billion annually. Table 19.5 indicates this trend reversed during the 1970s, with the United States mounting large trade deficits every year except two. Chronic trade deficits continued into the 1980s, reaching $158 billion in 1986, then growing to $375 billion in 1999.

Several factors were responsible for the large U.S. trade deficits during this period. First, the United States underwent its longest sustained period of economic growth since World War II, stimulating strong demand for imported goods. Nearly two-thirds of the U.S. merchandise trade deficit was accounted for by oil/petroleum products and automobiles. Second, the U.S. dollar rose to record levels against other currencies, leading to less expensive imports and more expensive exports. Thus, the balance of trade declined. Third, a proliferation of nontariff barriers to trade such as import licenses, quotas, and export subsidies were used by key trading partners, such as Japan and the EU, to limit imports to protect domestic employment from foreign competition. Finally, because of a large domestic market and inward orientation, many U.S. firms lacked the necessary incentives and expertise to aggressively pursue international marketing opportunities.

U.S. agriculture has consistently generated a trade surplus every year since 1960 (Table 19.5). Although this surplus was relatively small during the 1950s and 1960s, it grew rapidly during the 1970s and 1980s, reaching a record $26.6 billion in 1981. Further, agriculture exhibited strong trade performance despite rapid growth in imports. Although agricultural exports as a share of total U.S. exports

---

[3]Merchandise trade is normally reported as the nation's trade balance, in contrast to trade in services, such as shipping, insurance, banking, and tourism, which are referred to as trade in invisibles.

have declined steadily from 30% in 1950 to 11% in 1990, agriculture has managed to remain one of the most competitive sectors in the U.S. economy. In 1999, the U.S. agricultural trade surplus of $10.0 billion made a major contribution to the U.S. trade balance.

The balance of trade has two important implications for the economy as a whole and for agriculture in particular. First, a trade surplus provides capital to a nation because the nation is selling more than it is buying. This capital surplus may be used to satisfy foreign debt or to increase both domestic and foreign investment, enhance consumption by increasing imports of foreign goods, or be saved for future consumption and investment. For agriculture, the trade surplus has been an important means of partially offsetting the large U.S. trade deficit and stimulating the import of key food and fiber products at a lower cost to consumers. Second, trade deficits, particularly if chronic, may reduce a nation's capital reserves, leading to policies to restrict or limit imports or necessitating the need to borrow foreign capital. The trade balance also has some important impacts on the exchange value of currencies, which is discussed in Chapter 20.

Remember that a trade deficit is not necessarily bad and that a trade surplus is not necessarily good. When trade is in surplus, a country is giving up more than it receives. A trade deficit indicates that a country is receiving more goods than it is giving up. Neither condition can be sustained indefinitely without serious undermining of the nation's economy. Each case of surplus or deficit must be analyzed based on the particular circumstances causing the underlying trade position. Neither chronic surpluses nor chronic deficits are in the best interests of any nation. Care should be taken not to overemphasize any short-term condition or report of a partial measure of trade or performance.

SUMMARY

This chapter reviewed recent trends in agricultural trade, focusing on the importance of those trends for U.S. farms and agribusinesses. The importance of trade to the overall economy and the changing product mix for U.S. agricultural trade were discussed. Changing markets for U.S. agricultural exports were highlighted, followed by a discussion of trade performance and its importance to agriculture. The major points made in this chapter may be summarized as follows:

1. U.S. agricultural exports experienced record growth during the 1970s, but declined during the 1980s. Increasingly unstable trade led to greater uncertainty for farm and agribusiness managers.

2. Distortions to agricultural trade worldwide have prompted the major trading nations to complete an eighth round of negotiations in the General Agreement on Tariffs and Trade (GATT). These negotiations will liberalize trade in agriculture by reducing some elements of government support that distort trade.

3. U.S. agriculture has become increasingly export dependent, and large shares of major crops are currently sold to other countries. Greater export dependence

means that farmers and agribusinesses now have to rely more on transnational factors and events for price determination. Effective decision making will require improved understanding of global forces and better management of international risks.

4. Imported food products are supplying a growing share of domestic food consumption. Food imports are either competitive or noncompetitive. Imports can be a source of competition for some producers. Consumers depend on food imports for variety, quality, and as a source of low cost products. The growth in imported foods has heightened consumer awareness of food safety issues and international environmental concerns.

5. Agricultural exports represent important employment opportunities and economic activity for both the farm and nonfarm sectors of the U.S. economy. Value added exports generate more jobs and have a larger economic impact than bulk commodity exports. Value added exports are becoming more important to the economy in terms of value, but bulk commodity exports are still important in terms of tonnage.

6. Major markets for U.S. agricultural exports are changing. Japan is the single largest country market for U.S. food and fiber exports. However, trends are changing and customers in the developing countries of Asia and Latin America are becoming more important as customers in Western Europe are becoming less important. Developing countries represent the important growth markets for the future.

7. Developed countries are supplying a growing share of food imports to the U.S. market. Competitive imports account for three-fourths of all U.S. agricultural imports. The growth rate of competitive imports far exceeds growth of noncompetitive imports.

8. U.S. agricultural trade has exhibited strong performance over the last 30 years. Although agricultural trade has declined as a share of total trade, agriculture has generated a trade surplus each year, partially offsetting the total trade deficit. The agricultural trade surplus is important to the nation because it provides additional funds for consumption, investment, or savings.

DEFINITION OF KEY TERMS

Balance of trade: the monetary value of a nation's merchandise exports minus the value of merchandise imports for a given period of time.

Export dependence: occurs when a relatively large share of a country's agricultural production is exported.

Exports: the quantity and value of goods or services sold and shipped to other countries.

Food security: physical and economic access to minimum food requirements.

Import dependence: occurs when a relatively large share of a country's domestic food or fiber consumption is accounted for by imported products.

Import quota: a quantitative restriction on the amount of a product that may enter a country during a given time period.

Imports: the quantity and value of goods or services purchased from other countries and shipped to the United States.

Import tariff (duty): a government tax imposed on goods or products entering a country.

Petrodollars: U.S. dollars remitted to oil-producing countries for payment of oil and petroleum-related purchases by other nonoil producing countries.

Section 22 of the Agricultural Adjustment Act of 1933: authorizes the president of the United States to limit by tariff and/or quota the import of products that would interfere with the operation of farm income or price support programs. Section 22 authority has been used to limit imports of dairy, cotton, peanuts, and sugar.

Self-sufficiency: the capability of a nation to meet domestic food consumption requirements from domestic production capacity without importing.

Strong dollar: an increase in the value of the U.S. dollar relative to other currencies so that the dollar buys more foreign currency than before.

Tonnage: metric tons equal to 2204.62 pounds per ton.

Trade deficit: the monetary amount by which a nation's merchandise imports exceed exports during a given time period, usually one year.

Trade surplus: the monetary amount by which a nation's merchandise exports exceed imports during a given time period, usually one year.

U.S. Meat Import Act of 1964: a law authorizing the imposition of quantitative poundage quotas on imports of fresh, chilled, and frozen beef, veal, mutton, and goat if imports are expected to exceed certain prespecified minimum quantities.

Value: million or billion U.S. dollars.

Voluntary Export Restraint (VER): a voluntary agreement, in lieu of more restrictive measures such as higher duties or lower quotas, between two or more countries to limit the export of a certain product or commodity to a prespecified national market.

Weak dollar: a decline in the value of the U.S. dollar relative to other currencies so that the dollar buys less foreign currency than before.

REFERENCES

Edmondson W, Robinson M: U.S. agricultural trade boosts overall economy, *Foreign Agricultural Trade of the United States,* September/October 1991.

Elliott M: A survey of America, *The Economist,* October 26, 1991.

Jurenas R: *U.S. agricultural import protection and GATT Negotiations,* CRS Issue Brief IB92029, Congressional Research Service, The Library of Congress, January 17, 1992.

Miller WJ: *Encyclopedia of international commerce,* Centerville, Md, 1985, Cornell Maritime Press.

St. Clair T, Miyasaka E: *Desk reference guide to U.S. agricultural trade,* United States Department of Agriculture, Foreign Agricultural Service, Agriculture Handbook No. 683, Revised August 1991.

U.S. Department of Agriculture: *Agricultural trade highlights,* Washington, D.C., January 1992, U.S. Government Printing Office.

U.S. Department of Agriculture: *Effects of the Uruguay round agreement on U.S. agricultural commodities,* Washington, D.C., 1994, U.S. Government Printing Office.

U.S. Department of Agriculture: *Foreign agricultural trade of the United States,* Washington, D.C., 1980-1994, U.S. Government Printing Office.

U.S. Department of Agriculture: *U.S. foreign agricultural trade statistical report,* Washington, D.C., 1970, U.S. Government Printing Office.

Wisner RN, Wang W: *World food trade and U.S. agriculture, 1960–1989,* ed 10, Ames, Iowa, 1990, Midwest Agribusiness Trade Research and Information Center.

EXERCISES

1. During the past 30 years, agriculture has generated a trade surplus that provides additional funds for investment, consumption, or savings. T F

2. Expanded U.S. involvement in world trade resulted in higher farm prices, greater market stability, and increased farm incomes. T F

3. The "Uruguay Round" of the GATT negotiations is the only round that attempts major reform of agricultural trade. T F

4. World trade in value added products has declined relative to world trade in bulk commodities. T F

5. Consumers depend on imported foods for quality, variety, and as a source of low-cost products. T F

6. Which group of countries most likely represents future markets for U.S. agricultural products?

   a. developing countries.

   b. developed countries.

   c. centrally planned countries.

   d. all of the above.

7. Competitive imports substitute for domestic production causing:

    a. higher prices.

    b. increased competition for U.S. farmers.

    c. increased grower returns.

    d. all of the above.

8. Section 22 of the Agricultural Adjustment Act of 1933 was used to restrict imports of:

    a. corn.

    b. peanuts.

    c. fresh vegetables.

    d. live animals (mainly feeder cattle).

9. The North American Free Trade Agreement will eliminate _____ , convert quotas to _____ , and institute _____ over a 10-year period.

10. U.S. agricultural exports generate _____ , _____ , and _____ throughout the economy.

11. Imports are classified under the two broad categories of _____ and _____ . Imports that complement U.S. production are considered _____ .

12. What is the difference between competitive and noncompetitive imports? Give two examples of each type.

13. Define "balance of trade" and discuss the two important implications for the economy as a whole and for agriculture in particular.

20

EXCHANGE RATES AND AGRICULTURAL TRADE

Exchange can . . . bring about coordination without coercion.

Milton Friedman

TOPICS OF DISCUSSION

International capital markets have grown to surpass world trade in goods and services. International capital flows not only exceed funds needed to finance trade, but have become largely independent of trade. The uncoupling of international capital movements and trade flows has major implications for industries that are dependent on exports, compete with foreign imports, and tend to be capital intensive. U.S. agriculture is one such industry. The interactions among U.S. monetary and fiscal policies, agricultural policy, and trade policies will become increasingly important for U.S. agricultural trade, and consequently agribusinesses and rural communities.

The purpose of this chapter is to provide an overview of the international monetary environment affecting U.S. agricultural trade. Exchange rates and the setting in which they function—the international monetary system—will be discussed. The balance of payments will be defined and its relationship to exchange rates examined. An overview of recent developments in the international monetary system will be provided, followed by a review of the basic factors that determine the value of currency on the world market. The final section will emphasize important exchange rate impacts on agriculture, the role of national economic policies, and key considerations for policy coordination.

EXCHANGE RATES AND THE FOREIGN EXCHANGE MARKET

In Chapter 19, we learned that many factors influence the level and direction of agricultural trade. The **exchange rate,** the number of units of foreign currency that can be exchanged for one unit of domestic currency, is one of the most important factors affecting the level and the destination of U.S. agricultural exports. Relative prices determine which goods are traded and where they are shipped. For the business person making the decision of where to buy and where to sell, the process is one of converting one currency to another at the prevailing rate of exchange and comparing the prices. The difference in relative prices determines the flow of goods and services and, therefore, the patterns of international trade.

Exchange Rates Defined

Trade is affected by exchange rates through the price linkage. Exchange rates are used to convert international market prices to domestic currency equivalents. If the exporter's currency declines in value, for example, the importer's cost of foreign exchange will decrease, thereby lowering the commodity price in the import market and increasing the quantity demanded. Consequently, world prices will rise, inducing exporters to increase sales on the world market. If the exporter's currency increases in value, the foreign currency cost to the importer will rise, resulting in higher prices and a decline in quantity demanded of imported goods.

Today, the currencies of most developed nations are fully convertible to one another at market determined rates. As such, the exchange rate may be thought of as the value, or price, of one currency in terms of another currency. For example, on May 16, 1995, the German mark (DM) was valued at $.69 and the British pound sterling (£) at $1.56. In this case, the exchange rate is equivalent to the number of U.S. cents, or dollars, one German mark or one British pound sterling will buy. The reciprocal of this relationship reflects the number of foreign currency units per one U.S. dollar. For instance, if DM 1 = $.69, then $1 = DM 1.44, which can be calculated as 1 ÷ $.69. The U.S. dollar, in effect, will buy DM 1.44. The relationship between the mark and the pound can also be expressed as the **cross rate,** which is the exchange rate between two currencies as calculated from the value of a third currency. For example, the cross rate between the mark and pound is calculated by multiplying the number of dollars per one pound by one mark per number of U.S. cents as:

$$(\$1.56/£) \times (DM/\$.69) = (DM\ 1.56/£69) = DM\ 2.26/£$$

The Foreign Exchange Market

Currency values are determined on the foreign exchange market where they are traded. In early 1989, global currency trading exceeded $430 billion each business day, far surpassing world trade in goods and services. In fact, international financial transactions account for about 90% of the foreign exchange market activity around the world. Major international centers and their daily currency transaction amounts include London ($187 billion), New York ($130 billion), and Tokyo ($115 billion).

In reality, there is no "Big Board" on which currencies are traded or any one foreign exchange market. There are, however, a large number of private and institutional users of foreign currency who have foreign exchange departments and are in continuous communication regarding currency values and quantities traded. The foreign exchange market resembles a network of telephone lines, cables, facsimile machines, and on-line computer services, linking all the major money centers of the world. Important participants in foreign exchange markets are: (1) traditional users that include importers, exporters, tourists, speculators, traders, individual firms, and investors who buy and sell currencies to settle international accounts, (2) commercial banks that buy and sell currencies to traditional users and that maintain correspondent relationships with banks in other countries, (3) foreign exchange brokers who act as market intermediaries between banks for the purpose of settling commercial accounts by buying or selling currency, and (4) central banks, which act as the lenders of last resort by intervening in foreign exchange markets to balance national foreign exchange earnings and expenditures.

The primary function of foreign exchange markets is the transfer of funds among countries and currencies. Much of the activity on foreign exchange markets is related to the exchange of assets such as stocks, bonds, or bank accounts. Currencies are traded to maximize the expected returns on interest rates or dividends in various countries. Many institutions and individuals now maintain their wealth in both domestic and foreign assets to reduce the risk of large swings in asset values.

The volume of foreign exchange transactions has increased in recent years. In fact, some estimates indicate that global financial transactions now exceed world trade value by 40 times. Changes in the world economy are now much more important to the United States than ever before. Much of the growth in global financial flows has been aided by the development of a fully integrated capital market in which financial assets can be transferred among countries almost instantaneously. Improvements in telecommunications, such as electronic funds transfer and facsimile machines, along with the existence of a 24-hour foreign exchange market, have facilitated the growth in financial transactions. Also related to the growth of global capital flows has been the development of the **Eurodollar** market, which refers to U.S. dollars on deposit outside the United States.

Since 1973, exchange rates of most major currencies have been allowed to fluctuate in response to changes in supply and demand conditions on foreign exchange markets. This type of market mechanism is referred to as a flexible exchange rate system, which is synonymous with a floating exchange rate system. The current flexible rate system is in contrast to the fixed exchange rate system, which was established in 1944 at Bretton Woods, New Hampshire, and existed until 1973. The Bretton Woods Agreement required government central banks to intervene in foreign exchange markets in order to maintain the relative values of their currencies at fixed, agreed upon rates.

Exchange rates alone give little or no indication about the relative strength of currencies or the economies they represent. For example, the fact that the German mark is trading at $0.69 and the Japanese yen at $0.0115 does not indicate that the mark is 80 times stronger than the yen or that the German economy is stronger than that of Japan. What is important, however, is the relative change in currency value over time. Exchange rates are highly sensitive, fluctuating in many cases on a daily or even an hourly basis. Currency variability is important because exchange rates can increase and decrease in value. For example, Figure 20.1 shows that from early 1996 to mid-1997, the U.S. dollar was stable and then increased when compared with currencies in 19 other major trading countries.

Although exchange rate indices are useful in understanding the relationship between the dollar and a group of other currencies, nothing is revealed about the movements of individual currencies relative to the dollar. To determine the actual exchange rate change between two currencies requires specific information about each currency. Figure 20.1 contains the exchange rates of currencies traded among banks in New York. For example, the Swiss franc (SF) was quoted at SF 1.48/$ on May 16, 1995. When this rate of exchange is compared to SF 1.59/$, (the 12-month low), it can be determined that the U.S. dollar actually declined in value by 7.4%. This decline in the value of the U.S. dollar is called a currency depreciation and is calculated as:

$$(SF\ 1.59 - SF\ 1.48) = SF\ .11 \div SF\ 1.48 = (.074 \times 100) = 7.4\%$$

Currency depreciation refers to a decrease in the foreign price of the domestic currency. For instance, the U.S. dollar would buy SF 1.59 on May 16, 1994, and

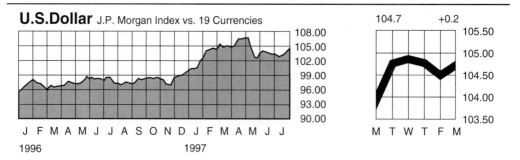

U.S.Dollar J.P. Morgan Index vs. 19 Currencies

CURRENCY	LATE NY	LATE FRI	DAY'S HIGH	DAY'S LOW	12-MO HIGH — LATE NY —	12-MO LOW
British pound (in U.S. dollars)	1.6787	1.6805	1.6813	1.6725	1.7125	1.5377
Canadian dollar (in U.S. dollars)	0.7260	0.7274	0.7277	0.7258	0.7520	0.7139
Swiss franc (per U.S. dollar)	1.4750	1.4725	1.4717	1.4790	1.1926	1.4870
Japanese yen (per U.S. dollar)	116.15	115.47	115.43	116.22	106.63	127.09
German mark (per U.S. dollar)	1.7964	1.7910	1.7867	1.7985	1.4725	1.7970

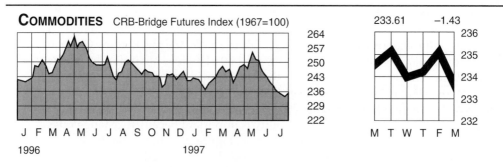

COMMODITIES CRB-Bridge Futures Index (1967=100)

COMMODITY	CLOSE	CHANGE	FRI	YR AGO	12-MO HIGH — AT CLOSE —	12-MO LOW
Gold (Comex spot), troy oz.	$325.50	$– 3.40	$328.90	$385.20	$389.80	$318.00
Oil (W. Tex. int. crude), bbl.	19.18	– 0.10	19.28	21.40	28.10	18.53
Wheat (#2 hard KC), bu.	3.58	+ 0.04	3.54	5.21	5.21	3.27
Steers (Omaha choice), 100 lb.	65.00	unch	65.00	64.00	75.50	61.75

NOTE: Monthly charts based on Friday close, except for Federal Funds, which are weekly average rates.

FIGURE 20.1 The exchange value of the U.S. dollar, as measured relative to 19 other currencies, depreciated in 1994 and 1995. When the dollar "appreciates" in value, the index rises; when the dollar "depreciates," the index falls. The change in the dollar's value over the previous week is indicated by the solid black line in the chart to the right. In this example, the exchange value of the dollar increased 2.7% from Thursday through the following Thursday, with a weekly index high of 85.5. (*Source: Wall Street Journal,* July 22, 1997.)

one year later one dollar would buy only SF 1.48. While the U.S. dollar depreciated in value, the Swiss franc increased in value.

Currency appreciation refers to an increase in the foreign price of the domestic currency. Suppose that one year ago the exchange rate between the U.S. dollar and Mexican peso (P) was P3.35/$ and that today the rate is P5.92/$. To determine the appreciation in the value of the dollar, calculate the percent change as

$$(P5.92 - P3.35) \div P3.35 = 76.7\%$$

The U.S. dollar buys 76.7% more Mexican pesos today than it did one year ago.

THE BALANCE OF PAYMENTS

The balance of payments position of a nation links trade in merchandise and services to international capital flows and exchange rates. It has been argued in recent years that macroeconomic policies are likely to have more influence on agricultural trade than commodity or farm policy (Schuh, 1990; Denbaly, 1988). Therefore, to correct for an imbalance in agricultural trade, the appropriate policy action might be to influence interest rates and exchange rates through macroeconomic policies rather than change farm or commodity policy. When the United States has experienced large trade deficits and a weak balance of payments position, popular policy prescriptions have attempted to address the specific commodity or sector in question, rather than dealing with the more important underlying cause of the problem—international capital markets, interest rates, and macroeconomic policies. Most of these efforts have had little success, supporting the arguments of Schuh and others.

The **balance of payments (BOP)** is an accounting record of all foreign transactions by U.S. private and public entities during a specific period of time, which is usually one year. Because of the growing importance of fully integrated international capital markets and exchange rates, international trade in goods cannot adjust independently of trade in international financial assets. Debit items, such as imports, are recorded as negative entries because they represent payments to other nations. Credit items, such as exports or foreign investment in the United States, are recorded as positive entries because they represent receipts from other nations. The balance of payments reflects the net international position of a nation relative to its trading partners and foreign investors. The main components of the balance of payments indicate trade in goods and services, U.S. purchases of foreign assets, and foreign purchases of U.S. assets.

The Current Account

The BOP is made up of three main accounts: the current account, the private capital account, and the official settlements account. The **current account** is composed of the merchandise trade account, the services trade account, and transfer payments. Merchandise trade is composed of exports and imports as discussed in Chapter 19. The services account reflects income from international capital investment, tourism, transportation, insurance, and other services. Most of the transactions in the services account are generated from investments in international real and financial assets. Interest income from a previously purchased foreign security is recorded as a service export. Conversely, interest income payments to foreign holders of U.S. securities are recorded as service imports. Transfer payments reflect private gifts and government expenditures, such as foreign aid.

To summarize, the current account is primarily composed of merchandise trade, services trade, and investment income. As shown in Table 20.1, merchandise trade has dominated the current account balance, representing three-fourths of all transactions in recent years. Therefore, important factors influencing the current account balance include differences in domestic and foreign income growth, prices, and exchange rates. The current account indicates the foreign contribution to domestic aggregate demand and, as such, receives much attention from policymakers and analysts.

The Private Capital Account

The **private capital account** summarizes all transactions in international real and financial assets and all international activities of U.S. private banking institutions. Government-to-government long-term loans are also included. The capital account mainly reflects the purchase and sale of foreign and U.S. assets. For example, U.S. **direct foreign investment (DFI),** which is the acquisition of at least 10% control of any foreign productive asset, such as farms, mines, factories, or facilities, represents an outflow of dollars from the United States and, therefore, is recorded as an import in the capital account. However, the actual income earned from DFI, such as a farm in Mexico, is recorded as a service export in the current account. Investments in foreign assets where control is limited (less than 10% ownership) are termed **portfolio investments** and represent dollar inflows into the United States and are, therefore, recorded as an export in the capital account. The capital account balance is most influenced by expectations regarding differentials in economic growth, interest rates, prices, and exchange rates.

Official Settlements Account

The **official settlements account** summarizes net changes in official holdings of international reserve assets, such as foreign currencies, gold, or special drawing rights of the International Monetary Fund. This figure was significant in 1989 when it reached $26 billion, but it declined to $2.2 billion in 1990.

The United States as a Debtor Nation

The current account and capital account balances are important for two primary reasons. First, the current account balance represents the contribution of total trade to aggregate demand. A current account surplus, for example, implies that trade

TABLE 20.1 International Transactions of the United States, 1980–1998

Year	Net Merchandise Trade	Net Services	Receipts on U.S. Assets Abroad	Payments on Foreign Assets Abroad	Net Investment Income	Balance on Goods, Services, and Income	Net Transfers	Balance on Current Account	U.S. Assets Abroad	Foreign Assets in the U.S.
1980	−25,500	7,915	72,606	−42,532	30,074	12,489	−8,349	2,317	−85,815	62,612
1981	−28,023	12,696	86,529	−53,626	32,903	17,576	−11,702	5,030	−113,054	86,232
1982	−36,485	12,217	91,690	−56,572	35,118	10,850	−17,139	−6,177	−127,825	96,578
1983	−67,102	9,897	90,050	−53,703	36,347	−20,858	−17,778	−39,198	−66,423	88,783
1984	−112,492	5,966	108,958	−73,977	34,981	−71,545	−20,661	−94,753	−40,515	117,973
1985	−122,173	4,685	98,736,	−73,156	25,580	−91,908	−22,762	−119,062	−44,946	146,452
1986	−145,081	10,475	97,274	−81,907	15,367	−119,239	−24,818	−149,236	−111,933	230,345
1987	−159,557	10,648	108,428	−94,273	14,155	−134,754	−24,047	−162,645	−79,540	249,016
1988	−126,959	17,823	137,000	−118,452	18,548	−90,588	−26,139	−123,046	−106,860	246,948
1989	−115,245	30,485	161,566	−141,842	19,724	−65,036	−27,116	−98,900	−175,662	225,307
1990	−109,030	36,690	172,078	−143,649	28,429	−43,911	−27,821	−79,332	−81,570	142,028
1991	−74,068	49,860	149,558	−125,608	23,950	−258	9,819	4,284	−64,732	111,332
1992	−96,106	60,528	132,523	−110,253	22,270	−13,308	−35,873	−50,629	−74,877	171,815
1993	−132,609	61,285	134,621	−111,445	23,176	−48,148	−38,522	−85,286	−201,014	283,230
1994	−166,192	65,227	165,968	−150,061	15,907	−85,058	−39,192	−121,680	−176,586	307,306
1995	−173,729	71,590	212,233	−192,823	19,410	−82,729	−35,437	−113,566	−330,675	467,552
1996	−191,270	82,245	224,619	−207,409	17,210	−91,815	−42,187	−129,295	−380,762	574,847
1997	−196,651	86,058	258,663	−255,432	3,231	−107,362	−41,966	−143,465	−465,296	751,661
1998	−246,932	78,336	258,324	−270,529	−12,205	−180,801	−44,075	−220,562	−292,818	502,637

Source: Created from *Economic report of the president*, February 2000.

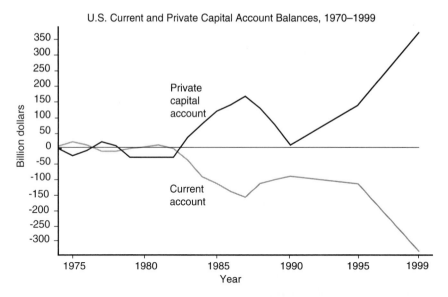

FIGURE 20.2   Large U.S. current account deficits began to appear in 1983, due mainly to a growing merchandise trade deficit. Inflows of foreign capital, measured by the private capital account, were used to pay for deficits in merchandise trade. (*Source: Economic report of the president*, February 1992.)

adds to aggregate demand because other nations are buying more U.S. output than U.S. residents are buying from foreign sources.

Second, as pointed out by Baker in 1990, the current account measures the extent to which a nation lives within its means. A current account deficit indicates that a nation is spending more on imports of goods and services than it is earning from exports of goods and services. For example, if the United States experiences chronic current account deficits, similar to that which occurred during most of the 1980s, then it must be liquidating its stock of foreign assets or increasing its borrowings from other nations. Figure 20.2 indicates that since 1983, the United States has been running increasingly large deficits on current account, financed by a growing inflow of foreign capital. However, this position may not be sustainable indefinitely because the U.S. stock of foreign capital is limited and there is probably some limit to how much foreigners are willing to lend to the United States.

As the United States has continued to sell more of its assets than it is accumulating abroad, much concern has been expressed about the United States becoming a net debtor nation. There is little doubt, however, that so long as U.S. macroeconomic policies result in high real interest rates, U.S. international debt will continue to rise and erosion of the current account balance will continue. In addition to economic factors, several other reasons exist for the expected accumulation of international debt. The United States is viewed by foreign investors as an excellent place to put capital. With a large, politically stable free market economy, the United States represents one of the most attractive investments in the Western Hemisphere. The high U.S. dollar of the early 1980s encouraged U.S. imports, yielding foreigners more funds to invest.

Many policymakers, the public, and some analysts are troubled by the decline of the United States as a net creditor to the rest of the world. The importance of the U.S. being a net debtor is that servicing foreign debt can be a drain on national income. Many have argued that the United States is leveraging future prosperity for present consumption. However, whether or not this is true depends crucially upon how the foreign funds invested in the United States are used. For example, if foreign capital is used to finance productive investment, then increases in future income should exceed expansion of debt. However, if foreign funds are used for consumption, especially frivolous consumption, then indeed, the United States may be trading away future prosperity for current consumption.

The net inflow of capital into the United States has had other important impacts that are often overlooked. As a result of large foreign borrowing by the United States, interest rates have been lower than they otherwise might have been. Foreign capital has provided credit for business expansion and consumer purchases, and has effectively maintained interest rates below where they would have been without this important source of funds. Conversely, it has been argued that these large inflows of capital have reduced available funds in other countries, slowing investment and economic growth in parts of the developing world.

Finally, under the current system of managed exchange rates, the reduction of official reserve assets only reflects the degree of government intervention required to influence currency values. The effects of currency depreciations (appreciations) are impossible to measure; therefore, the balance of payments position is no longer accurately measurable. Because of this, the United States suspended calculation of the deficit or surplus on BOP in 1976. In fact, the statement of U.S. international transactions no longer shows the net balance on the official reserve account. However, calculation and analysis of selected accounts, such as the current account and private capital account, still provide useful and meaningful information for policymakers.

Although large trade deficits and the transition of the United States from a creditor to a debtor nation have increased the calls for protection of domestic industries negatively impacted by import competition, they do offer several benefits, which should not be overlooked. First, resource reallocation has occurred, making the economy more efficient. Second, inflation has declined as less expensive foreign imports have become available to meet the needs of U.S. consumers. Third, more foreign competition will increase the productivity of U.S. businesses over the long run. Finally, the large U.S. trade deficit has stimulated economic growth worldwide over the last decade, allowing other nations access to one of the largest global markets, and in turn has stimulated their economic growth, allowing more imports from the United States.

THE INTERNATIONAL MONETARY SYSTEM

International trade and exchange rates are heavily influenced by the international monetary system, designed to allow for the maximum flow of international trade and investments and provide an equitable distribution of the gains from trade among nations. The main purpose of the international monetary system is to provide for the orderly and timely settlement of international financial obligations. Operations of the current system are controlled by rules and regulations, customs, and institutions agreed to and supported monetarily by the member nations. Understanding of the international monetary system is important because it sets rules and guidelines for exchange rate operations related to international agricultural trade.

The Gold Standard and the Interwar Years

From 1880 to 1914, world trade operated on the gold standard under which each nation's currency was defined relative to a predetermined value of gold. Exchange rates between nations were essentially fixed. For example, 1 oz. gold = $20.67 = £4.25 therefore, £1 = $4.87. The importance of gold as a medium of exchange should not be overemphasized, because gold had only a reserve currency role and most commercial transactions were conducted in paper currency. However, this period was characterized by economic prosperity and stability due mainly to free trade and income growth, so there is a tendency to give undue credit to the role that gold had during this period. The gold standard ended with the beginning of World War I in 1914.

The interwar years (1914–1939) were characterized by flexible exchange rates and a high degree of exchange rate instability, competitive currency devaluations in an attempt to export unemployment, and high tariff protection. This was one of the most chaotic periods in world monetary history. During this period, the United States passed the Smoot-Hawley Tariff Act (1930), which effectively raised duties of goods entering the United States to 59%, one of the highest levels in history. Other countries passed similar protective measures, leading to a precipitous decline in world trade. Many analysts attribute the onset and worsening of the Great Depression to the restrictive trade policies of this period.

The Bretton Woods System

Contemporary international monetary policy had its beginnings at Bretton Woods, New Hampshire, in July 1944, when the United States, the United Kingdom, and 42 other countries met to determine the structure of the new international monetary system, which would emerge from the chaos of World War II. The primary purpose of Bretton Woods was to devise a plan for postwar reconstruction and economic stability. Several important lessons had been learned from the time the gold standard ended in 1914 until the beginning of World War II. First, according to some analysts, when nations pursue their own interests without regard for the interests of other nations, all may suffer. Second, a stable monetary system, coupled with international financial liquidity, is critical for achieving economic growth and development worldwide. Finally, currency instability, excessive use of exchange controls, and policies that restricted international trade would undermine economic growth, the achievement of full employment, and the hopes for political stability following the war. Institutions, mechanisms, and policies established by the Bretton Woods System provide the basis for today's international monetary system.

The Bretton Woods System was essentially a return to a gold standard, but with some currency flexibility. From 1945 to 1971, exchange rates operated under a crawling peg system in which currencies were pegged to each other, but allowed to adjust within a limited range. The United States played a pivotal role under Bretton Woods. The exchange value of the dollar was set relative to gold at $35 per ounce. The United States agreed to exchange dollars for gold with other monetary authorities at the fixed price, but was not obligated to sell gold on the private market. For purposes of commercial trade and investment, foreign currency convertibility directly into dollars was limited. Other nations fixed the value of their currencies relative to the dollar and therefore implicitly to the price of gold. Because of this strong linkage to gold, the Bretton Woods System is sometimes referred to as the gold-exchange standard. Further, other countries agreed to intervene in foreign exchange markets to prevent the exchange rate from varying by more than 1% above or below its fixed price. In essence, currencies were fixed relative to each other and the dollar and were free to fluctuate only within a limited range.

The United States was firmly established as the world's central banker under the Bretton Woods System. Consequently, tremendous financial and political responsibility was placed on the United States. Most other currencies did not become fully convertible into dollars until the late 1950s and early 1960s. During this time, the U.S. dollar was the only **intervention currency,** creating, in effect, a dollar gold standard maintained by the United States. The role of intervention currency meant that the U.S. dollar was the only currency used by other nations' monetary authorities to keep currencies within their predefined ranges. The expanded role for the dollar resulted in many private international transactions being financed in dollars. The emergence of the Eurodollar market in the 1960s was a direct result of these events. New York became a major international financial center. Both of these events would have major consequences for global monetary stability by the early 1960s.

In the early 1960s, the United States began running a relatively large balance of payments deficits, which were financed by dollars. High capital outflows for investment in other countries, coupled with excessive increases in the money to fund the Vietnam War, acted to increase U.S. inflation and worsen the trade deficit. By 1970, foreign dollar holdings exceeded $40 billion, representing potential claims against U.S. gold reserves, which had declined by 50% to only $11 billion. **Currency devaluation,** or an action taken by a monetary authority to deliberately decrease the value of the dollar from a fixed level relative to other currencies, was not possible because of the reserve role of the dollar.

When U.S. balance of payments deficits persisted in the face of sharply lower gold reserves, it became apparent that in order for the United States to stimulate exports in an attempt to reduce its mounting trade deficit, currencies would have to be realigned. Attempts to convince West Germany and Japan to revalue their currencies failed because neither nation was willing to sacrifice its export competitiveness. Expectations that a U.S. dollar devaluation was imminent led to massive capital flight when foreign investors and residents alike sought to relocate their funds in markets with higher potential returns and more stability. On August 15, 1971, President Richard Nixon suspended convertibility of the dollar and imposed a 10% import duty. The Bretton Woods era had ended. Attempts to revive the system failed. The dollar was devalued to $38 per ounce of gold in 1971 and to $42 per ounce in 1973, but more and more nations severed their ties to the fixed exchange rate system. Ironically, however, the dollar remained an international currency, even increasing in importance after losing value relative to other currencies.

The Bretton Woods System offered several advantages and disadvantages. Exchange rate stability was established, in that currency values could be changed only after consultation with the International Monetary Fund and then only by a maximum of 10% of the initial fixed value. The United States began to accumulate dollar liabilities to other nations, while foreign countries acquired dollar assets. Dollar reserves became preferable to gold. For the United States, large balance of payments deficits could be financed with a reserve currency accepted worldwide. However, foreign individuals and institutions held dollar balances in interest bearing accounts, which had to be settled with dollars paid by the United States. The primary drawback of this system, however, was the inability of the U.S. government to change the value of the dollar. The government's ability to pursue domestic economic policy objectives was severely limited because the value of the dollar was, in effect, determined by foreign monetary authorities as a residual of their own currency values.

The Present International Monetary System

Since March 1973, world trade and investment have operated under a managed float exchange rate system. Under a **managed float,** or "dirty" float as it is sometimes called, the government intervenes periodically to change currency values. This is in contrast to a **floating exchange rate** system in which currency values are determined by market supply and demand conditions, with minimal government intervention. The managed float is especially important for U.S. agriculture because many countries manage exchange rates to give themselves a competitive advantage in the export of food and fiber products, thus eroding the U.S. competitive advantage on the international market.

Exchange Rate Arrangements.   Current exchange rate arrangements may be classified in three basic categories: pegged, managed float, and independent float. Figure 20.3 shows the major exchange rate arrangements and number of countries subscribing to each arrangement. In 1994 the International Monetary Fund reported that 75 currencies, mostly those of developing nations, were pegged (fixed)

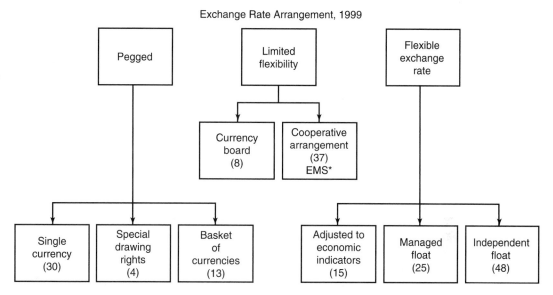

FIGURE 20.3 Only 48 countries, including the United States, Japan, Canada, and Australia, allow their currencies to float independently on the foreign exchange market. Most countries peg, or fix, their currencies relative to another currency, a basket of currencies, or the special drawing rights. (*Source:* International Monetary Fund, International Financial Statistics, August 1999.)
*EMS (Exchange Rate Mechanism II replaced EMS on January 1, 1999)

to one other major currency; 24 of these were pegged to the U.S. dollar. Therefore, when the U.S. dollar appreciates, each of these 24 currencies appreciates also. Others were pegged to the French franc (14), some other currency (4), the special drawing rights (6), or a composite basket of currencies (25). The currencies of 14 countries were defined to have limited flexibility. Four of these were managed relative to some other single currency. The 10 currencies of the European Monetary System (EMS) operated through a cooperative arrangement whereby each single country currency is pegged to each of the others, but allowed to fluctuate within a specified narrow range. Of the 182 members of the International Monetary Fund, only 48 allowed their currencies to operate under an independent float, with minimal government intervention. Among the most important were the United States, Japan, Canada, Australia, New Zealand, Brazil, Uruguay, and Venezuela.

International Monetary Institutions. Two important institutions were created as a result of Bretton Woods: the International Monetary Fund (IMF) and the International Bank for Reconstruction and Development (IBRD), also known as the World Bank. The IMF was created to maintain exchange rate stability and facilitate settlement of temporary balance of payments difficulties. Today IMF member nations finance temporary balance of payment deficits by borrowing from the international reserve fund, which was established in 1947. Each nation member was allocated a quota based on volume of trade, level of income, and international reserves. This quota was paid to the IMF, 25% in gold or dollars and 75% in domestic currency. Countries could then borrow against this stock of reserve currency. By 1967 demand for international reserves far exceeded available supply.

To aid with the shortage of international reserve currency, the IMF created **special drawing rights (SDR),** an artificial currency in the form of accounting entries to supplement reserve assets of the IMF. SDRs are not backed by gold or any other international reserve currency, but represent authorized reserve assets of the IMF. Further, SDRs cannot be used in private transactions, but are reserved for use only by central banks to settle official balance of payments deficits or surpluses. All funds borrowed from the fund must be repaid within three to five years to avoid tying up reserves for long periods. The IMF remains the cornerstone of the present international monetary system.

In contrast to the IMF, the World Bank was established to provide long-term development assistance. After post–World War II reconstruction was complete, the World Bank expanded its role to include 20 to 30 development projects in developing countries. In addition, the International Development Association (IDA) was created to finance projects in the poorest countries for the development of infrastructure, agriculture, and the improvement of literacy. Later, the International Finance Corporation was established to stimulate private investment in developing countries.

The European Monetary System

One of the most significant developments in recent monetary history was the creation of the European Monetary System (EMS). In March 1979, eight members of the EC—France, Italy, West Germany, Spain, Belgium, Denmark, the Netherlands, and Ireland—formed the EMS to promote exchange rate stability, economic growth, and employment. Proponents claim that the EMS has been effective in fostering low inflation, low interest rates, and high levels of investment among members, and has reduced the gap between economic growth of member nations.

Major features of the EMS included more complete monetary integration and the creation of a common currency among members. To accomplish these goals, the EMS provided for: (1) the creation of a common accounting unit, the European Currency Unit (ECU), (2) a range of fluctuation for the currency of each country, and (3) short-term balance of payments assistance to member states through the European Monetary Cooperation Fund (EMCF). During the European monetary crisis in 1992, the United Kingdom and Italy withdrew from the EMS.

The European Union and the European Monetary System

Six European nations—Belgium, France, West Germany, Italy, Luxembourg, and the Netherlands—established the European Economic Community (EEC) through the Treaty of Rome in 1957. Complete economic integration among member countries was one of the major objectives of the EEC, with plans for a Common Market for products, people, and monies. The Single European Act (1987–1992) established the criteria for formal and more complete integration of members and led to the creation of the European Union. The EU now has 15 members. Britain, Ireland, and Denmark joined in 1973, followed by Greece in 1981, Spain and Portugal in 1986, and most recently Austria, Finland, and Sweden (1995). EU enlargement negotiations were started on March 30, 1998. Negotiations are currently being held with 12 countries: Bulgaria, Cyprus, the Czech Republic, Estonia, Hungary, Latvia, Lithuania, Malta, Poland, Romania, Slovakia, and Slovenia.

The European Monetary System (EMS) was based on a system of fixed but flexible exchange rates, with each member country pegging its currency to every other member's currency. EMS currencies were allowed to fluctuate relative to the U.S. dollar by 2.25% until 1993 when large depreciations forced member governments to widen the band to 30%. This type of exchange rate mechanism (ERM) was called an *adjustable peg*, allowing central banks to intervene on foreign exchange markets to affect exchange rates directly, or governments could alter monetary or fiscal policies.

The Treaty of Maastricht (1992) created the basis for the European Monetary Union (EMU). Five convergence criteria were used to allow member countries to adopt the **euro:** government budget balance, public debt, inflation, interest rates, and exchange rate stability. The treaty was ratified by all EU members and all joined the EMU except Britain, Denmark, Greece, and Sweden. The European Central Bank was also created to implement and coordinate the EMU and its policies. The euro replaced the ECU on January 1, 1999, and euro coins and notes will replace all national currencies of the 11 EMU members on January 1, 2002. The euro, however, is presently being used by consumers, retailers, businesses, and public authorities in noncash forms. Some bank deposits, loans, travelers checks, and bonds are denominated in the ECU.

EXCHANGE RATE DETERMINATION

Under the current flexible exchange rate system, the foreign exchange value of currencies is determined by the interaction of the supply and demand for currencies. Market forces combine to determine the equilibrium exchange rate, which clears the foreign exchange market. In the short run, interest rates, fiscal and monetary policies, and expectations about important economic variables affect movement in exchange rates. Over the long run, however, price changes and the balance of trade are the key determinants of change in currency values. Although both short-run and long-run factors are clearly important, a comprehensive exchange rate theory has not yet been fully developed. Therefore, an explanation of each main factor is necessary to gain a complete understanding of the forces affecting currency values.

Demand and Supply of Foreign Currencies

To simplify the analysis of exchange rate determination, it will be assumed that only two currencies are relevant, the U.S. dollar and the German mark. In reality,

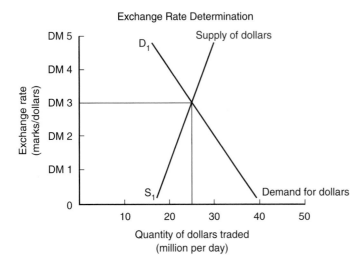

Exchange Rate Determination

FIGURE 20.4   Exchange rates are determined by the interaction of forces affecting the supply and demand for currencies on the foreign exchange market. Exchange rates influence how competitive industries such as agriculture are on the international market. If the exchange rate rises, the price of U.S. goods to German buyers increases.

there are more than 100 currencies traded worldwide, giving rise to numerous rates of exchange between any two.

Figure 20.4 illustrates the relationship between the dollar and the mark (DM) on the foreign exchange market. The foreign demand for U.S. dollars is negatively sloped, indicating that the higher the exchange rate, the lower the quantity of dollars demanded. The lower the exchange rate (the fewer number of marks required to buy dollars), the cheaper it is for Germany to import goods and services and to invest in the United States. When this occurs, the quantity of marks demanded by German residents will increase.

The supply of U.S. dollars is positively sloped, indicating that the higher the exchange rates, the greater the quantity of dollars supplied to the market. At higher rates of exchange, U.S. residents receive more marks for their dollars, making German goods, services, and investments cheaper and more attractive. The supply of marks to the United States therefore increases. The initial equilibrium conditions denoted by the supply and demand conditions in Figure 20.4 reflect a market exchange rate of DM 3 = $1 and $25 million traded each day. This relationship is similar to the commodity market equilibrium conditions explained in Chapter 10.

Relative Interest Rates

One of the major factors affecting the value of exchange rates in the short run is interest rate differentials between countries. Assume that U.S. interest rates increase relative to prevailing interest rates in Germany. This can be viewed as an increase in demand for U.S. assets. Foreign investors, speculators, businesses, and institutions would transfer funds to the United States to be placed in interest yielding securities and other financial assets. The net result of higher U.S. interest rates is an increase in demand for dollars.

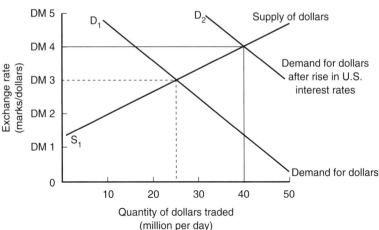

FIGURE 20.5   Any force or factor affecting the supply or demand for either the German mark or the U.S. dollar will influence the rate of exchange between the two currencies. For instance, higher U.S. interest rates increase the demand for dollars in Germany, leading to an upward shift in the demand curve to D_2. The price of the dollar in German marks, its exchange rate, increases from *DM 3/$* to *DM 4/$*.

The effects of rising U.S. interest rates on the exchange rate and quantity of dollars traded are shown in Figure 20.5. Rising U.S. interest rates cause the demand for U.S. dollars to increase from D_1 to D_2. The exchange rate increases from DM 3 to DM 4 and the quantity of dollars traded expands from $25 million to $40 million per day. Because the exchange value of the dollar has increased, the value of the German mark has depreciated. By calculating the inverse of the *DM/$* exchange rate, it can be determined that the mark depreciated from $.33/DM to $.25/DM as a result of higher U.S. interest rates. It is important to remember that an increase in German interest rates relative to interest rates in the United States will increase the demand for German assets, driving up the value of the mark and driving down the exchange value of the U.S. dollar. Interest rate declines have the opposite effects.

Aside from market forces, exchange rates can be influenced by government intervention for economic or political reasons. For example, if government authorities believe that the exchange rate is too high, thereby damaging the competitiveness of exports, foreign exchange can be sold to increase its supply and force the value of the currency to decline. Similarly, if government thinks the exchange rate is too low, undermining the ability to import foreign goods and threatening to cause inflation, it can intervene in the market by purchasing foreign exchange and causing the exchange rate to appreciate. However, government actions usually have only short-term impacts on currency values, with market forces dominating over the longer term.

Governments have several options for obtaining the currency required for market intervention. One option is to draw upon the stocks of foreign currency held strictly for intervention purposes, called international reserves. In addition, domestic currency can be printed and used to purchase foreign exchange. Finally, government bonds can be sold to raise the necessary cash with which to purchase needed amounts of foreign currency. Which option(s) is used will depend upon the overall financial strength of the country and its underlying policy objectives.

Monetary and Fiscal Policies.   Monetary and fiscal policies affect interest rates and therefore have major implications for subsequent impacts on exchange rates. For example, contractionary monetary policy, other things constant, raises interest rates, thereby causing the exchange rate to appreciate. Monetary contraction can be accomplished either directly by raising the discount rate, or indirectly by decreasing the money supply in the economy. Expansionary monetary policy has the opposite effects of lowering the interest rate and causing the exchange rate to depreciate. It is important to note that even expectations of changes in the discount rate or money supply can have the same effects on changes in the exchange rate. Fiscal contraction, all else constant, decreases aggregate demand in the economy, leading to reduced demand for money and lower interest rates. The exchange rate subsequently declines in value relative to other currencies. Fiscal expansion has the opposite effects of increasing the demand for money and raising interest rates, resulting in exchange rate appreciation.

Changes in Relative Prices

Differing rates of inflation between countries are a major cause of exchange rate changes. In fact, any factor that changes the general level of price within a nation can have major effects on the exchange rate. For example, if the level of inflation in the United States rises relative to inflation in other countries, U.S. goods and services would become more expensive. In response to higher U.S. prices, other countries would export more to the United States. As U.S. foreign purchases increase, the supply of dollars on the world market would expand. Other countries would also import less from the United States, thereby reducing the demand for dollars. When the supply of dollars increased and the demand for dollars declined, the value of the dollar, or its exchange rate, would fall.

Inflation rates can also affect exchange rates through monetary policy. For example, expansionary monetary policy would increase the general level of prices in the economy. When inflation increases, the exchange rate would depreciate. Contractionary monetary policy would have the opposite effect, such as currency appreciation due to lower inflation. Therefore, monetary policy affects exchange rates through mutually reinforcing interest rate and inflation differentials.

Balance of Trade Impacts

The balance of trade is one of the most important long-run factors affecting the level of exchange rates. However, since moving to a flexible exchange rate system in the early 1970s, it has become less clear whether exchange rates cause changes

Effects of U.S. Trade Deficits on Exchange Rates

FIGURE 20.6   When U.S. imports exceed U.S. exports, U.S. dollars in the foreign exchange market exceed other foreign currency. As a result, the supply of dollars increases to S_2, causing the exchange value of the dollar to decline from *DM 3/\$* to *DM 2/\$*. At the new exchange rate, exports are less expensive to foreign buyers and foreign imports are more expensive to U.S. consumers. Consequently, U.S. imports will fall and U.S. exports will increase.

in the balance of trade, respond to the trade balance, or both. The consensus is that large, chronic trade deficits and surpluses *do* have a major impact on the value of currencies, but there appears to be a significant lag time for exchange rate changes to occur.

Based on the balance of trade approach to exchange rate determination, the exchange rate balances the value of a country's exports and imports. If a nation experiences a trade deficit, the value of the exchange rate in domestic currency will depreciate. Exports will become cheaper in other countries, while imports will become more expensive domestically. The net result is that exports rise and imports decline to balance trade. Figure 20.6 illustrates that a U.S. trade deficit with Germany would increase dollars available on the world market, shifting the supply curve for dollars to the right to S_2. The supply of dollars increases because as goods are imported from Germany, they are paid for in U.S. currency, thereby raising the number of dollars available in foreign exchange markets. As a result, the exchange value of the dollar depreciates relative to the German mark. Following this analysis, we may conclude that when world demand for U.S. goods and services increases, the exchange value of the dollar would appreciate. Conversely, a decrease in demand for U.S. goods and services leads to dollar depreciation.

The rate of adjustment in the trade balance depends on how responsive, or elastic, imports and exports are to exchange rate changes. The more elastic are import demand and export supply, the faster the rate of adjustment in the balance of trade. Although important, the balance of trade approach to exchange rate determination fails to explain how the U.S. dollar appreciated during the early 1980s when the U.S. trade deficit reached record levels. Neither is it useful in explaining why the U.S. trade deficit did not fall when the dollar depreciated in the mid-1980s. It is obvious that other, more important short-run determinants of exchange rates, such as interest rates, were offsetting the effects of the trade balance during these periods.

The Role of Expectations

With the advent of information technology such as cable and facsimile, the instantaneous electronic transfer of capital, and almost 24-hour money center operation, private and public decision makers are constantly aware of global events affecting their returns on assets invested in other countries. Consequently, the role that expectations play in determining exchange rates has taken on added importance in recent years. For example, if Japanese investors learn that U.S. monetary policy has tightened, their expectations would be for higher U.S. interest rates and lower inflation, both of which reinforce an appreciation of the U.S. dollar in foreign exchange markets. Their reaction may be to convert large amounts of yen to dollars, with the *expectation* that the dollar will appreciate. By purchasing dollars, the Japanese actually increase the demand for dollars, thereby causing the exchange value of the dollar to appreciate even before the monetary policy is implemented.

Market psychology also is important in determining relative currency values. For example, if the dollar has been appreciating in value relative to the currencies of other major developed nations, investor speculation about continued appreciation can have an important impact. If fundamental factors, such as a large U.S. trade deficit, suggest that the dollar should be declining, then investors may refrain from investing in dollar assets until after the dollar falls. Inaction on the part of market participants may lead to a shortage of dollars on foreign exchange markets, causing the value of the dollar to actually increase when it was expected to decline. Conversely, if investor confidence is eroded by market factors or political action, a dollar selloff could follow, resulting in even further depreciation of the dollar relative to other currencies.

EXCHANGE RATES AND U.S. AGRICULTURAL TRADE

Exchange rates are key in determining the international prices for agricultural products, thereby determining how much other countries purchase. Although two-thirds of total world trade is conducted in U.S. dollars, every international sale involves two basic transactions: (1) an exchange of currency and (2) an exchange of goods or services. Therefore, even if U.S. wheat is sold to Mexico and the sale is in U.S. dollars, someone, usually the buyer, must convert Mexican pesos to U.S. dollars to make the purchase.

Exchange rates are especially important to agriculture in the United States, where exports account for a large share of agricultural production. In addition, exchange rates have become crucial variables in transmitting macroeconomic policies to the trade sector and in influencing the final outcome of U.S. farm policies.

The exchange rate is one of the most important prices in the economy due to its influence on exports and imports. When the dollar increases in value relative to other currencies, the price of U.S. goods valued in those currencies rises. It takes

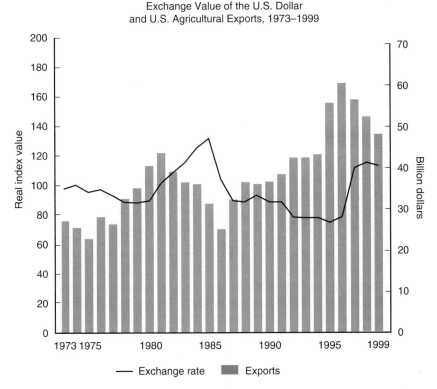

FIGURE 20.7 The real value of the U.S. dollar and agricultural exports have a strong negative correlation. When the dollar rises, exports fall, and when the dollar declines, exports rise. U.S. agricultural exports increased more than tenfold in real terms during the decade of the 1970s. The exchange value of the U.S. dollar fell by 30%. When the dollar declined after 1985, agricultural exports increased.

more foreign currency to buy the same quantity of goods, even if U.S. commodity prices have remained constant. When the dollar declines in value, the prices of U.S. goods decline in terms of foreign currency.

As shown in Figure 20.7, changes in the exchange value of the U.S. dollar have been very closely correlated with changes in U.S. agricultural exports over the last two decades. The exchange value of the dollar relative to other currencies has important effects on how competitive U.S. agriculture actually is at any particular time. For example, the real value of the U.S. dollar decreased by one-third from 1970 to 1979, and U.S. agricultural exports increased in real terms more than eightfold. From 1981 through 1985, the dollar increased by 42% when agricultural exports fell by 31%. Recent USDA estimates indicate that more than 25% of the increase in U.S. agricultural exports since 1985 can be accounted for by changes in the exchange rate. During this period, the dollar declined in value by 21%, and exports increased 15%. The exchange rate has become one of the critical links between U.S. agriculture and international trade.

Exchange Rate Indices

The exchange rate index referred to in Figure 20.7 is a real trade-weighted exchange rate index. **Real exchange rate** indices show the change in the value of one currency relative to other currencies, accounting for different inflation rates between countries. This type of index is useful to assess the effects of currency changes on trade flows. A trade-weighted index is necessary because a currency may appreciate against some currencies and depreciate against others during the same time period. Further, average indices may not be reflective of trade with any particular group of countries. For example, to account for agricultural trade patterns, a realistic index would need to take into account the relative importance of Japan, the EU, and developing countries as trading partners. The index used above is a **bilateral trade-weighted exchange rate index** and is determined by using U.S. agricultural trade volumes with major trading partners to calculate weights, then multiplying those weights by the real exchange rate of each country. This is sometimes referred to as an effective exchange rate index.

Exchange Rate Impacts on Prices

The exchange rate link between domestic market prices and international prices has become especially important. For example, the hypothetical case shown in Table 20.2 illustrates how exchange rate changes can affect the price of domestic goods on the international market. The initial situation is as follows: the U.S. price of wheat is $100 per ton and the exchange rate between the U.S. dollar and the Japanese yen is ¥(yen)120 = $1.00. The international price of U.S. wheat is then ¥12,000, calculated as ¥120 × $100/ton.

Case 1.   If the United States experiences a shortfall in wheat production, for example, the domestic wheat price could rise to $110 per ton. If the exchange rate remains the same at ¥120/$, then the international price of U.S. wheat becomes

TABLE 20.2   Example of Exchange Rate Effects on Prices

Case	U.S. Wheat Price $/ton	Exchange Rate ¥/$[2]*	International Wheat Price ¥/ton
Initial situation	$100	¥120 = $1	¥12,000
Case 1 U.S. price increase	$110	¥120 = $1	¥13,200
Case 2 Exchange rate increase	$100	¥132 = $1	¥13,200

*¥ measured as number of Japanese yen per one U.S. dollar

¥13,200. U.S. wheat has become more expensive to Japanese buyers due solely to a reduction in supply and the subsequent rise in prices.

Case 2.   Now assume that the U.S. wheat price is again $100 per ton, but the exchange rate increases to ¥132 = $1. The cost to the Japanese buyer increases to ¥13,200 due solely to the increase in the value of the dollar relative to the yen. In this case, the U.S. dollar has appreciated relative to the yen, and therefore buys more yen. Conversely, the yen has depreciated in value relative to the dollar and will buy only $0.0076 at ¥132/$ compared to $.0083 at ¥120/$. The dollar per yen exchange rate is found by taking the inverse of 132 or 1/132 = 0.0076. When yen are converted to dollars by Japanese buyers, each yen is able to purchase fewer dollars than before. As a result, the cost of U.S. wheat in Japan has risen due to the increased value of the dollar, or conversely, the lower value of the yen. Exchange rate changes, therefore, have had the same ultimate effect on the international price of U.S. wheat as an increase in domestic wheat prices.

CONSIDERATIONS FOR POLICY COORDINATION

Macroeconomic and agricultural policies combine to influence the performance of agricultural trade. Monetary and fiscal policies represent the major elements of macroeconomic policy, which affects interest rates, inflation rates, and the exchange value of the dollar. Macroeconomic policies also influence the overall level of economic activity and incentives to import and export. U.S. agriculture, the broader U.S. economy, and the rest of the world are linked through the macroeconomic policies of the United States.

Macroeconomic Policy Coordination

Monetary policy influences the supply of money in the economy by changing the amount of money or credit available to the banking system. Expansionary monetary policy lowers the prevailing interest rate, causing the exchange rate to depreciate. Consequently, net exports increase due to the lower exchange rate, while restrictive monetary policy, a smaller money supply, tends to strengthen the value of the dollar by raising interest rates. As a result, exports decline and imports rise.

Fiscal policy adjusts government spending through government expenditure and tax regulations. Expansionary fiscal policy increases interest rates and exchange rates and, therefore, lowers exports. Contractionary fiscal policy has the opposite effects.

Although the short-run, direct effects of monetary and fiscal policies are fairly clear, their interaction is less clear but critical to agricultural exports. Expansionary fiscal policy puts upward pressure on the exchange value of the dollar because real interest rates rise. If, however, the money supply is expanded by purchasing bonds, the demand for money also increases, putting upward pressure on inflation and downward pressure on the value of the dollar. Although interest rates rise, the appreciation of the dollar would be partially offset. This policy mix would benefit agriculture in the short run because it would hold down interest rates and the exchange value of the dollar while stimulating exports.

Restrictive monetary policy accompanied by expansionary fiscal policy would cause real interest rates and the value of the dollar to rise. The failure to coordinate these two key elements of macroeconomic policy almost certainly contributed to the downturn in U.S. agricultural exports in the early 1980s. Fears of rampant inflation prompted the Federal Reserve Board to adopt a restrictive policy of money supply growth in the early 1980s. As a result, inflation was reduced from the double-digit levels of the late 1970s to less than 4% in 1983.

Expansionary fiscal policy by the federal government pushed the budget deficit over $100 billion in the early 1980s, and it exceeded $200 billion by 1983. Credit markets were squeezed by federal borrowing, causing interest rates to increase dramatically. The combination of high interest rates and low inflation resulted in a record high level of real (inflation adjusted) interest rates. When foreign investors responded to record rates of return on dollar valued assets, the value of the U.S. dollar, measured in its exchange rate relative to other currencies, increased 40% in real terms. The 40% dollar appreciation almost certainly contributed to a 32% drop in farm exports between 1981 and 1983.

To summarize, the soaring U.S. federal budget deficit has been financed by borrowing liquid funds from all over the world. Insofar as the trade deficit is a loan from the sellers of goods and services to buyers of goods and services, the federal deficit is, in effect, a fundamental cause of the U.S. trade imbalance. To avoid policy conflict in the future will require careful coordination of both monetary and fiscal policies.

Domestic Agricultural Policy Coordination

The 1996 farm bill resulted in a more flexible policy for U.S. farm products and should enhance agriculture's competitiveness. U.S. agricultural policy played a major role in eroding the competitive position of farm exports in the early 1980s, however. Under the 1981 Farm Bill, loan rates and target prices were set to escalate with inflation. There were no provisions to adjust loan rates downward to remain competitive on the world market during a period of falling world commodity prices. When the dollar appreciated, high loan rates made U.S. farm products less price competitive.

High loan rates acted as a price umbrella under which competitors could increase production. Instead of purchases being made by foreign customers, the U.S. government became the buyer of last resort. Government stocks of agricultural commodities rose when exports fell; the U.S. share of the world export market for food and fiber products declined.

When the dollar gained strength, foreign customers were forced to spend more of their currency to pay for a given amount of U.S. farm imports. In 1981 soybeans were $7.98/bu in Chicago. The same bushel sold for DM 15.8 in West Germany. By 1985 soybeans were $5.97/bu in Chicago, and the price in West Germany was DM 18.4. The 25% decline in the U.S. price was offset by a 16% appreciation in the U.S. dollar relative to the German mark. The dollar also gained strength relative to major competitors such as Canada, Australia, Argentina, and France.

The strong dollar also changed the U.S. cost position relative to other countries, making U.S. farm production less competitive. When the dollar gained strength, the U.S. cost position eroded. While the dollar was valued at 7.0 French francs, U.S. wheat production costs were $115/acre compared with Fr 810 or $116/acre for French wheat. When the dollar reached almost Fr 10, U.S. costs were still $115 but French wheat costs had fallen to $81/acre. This change in relative cost position was due strictly to currency appreciation.

In summary, for U.S. agriculture to remain competitive on the world market will require careful coordination of agriculture and macroeconomic policies. In fact, it has been argued that macroeconomic policies now have more influence on agriculture than do farm policies. This has occurred primarily because of the increased interdependence between agriculture and the international market. Exchange rates have become one of the most important variables linking agriculture to the world economy. Currency values influence not only agricultural exports, but international competition and investment decisions. Exchange rates are especially important to U.S. agriculture because such a large share of production is exported and a large share of consumption is imported. For U.S. agriculture to maintain its competitive position on the world market will require agricultural policies that ensure flexible commodity prices and well-coordinated macroeconomic policies so that large swings in exchange rates do not undermine international competitiveness or discourage the importation of U.S. food and fiber products.

SUMMARY

The purpose of this chapter was to provide an overview of the international monetary environment affecting U.S. agricultural trade, emphasizing exchange rates, the foreign exchange market, and the international monetary system. The balance of payments was discussed and the relationship to exchange rates examined. Exchange rate determination was explored, with emphasis on important links to agriculture, including interest rates, prices, the trade balance, and expectations. The importance of exchange rate impacts on U.S. agricultural trade was examined, with special attention focused on prices, international competition, and agricultural policy. Finally, some considerations for the coordination of macroeconomic and agricultural policy were discussed, emphasizing the importance of maintaining the competitive position of U.S. agriculture. The major points in this chapter may be summarized as follows:

1. Exchange rates, the number of foreign currency units per unit of another currency, have become one of the most important economic variables affecting U.S. agricultural trade. International financial flows—capital and foreign

exchange—have grown to exceed world trade by 40 times, reaching $430 billion per day. The value of foreign currencies is determined in the foreign exchange market where supply and demand interact to determine the exchange rate. Currency values can appreciate (increase in value) or depreciate (decrease in value), determining the level and direction of U.S. agricultural trade.

2. The balance of payments (BOP) provides an important link between international trade, capital flows, and exchange rates. The current account, which is made up of trade and investment income, along with the capital account, which is composed of transfers of international assets, and official settlements are the primary components of the BOP. Large inflows of foreign capital since 1983 have transformed the United States into a net debtor nation. By importing foreign capital, the United States has run chronic, large deficits in the current account, while simultaneously providing capital for investment and consumption, and keeping interest rates lower than they otherwise might have been.

3. The international monetary system has evolved over the last 100 years from a system of exchange rates fixed to gold into a system called managed float, whereby exchange rates fluctuate, but within limits. These limits are controlled by the monetary authorities of the 156 members of the International Monetary Fund (IMF). Some countries peg their currencies to those of other nations, mainly the United States and France. Other nations have formed coordinated exchange rate arrangements, such as the European Monetary System of the European Union, for the purpose of stabilizing exchange rate movements.

4. Exchange rates are determined by the interaction of supply and demand for foreign currencies. In the short run, real interest rates, fiscal and monetary policies, and expectations about key economic variables are important in determining the value of currencies. The balance of trade, differentials in rates of economic growth between countries, and price levels are important in determining exchange rate movements over the long run. A comprehensive theory explaining all of the aspects of exchange rate determination has not been developed.

5. Exchange rates are especially important to U.S. agriculture because exports account for a large share of production. When the U.S. dollar appreciates in value, the price of goods in foreign currency increases, leading to a lower demand for U.S. goods, reduced prices, declining exports, and a loss of market share. U.S. agricultural policies that prevent U.S. prices from declining when the exchange rate appreciates can result in a loss of competitive position and market share.

6. Macroeconomic and agricultural policies combine to influence the performance of U.S. agricultural trade. Restrictive monetary policy raises interest rates and the value of the dollar. Expansionary fiscal policy also raises interest rates and causes dollar appreciation. This policy combination, combined with inflexible farm prices, can cause U.S. exports to decline. Macroeconomic poli-

cies that lead to higher real U.S. interest rates also place U.S. agriculture at a competitive disadvantage. For U.S. agriculture to maintain its competitiveness, it must coordinate both macroeconomic and agricultural policies.

DEFINITION OF KEY TERMS

Balance of payments (BOP): a record of all foreign transactions by private and public entities over a specific period of time, usually one year.

Bilateral trade-weighted exchange rate index: determined by using U.S. agricultural trade volumes with major trading partners to calculate weights, then multiplying those weights by the real exchange rate of each country.

Cross rate: the exchange rate between two currencies as calculated from the value of a third currency.

Currency appreciation: an increase in the foreign price of the domestic currency.

Currency depreciation: a decrease in the foreign price of the domestic currency.

Currency devaluation: action taken by monetary authority to deliberately decrease the value of currency from a fixed level relative to other currencies.

Current account: account composed of the merchandise trade account, the services trade account, and transfer payments.

Direct foreign investment (DFI): the acquisition of at least 10% ownership and control of any foreign production asset.

Euro: common currency of 11 EU member countries that participate in the European Monetary System.

Eurodollars: U.S. dollars on deposit outside the United States.

Exchange rate: the number of units of foreign currency that can be exchanged for one unit of domestic currency.

Floating exchange rate: system whereby currency values are determined by market supply and demand conditions, with minimal government intervention.

Intervention currency: currency used by other nations' monetary authorities to keep currencies within their predefined ranges.

Managed float: the government intervenes periodically to change currency values.

Official settlements account: summarizes net changes in official holdings of international reserve assets, such as foreign currencies, gold, or SDRs.

Portfolio investments: the acquisition of less than 10% ownership and control of foreign production assets, stocks, or bonds.

Private capital account: summarizes all transactions in international real and financial assets and all international activities of U.S. private banking institutions.

Real exchange rate: the value of one currency relative to other countries, accounting for different inflation rates between countries.

Special drawing rights (SDR): an artificial currency in the form of accounting entries to supplement reserve assets of the International Monetary Fund.

REFERENCES

Baker SA: *An introduction to international economics,* New York, 1990, Harcourt.

Denbaly M: *Balance of payments and macroeconomic policies,* Staff Report, AGES880815, Washington, D.C., 1988, USDA.

Economic report of the president, Washington, D.C., 1992, U.S. Government Printing Office.

International Monetary Fund: *International financial statistics,* Brussels, 1992, IMF.

Kreinin ME: *International economics: a policy approach,* 6th ed., New York, 1991, Harcourt.

Salvatore D: *International economics,* 3rd ed., New York, 1990, Macmillan.

Schuh GE: The exchange rate and U.S. agriculture, *American Journal of Agricultural Economics* 56(1): 1974.

Schuh GE: The evolution of the global economy, *Agrichemical Age,* February 1990.

Shane MD: *Exchange rates and U.S. agricultural trade* (Bulletin Number 585), Washington, D.C., 1990, USDA.

EXERCISES

1. An exchange rate can be thought of as the value, or price, of one currency in terms of another currency. T F

2. The balance of payments is a record of all transactions by U.S. private and public entities over a specific period of time, usually one year. T F

3. Currency devaluation is used to deliberately decrease the value of a currency from a fixed level relative to other currencies. T F

4. An increase in U.S. interest rates would cause a decrease in demand for U.S. dollars. T F

5. An expansionary monetary policy will cause the interest rate to decline and the exchange rate to depreciate. T F

6. Monetary contraction can be accomplished directly in the U.S. by raising the _____ or indirectly by decreasing the _____ in the economy.

7. If the level of inflation in the U.S. _____ relative to other countries, U.S. goods and services would become more _____ .

8. If a nation experiences a trade deficit, the value of the exchange rate in domestic currency will _____ .

9. In order to accomplish more complete monetary integration and the creation

of a common currency among members, the European Monetary System provided for:

a. creation of a common accounting unit, the European Currency Unit.

b. no currency fluctuations among countries in the E.C.

c. long-term balance of payments assistance to member states throughout the European Monetary Cooperation Fund.

d. none of the above.

10. Expansionary fiscal policy:

a. decreases exchange rates.

b. increases interest rates.

c. increases exports.

d. decreases imports.

11. A restrictive monetary policy accompanied by expansionary fiscal policy would cause:

a. real interest rates to decline and the value of the dollar to rise.

b. real interest rates to rise and the value of the dollar to decline.

c. real interest rates to rise and the value of the dollar to rise.

d. real interest rates to decline and the value of the dollar to decline.

12. As the U.S. dollar appreciates in value, the price of goods in foreign currency increases resulting in:

a. the demand for U.S. goods to decline.

b. the decline of prices.

c. loss of market share.

d. all of the above.

13. Identify and describe the important participants in foreign exchange markets.

14. The U.S. price of corn is $100 per ton and the exchange rate between the U.S. dollar and the Japanese yen is ¥120 = $1.00.

a. Calculate the international price of U.S. wheat for Japan.

b. If the exchange rate changes to ¥145 = $1.00, calculate the new international wheat price for Japan.

21

WHY NATIONS TRADE

It is the maxim of every prudent master of a family, never to attempt to make at home what it will cost him more to make than to buy. . . .

Adam Smith
(1723–1790)

TOPICS OF DISCUSSION

Trade occurs because individuals, firms, governments, or nations anticipate economic gains from the exchange of goods or services. Simply stated, people make money in international commerce by purchasing goods or services in the country where they can be produced the cheapest and selling those goods or services in the country where they are worth the most. Each nation is endowed with a unique set of human, institutional, and natural resources that give rise to different prices and trade relationships among countries.

The purpose of this chapter is to examine the underlying factors affecting trade patterns and to focus on why nations trade. The concepts of absolute and comparative advantage will be explored, along with an example of each. Gains from trade will be examined, with explanations of the importance of exchange, and specialization, and a discussion of the distribution of the gains from trade.

WHY TRADE?

The question of why nations trade has been at the crux of economic theory and analysis for nearly four centuries. One of the earliest attempts to explain why trade occurs was made by Thomas Munn (1571–1641) in his treatise on *England's Treasure by Foreign Trade.* Munn argued that a nation could become rich and powerful by exporting more than it imported. This philosophy, commonly referred to as **mercantilism,** became the accepted practice among trading nations. Highly protectionist trade policies, which were designed to restrict imports, were implemented by major traders such as England. The ensuing trade surplus with other countries could then be settled by an inflow of precious metals, namely gold and silver. In response to this policy, governments attempted to stimulate exports and restrict imports. When stocks of gold and silver were accumulated, the government became richer and more powerful. Governments realized that restricting imports and expanding exports would lead to additional economic output and employment. Pursuits of mercantilism led to government control of most major economic activity and strong movements toward economic nationalism.

The primary flaw in Munn's argument, and in the practice of mercantilism, was that all nations could not have a trade surplus at the same time. Mercantilism implies a zero-sum game. The only way for a nation to practice mercantilist principles was for it to gain and maintain a trade surplus at the expense of other nations. Recall that these were times of rule by monarch in most parts of the world. The obvious result of attempting to pursue these mercantilist policies was almost constant conflict among nations. With more gold, rulers could maintain larger armies and navies, leading to tremendous military power. Greater military strength was necessary for consolidation of power at home and to the acquisition of additional colonies abroad.

Although these events occurred in the seventeenth century, they are highly relevant to any discussion of international trade today. First, the body of knowledge brought forward by classical economists, such as Adam Smith and David Ricardo, was in response to, and an attack on, the mercantilist policies and the role of government in international commerce. Second, except for a brief period during the nineteenth century, no Western nation has ever been entirely free of mercantilist philosophy and policies. In fact, since the mid-1950s, there has been a resurgence of mercantilist philosophy in the United States known as neo-mercantilism. Neo-mercantilism led nations with high unemployment in import sensitive industries to restrict imports to recover lost jobs and increase sagging domestic production.

ABSOLUTE ADVANTAGE

One of the most convincing attacks on mercantilism was offered by Adam Smith in his classic book *An Inquiry into the Nature and Causes of the Wealth of Nations*, published in 1776. The key premise underlying Smith's attack on the mercantilist view was that for nations to trade, each had to gain something in the exchange of goods. If there were no gains, there was no incentive to trade. Smith's premise was based on the concept of **absolute advantage.** The concept of absolute advantage states that a nation will export those goods that it can produce more cheaply than others and import those goods that other nations can produce more cheaply. A nation is said to have an absolute advantage in the production of a good if its costs of production are lower than other nations' at prevailing prices and exchange rates. Nations can gain economically by specializing in the production of the goods for which they are most efficient and by trading the excess production with other nations. In this way, resources are used most efficiently and the output of nations will rise, leading to an increase in global economic welfare.

The basis for international trade is the efficiency with which nations combine productive resources to produce goods and services. For example, a farmer may trade grain for dollars and dollars for medical care. The doctor trades medical skills for dollars and dollars for food. Each individual is better off than before: the farmer is healthy and the doctor is well fed. Both individuals are endowed with, or have developed, a unique set of skills that allow them to specialize and practice what they do best, or what they do most efficiently.

Trade among nations is similar to trade between individuals. For example, because of its favorable climate, the United States is an efficient producer of wheat but an inefficient, or high cost, producer of coffee. Conversely, Mexico is an efficient producer of coffee, but an inefficient producer of wheat. The United States imports coffee from Mexico, which has been produced and marketed at prices below those in the United States. Mexico imports wheat from the United States at less cost than if it were produced in Mexico. With a given amount of income, consumers in both nations can purchase a greater variety of goods, in larger amounts, than if everything were produced domestically. In this way, nations can achieve a higher level of satisfaction from a given income with trade than without it.

The contrasts between the arguments of Munn and Smith are important to understand. The mercantilist doctrine held that one nation could gain from trade only at the expense of other nations and that strict government control of all economic activity and trade was necessary. This was most often accomplished by restricting imports and stimulating exports. Adam Smith, on the other hand, wrote that all nations would gain from free trade and advocated the **laissez-faire** philosophy, which emphasized little or no government control of a nation's economy. Smith further offered that free trade would lead to efficient resource use, thereby maximizing world welfare.

A numerical example of Adam Smith's absolute advantage should help solidify the concept and will serve to develop a conceptual framework for the discussion of comparative advantage. To simplify the analysis, the nation will be assumed to be the relevant economic unit. The use of nations as the economic unit facilitates the analysis of trade. Economic resources can normally be reallocated more easily within nations than among nations. Languages, laws, customs, culture, and institutions are often more alike within a nation. Policies and trade restrictions are generally not as important for domestic transactions as they are for international trade. The nation will be assumed to employ all its resources fully in the production of two goods, wheat and coffee. Further, resources are assumed to adjust efficiently and completely to changing economic conditions within countries. However, resources such as labor are assumed to have differing degrees of mobility among countries. Otherwise the major incentive for trade, different costs of production among countries, would disappear.

Table 21.1 shows that 1 hour of labor is required to produce 10 tons of wheat in the United States and 2 tons of wheat in Mexico. On the other hand, 1 hour of labor produces 6 tons of coffee in Mexico, but only 4 tons of coffee in the United States. Based on the results of this analysis, the United States is the most efficient, or least cost, producer of wheat, and Mexico is the most efficient producer of coffee. With trade, the United States would specialize in wheat production and exchange part of the surplus for coffee. Mexico would specialize in coffee production and trade part of the surplus for wheat.

The United States could trade 10 tons of wheat (10W) for 10 tons of coffee (10C) and gain 6C, saving 1.5 man-hours of labor. The 10W that Mexico receives from the United States would require 5 man-hours to produce in Mexico. With the 5 man-hours saved, Mexico can produce 30 tons of coffee, or 30C (5 hours times 6 tons per man-hour). By trading 10C for 10W, Mexico has gained an additional 20C, thereby saving 5 man-hours.

TABLE 21.1   Example of Absolute Advantage Measured as Units of Output per Unit of Labor

	United States	Mexico
	Tons/Man-Hour	
Wheat	10	2
Coffee	4	6

Although the concept of absolute advantage has intuitive appeal, it is not very useful in explaining the real-world relationships of trade that are prevalent today. For example, what if a nation has an absolute advantage in the production of all goods due to low cost labor, abundant natural resources, advanced technology, or superior management skills? Absolute advantage would dictate that this nation would export goods, but import nothing. In fact, it can be shown that absolute advantage is only a very special and limited case of the more general principle of comparative advantage. In the following section, we will learn that nations do not need an absolute advantage to gain from trade.

Comparative Advantage

The limitations of absolute advantage to explain trade among nations accurately and realistically led David Ricardo to develop more fully the principle of **comparative advantage.** The law of comparative advantage was first presented by Ricardo in his book *On the Principles of Political Economy and Taxation* published in 1817. One of the most important and still unchallenged laws of economics, comparative advantage states that even if one nation is less efficient in the production of both goods than another nation, there is still a basis for gains from trade. A nation should export the good for which its relative, or comparative, advantage is greatest and import the good for which its relative, or comparative, advantage is least. Conversely stated, a nation should specialize in the production and export of the good for which it has the least relative disadvantage and import the good for which its relative disadvantage is greatest. The key point is that trade is based on relative or comparative cost relationships rather than absolute costs.

A number of simplifying, yet important, assumptions were made by Ricardo in his explanation of comparative advantage: (1) a two-nation, two-good world, (2) completely open and free trade, (3) free movement of labor within nations, but no movement of labor between nations, (4) constant production costs, (5) absence of costs of transport and transfer of goods, (6) constant technology, and (7) that labor is the only factor of production or that it is used in the same fixed proportion in the production of both goods and that labor is homogeneous. This final assumption is called the labor theory of value and is often not used to explain the law of comparative advantage because its underlying precepts are unrealistic.

The principle of comparative advantage can be explained by the numerical example in Table 21.2. In this case, Mexico has an absolute disadvantage in the production of both goods, because it now produces only 3 tons of coffee instead of 6, indicating that productivity has fallen by 50% from the previous example. The United States can produce 10W compared with 2W for Mexico and 4C compared with 3C for Mexico. To find the comparative advantage of each nation, the relative productivity of labor must be determined. The United States can produce 10 tons of wheat with one unit of labor, compared with only 2 tons for Mexico, yielding a ratio of 10:2 or 5. The comparable ratio for coffee is 4:3 or 1.33. Although the United States is more productive in the production of both, its relative advantage is greatest in wheat, be-

TABLE 21.2   Example of Comparative Advantage Measured as Units of Output per Unit of Labor

	United States	Mexico
	Tons/Man-Hour	
Wheat	10	2
Coffee	4	3

cause 5 > 1.33. Therefore, the United States has a comparative advantage in wheat, and Mexico has the least comparative disadvantage in coffee. Conversely, Mexico's labor productivity yields output to labor input ratios of 0.2 and 0.75 for wheat and coffee, respectively. Based on this example, we may conclude that Mexico has a smaller relative disadvantage in the production of coffee than in the production of wheat. Or conversely, Mexico has a comparative advantage in coffee production.

To examine the gains from this trade relationship, the same example can be expanded. If the United States traded 10W for 10C, it would gain 6C because the domestic rate of exchange in the United States is 10W for 4C. The 6C gained represents a labor savings of 1.5 man-hours. The 10W that Mexico receives would require 5 man-hours to produce. These 5 man-hours could be used to produce 15C. By trading 10C for 10W, Mexico gains 5 tons of coffee. Because the United States could exchange 10W for 4C internally, it would gain from trade if it could get more than 4C for 10W. For Mexico to gain from trade, it must receive more than 2W for 3C. By exchanging 10W and 10C, both countries gain from trade.

A further extension of comparative advantage was made in 1936 by G. Haberler in *The Theory of International Trade.* In relaxing the value theory of labor assumption, Haberler was able to demonstrate that the law of comparative advantage was still valid and more universally applicable by employing the concept of **opportunity cost.** According to Haberler, opportunity cost reflects the cost of a good as measured by the amount of a second good that must be given up to release just enough resources to produce one additional unit of the first good. No underlying assumptions are made about the homogeneity of labor or that labor is the only factor of production. Haberler's extension is often called the **opportunity cost theory,** which states that a nation has a comparative advantage in the production of the good with the lowest opportunity cost. For example, the United States must give up 0.4 units of coffee to release just enough resources to produce one additional unit of wheat. In this case, the opportunity cost of wheat is 0.4 ton of coffee, calculated as 10W = 4C or 1W = .4C. In Mexico, the opportunity cost of wheat in terms of the coffee given up to produce one additional ton is 1.5 because 2W = 3C, then 1W = 1.5C. Because the United States has the lowest opportunity cost for wheat (.4<1.5), it has a comparative advantage in the production of wheat.

The same analysis can be conducted to determine the opportunity cost of coffee in Mexico. The amount of wheat that must be given up by Mexico to produce one additional unit of coffee is 0.67 tons. For the United States, the amount of wheat forgone to produce one additional ton of coffee is 2.5 tons, or 4C = 10W. Therefore, Mexico has the lowest opportunity cost of producing coffee, 0.67 tons compared to 2.5 tons in the United States. We may conclude that Mexico has a comparative advantage in coffee production, and the United States has a comparative disadvantage in coffee production. According to the law of comparative advantage, the United States should specialize in wheat production, exporting the surplus to Mexico in return for coffee. This is the identical result found when the analysis was based on the labor theory of value, but now it is based upon the less restrictive and widely applicable concept of opportunity cost.

Factors Affecting Comparative Advantage

As noted by James Houck, comparative advantage is a real, not a monetary, concept. Because comparative advantage is based on the structure of relative opportunity costs among nations, it is not affected by changes in exchange rates or inflation. While exchange rates may vary among different currencies, they act to translate comparative advantage into absolute advantages, which are compared by individual buyers and sellers worldwide. If all resources could be readily transferred among nations, opportunity costs would be altered and incentives to trade would disappear. However, because productive resources such as land, climate, and labor are not mobile, comparative advantage still may be used to explain much of the world's trade in agricultural products. Due to the increased international mobility of capital and technology, however, opportunity costs and comparative advantage do change over time, leading to different trade patterns.

National differences in the opportunity cost of resources determine comparative advantages and the patterns of trade. Opportunity costs are dependent on several factors. First, the availability of productive resources varies among nations, leading to different costs. The cost of using plentiful resources will be lower than the cost of using scarce resources. Land, labor, capital, technology, and management skills are all affected by the abundance with which they exist in a given nation. Second, the production of different goods and services requires different resources in different proportions. Third, most products can be produced by more than one process, resulting in many different resource combinations. Finally, resource mobility differs among nations. Any of these factors, alone or combined, can affect the opportunity costs within nations and their comparative advantages.

Comparative Advantage and Competitive Advantage

In the long run, comparative advantage, or the relative efficiency with which resources are combined, is reflected in the economic incentives or disincentives for resource use. Agricultural exports represent additional demand for U.S. products and, therefore, depend on the level of economic development and growth in other countries. However, U.S. products must compete with goods from Europe, Australia, and Canada. How successfully those products compete depends on price, quality, timeliness of delivery, and many other factors. In competitive markets, however, international market prices reflect the long-run costs of the human, natural, and manufactured resources required to produce the goods.

In the short run, however, relative absolute costs among nations may not reflect comparative advantage. Michael Porter was the first to use the term competitive advantage. Market distortions brought about by government intervention have been especially prevalent in agricultural trade. **Competitive advantage,** which is defined as the pure economic competitiveness of a nation reflected by the absolute cost of a given good in a given market at a particular point in time, may provide a better understanding of historical trade patterns in agriculture than the more restrictive law of comparative advantage. Further, the existence of historical social ties, common ideology, common defense treaties, and agreement on basic human values may be as important in understanding trade patterns today as reliance on the law of comparative advantage. For example, Cuba is not allowed to ship sugar to the United States, regardless of its comparative advantage or the potential cost savings to U.S. consumers. This section highlights some of the most important factors and forces affecting the competitive advantage in international trade of agricultural products.

Domestic agricultural policies, in combination with the use of import duties and export subsidies, were instrumental in transforming the European Community from a net importer of grains into a net exporter. Subsequently, the United States lost large shares of the world grain market. Many analysts questioned whether the United States had lost its comparative advantage in grain production. The United States responded to these EC policy actions by implementing more competitive farm prices and export subsidies of its own to regain lost markets. In reality, the United States had not lost its comparative advantage, but had become less competitive relative to the EC, thereby losing its competitive advantage and the ability to compete for international markets at competitive world prices.

Import quotas can mask the true comparative advantage of a nation. An import quota forces the effective demand for a good to zero once the quota is filled. When the United States implemented the meat import quota in 1991, it made little difference that Australia had a comparative advantage in meat production. Once the United States had accepted all agreed-to tonnage, the market was closed.

Exchange rates have had major impacts on agricultural trade. For example, we may argue that U.S. borrowing in foreign money markets to fund growing budget deficits in the early 1980s drove up interest rates, and increased the demand for dollar valued assets, and, hence, the value of the U.S. dollar. The resultant higher prices of U.S. goods in international markets perpetuated the loss of competitive advantage for U.S. agriculture. Over time, however, exchange rates declined and the United States regained its competitive advantage. Although the United States maintained its comparative advantage in agriculture during this period, it became much less competitive relative to other countries, losing markets and export sales.

A lack of marketing infrastructure, such as storage and handling facilities, communications systems, and other logistical factors can alter the competitive advantage of nations, increasing the absolute cost of accessing world markets. High cost and lack of available transportation, along with other transfer costs, can result in short-run distortions in competitive advantage. In summary, competitive advantages that arise from nonmarket factors and forces such as tariffs, quotas, and subsidies are no less real and effective than those reflected by lower opportunity costs. However, competitive advantages based on nonmarket factors are dependent upon politics rather than economics for sustained existence. These basic differences between comparative and competitive advantage should be considered when analyzing trade and competition among nations.

GAINS FROM TRADE

For trade to occur, both nations must benefit from the transaction. Real incomes are raised by trading with other nations. By importing, goods can be purchased at a lower cost and in greater variety and quantity than without trade. Lower prices allow consumers to buy more with their limited resources, thereby increasing disposable incomes for each consumer individually and increasing the overall welfare of the nation. Prices of exported goods may be higher, with additional economic activity and income spread throughout the economy. Finally, consumers with a given amount of income are able to achieve more satisfaction from those incomes with trade than without trade. However, these gains from trade are dependent upon increased specialization in production and exchange of goods.

The Importance of Exchange and Specialization

The law of comparative advantage can be applied to demonstrate the gains from trade and the subsequent increase in world welfare. Table 21.3 is an extension of the previous hypothetical example. Combinations of wheat and coffee can be produced in the United States and Mexico by fully employing all available resources, the most productive technology, and a maximum of 20 man-hours of labor per year in each country. These production possibilities can be combined to form the complete **production possibility schedule** in Table 21.3. For instance, the United States can produce 200 tons of wheat and no coffee, 160 tons of wheat and 16 tons of coffee, 80 tons of wheat and 48 tons of coffee, or no wheat and 80 tons of coffee. For every 10 tons of wheat produced, the United States releases enough resources to produce 4 tons of coffee. If the United States increases wheat output from 120 tons to 160 tons, it must give up the opportunity to produce 16 tons of coffee, or coffee production must decline from 32 tons to 16 tons. The ratio of 40W to 16C determines the rate at which wheat production is substituted for coffee production for each combination of alternatives. The opportunity cost of wheat for coffee is constant at 10W = 4C for every combination of U.S. production possibilities.

The same exercise can be done for Mexico, revealing that the set of production combinations ranges from 40W = 0C, 32W = 12C, 16W = 36C, to 0W = 60C. The cost

TABLE 21.3   Production Possibility Schedule for Wheat and Coffee in the United States and Mexico

United States		Mexico	
Wheat (Tons/Year)	Coffee (Tons/Year)	Wheat (Tons/Year)	Coffee (Tons/Year)
200	0	40	0
160	16	32	12
120	32	24	24
100	40	20	30
80	48	16	36
40	64	8	48
0	80	0	60

of wheat production in Mexico, in terms of the amount of coffee production forgone, is constant at 2W = 3C for every combination of production possibilities. For Mexico to increase wheat output by 16 tons, from 8 tons to 24 tons, requires that the opportunity to produce 24 tons of coffee be forgone. The relationship of 2W = 3C, or 1W = 1.5C, is constant throughout the production possibility schedule for Mexico.

United States and Mexico production possibility schedules are graphed in Figure 21.1, with production possibility frontiers reflecting the combinations of wheat and coffee that can be produced by each nation. For example, at point *A*, the United States produces 100 tons of wheat and 40 tons of coffee. At *A'*, Mexico produces 32 tons of wheat and 12 tons of coffee. The negative sloping frontiers indicate that for either nation to increase output of one good, some output of the other good must be forgone. For example, if the United States increases wheat production by an additional 10 tons, it would need to give up the opportunity to produce 4 tons of coffee. For Mexico to increase coffee output by 6 tons, it must be willing to give up 4 tons of wheat production.

Distribution of the Gains from Trade

As was noted earlier, the basis for trade is the differing opportunity costs among nations. This concept is illustrated in Figure 21.2. In the absence of trade, assume the United States and Mexico produce at *A* and *A'*, respectively. With trade, the United States would specialize in wheat production (point *B*), the good for which it has a comparative advantage, and Mexico would specialize in coffee (point *B'*). The United States could trade 50W for 45C and reach point *C*, where 150W and 45C are consumed. Under these conditions, the United States is producing 200 tons of wheat, trading 50W to Mexico for 45 tons of coffee, and reaching a higher level of consumption through trade.

Mexico could produce 60 tons of coffee, trade 45C to the United States and import and consume 50 tons of wheat, thereby reaching point *C'*. Comparing

Production Possibility Frontier for the United States and Mexico

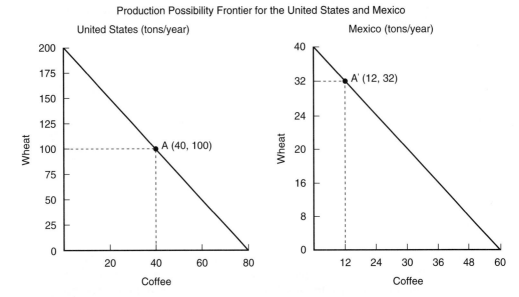

FIGURE 21.1   These relationships depict the combinations of wheat and coffee that can be produced in Mexico and the United States. At point *A*, the United States produces 100 tons of wheat and 40 tons of coffee. At point *A'*, Mexico produces 32 tons of wheat and 12 tons of coffee.

Gains from Trade

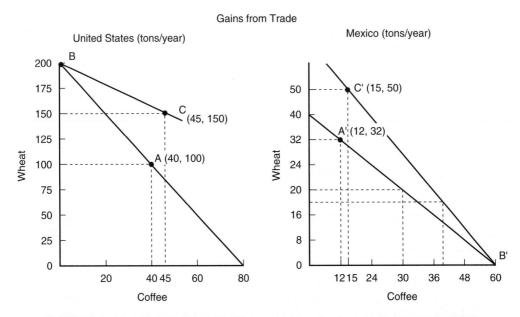

FIGURE 21.2   Through trade, the United States specializes in wheat production at point *B* (200 tons), trades 50 tons of wheat to Mexico for 45 tons of coffee, and consumes more of both goods at point *C* (45 tons of coffee and 150 tons of wheat). Mexico specializes in coffee production (60 tons) and produces no wheat (point *B'*), trades 45 tons of coffee for 50 tons of wheat, and consumes more of both goods (50 tons of wheat and 15 tons of coffee) than before trade occurred.

point *A* with point *C*, we may determine that the United States gained 50 tons of wheat and 5 tons of coffee through trade. Likewise, comparing *A'* and *C'*, Mexico gained 18 tons of wheat and 3 tons of coffee by trading with the United States.

By specializing in the production of the good for which it has a comparative advantage and trading with the other, each nation can increase consumption of both goods. Before trade, the United States produced 100W and Mexico produced 32W, a total of 132W. With specialization and trade, 200W is produced, all in the United States. Further, in the absence of trade, Mexico produced 12C and the United States produced 40C, for a total of 52C. After specialization and trade, 60C is produced, all in Mexico. The net increase in production of 68W and 8C from specialization in the goods for which each had a comparative advantage represents the gains from trade shared by each nation. These gains also reflect the increase in world welfare from increased specialization and trade.

SUMMARY

The purpose of this chapter was to examine the major factors affecting trade patterns and to determine why nations trade. The law of comparative advantage was reviewed and contrasted with absolute advantage and competitive advantage. The gains from trade were examined, with emphasis on how those gains were distributed between nations and the importance of specialization and exchange. The major points made in this chapter may be summarized as follows:

1. Trade occurs because individuals, firms, governments, and nations anticipate economic gains from the exchange of goods and services. International commerce is lucrative because goods can be purchased where they are abundant and cheapest and then sold where they are more scarce and highly valued. Differences in national resource endowments give rise to these differences in costs and prices, and provide the basis for trade.

2. The theory of international trade has evolved over the past four centuries. Mercantilist theory explained trade as a zero-sum game in which all nations strive to maintain a trade surplus at the expense of other nations. Next, the basic concept of absolute cost advantage was used to account for why nations traded with each other. Absolute advantage is not a necessary condition for trade, but the law of comparative advantage actually determines why nations trade and how the gains from trade are distributed.

3. Comparative advantage was contrasted with the concept of competitive advantage to explain how agricultural trade patterns are influenced by government intervention. Domestic agricultural policies, combined with export subsidies and import restrictions, have been important factors affecting the ability of nations to compete for world markets. Competitive advantages based on the role of government in trade are dependent upon political factors rather than economic forces for sustained existence.

4. The basis for trade is differing opportunity costs among nations. To receive gains from trade, nations must specialize in the production of goods for which they are most efficient and exchange those goods with other nations. Through increased specialization and exchange, all nations can benefit from trade, and world economic welfare will be increased.

DEFINITION OF KEY TERMS

Absolute advantage: exists when one nation can produce goods more cheaply than another nation.

Comparative advantage: ability of a nation to specialize in the production of the good for which it has the lowest opportunity cost.

Competitive advantage: economic competitiveness of a nation reflected by the absolute cost of a given good in a given market at a particular point in time.

Laissez-faire: economic philosophy of little or no government control of a nation's economy.

Mercantilism: the economic and political philosophy that national wealth and power were dependent upon a nation being able to export more than it imported.

Opportunity cost: the cost of producing a good as measured by the amount of a second good that must be forgone to release just enough resources to produce one additional unit of the first good.

Opportunity cost theory: a nation has a comparative advantage in the production of the good with the lowest opportunity cost.

Production possibility schedule: all combinations of national output that can be produced by fully employing all available resources and the most productive technology.

REFERENCES

Baker SA: *An introduction to international economics,* New York, 1990, Harcourt, Brace, Jovanovich.

Haberler G: *The theory of international trade,* London, 1936, W. Hodge and Company.

Houck JP: *Elements of agricultural trade policies,* Prospect Heights, Ill, 1986, Waveland Press.

Munn T: *England's treasure by foreign trade,* Oxford, 1928, Basil Blackwell.

Porter M: *Competitive Advantage: creating and sustaining superior performance,* New York, 1985, The Free Press.

Ricardo D: *On the principles of political economy and taxation,* Homewood, Ill, 1963, Richard D. Irwin.

Salvatore D: *International economics*, 3rd ed., New York, 1990, Macmillan.
Smith A: *An inquiry into the nature and causes of the wealth of nations*, New York, 1937, The Modern Library.

EXERCISES

1. The mercantilist philosophy argues that nations can become rich and powerful by exporting more than they import. T F

2. Comparative advantage states that a nation will export the foods that it can produce more cheaply than others and import the goods that other nations can produce more cheaply. T F

3. Opportunity costs reflect the cost of a good as measured by the amount of a second good that must be given up in order to produce one additional unit of the first good. T F

4. Comparative advantage is a monetary concept and is affected by changes in exchange rates or inflation. T F

5. Competitive advantages based on nonmarket factors are dependent on politics rather than economics for sustained existence. T F

6. The basis for trade is differing _____ among nations.

7. In order to achieve gains from trade, nations must specialize in the _____ of goods for which they are most efficient and _____ those goods with other nations.

8. The _____ shows all combinations of national output that can be produced by fully employing all available _____ and the most productive _____ .

9. The laissez-faire philosophy held that:

   a. strict government control of all economic activity was necessary.

   b. little or no government control of the economy was needed.

   c. nations need the government to establish export laws to stimulate the economy.

   d. none of the above.

10. According to comparative advantage, a nation should:

    a. not import any goods if the nation does not have a relative disadvantage.

    b. specialize in production of the good for which it has the least relative disadvantage.

c. import only the goods in which the nation has a disadvantage.

d. not produce any goods if the nation is at a relative disadvantage.

11. Opportunity cost theory states that a nation has:

   a. an absolute advantage in the production of the good with the lowest opportunity cost.

   b. no advantage in the production of any good with an opportunity cost.

   c. a comparative advantage in the production of the good with the lowest opportunity cost.

   d. none of the above.

12. Identify the factors affecting comparative advantage.

13. Discuss the difference between comparative advantage and competitive advantage.

14. If the United States produces 20 tons of grain and 10 tons of coffee per one man-hour and Mexico produces 15 tons of grain and 3 tons of coffee per one man-hour, what commodity should each country produce according to the law of competitive advantage?

22

AGRICULTURAL TRADE POLICY

*Despite strong theoretical arguments about the
benefits of freer trade, its application imposes
hardship on some industries and people.*

James P. Houck

TOPICS OF DISCUSSION

The formulation and implementation of effective trade policy poses a crucial, yet complex dilemma for policymakers. Large gains in income and employment can be achieved by reducing or eliminating policies that restrict the trade of food and agricultural products. As C. Ford Runge (1988) states, "Agriculture is emerging as a key issue in the politics of international trade." The difficult choices facing international trade policymakers are many. The first problem is how to disengage from trade-distorting domestic policies and still support farm incomes at home. The second problem is how to continue the process to significantly reduce trade barriers, which would foster large economic gains to all nations. Despite clear evidence that **trade liberalization** in agriculture would generate more than $4.0 billion in economic output worldwide, only slight progress has been made since World War II to seriously negotiate a global reduction in policies that restrict trade. The purpose of this chapter is to demonstrate that, although clear economic gains to trade may exist, it is often difficult to pursue those gains because of strong vested interests that pursue the status quo. Although the nation as a whole benefits from trade, there are both gainers and losers. It is the potential losers who effectively protect domestic markets using trade barriers.

TRADE AND WELFARE

In Chapter 21, we learned that in a two-nation, two-good world, each nation could specialize in the production of the good for which it had a comparative advantage and trade with the other nation, and the consumption of both nations was increased. Comparative advantage was used to explain why nations trade. In this chapter, we will use partial equilibrium analysis to determine the specific prices and quantities at which trade occurs. The gains to trade are illustrated using consumer and producer surplus, or welfare analysis, which demonstrates the importance of interdependence between nations resulting from trade.

Autarky or the Closed Economy

First, assume there are two nations: the United States and Japan. One commodity—for example, wheat—is produced by both nations. Figure 22.1 illustrates the hypothetical supply and demand relationships. Initially assume that each nation operates as a closed economy. Also, each nation is self-sufficient, or an **autarky,** and no trade occurs between it and any other nation. Initial market conditions are reflected by the intersection of supply and demand in each nation, in which P_{US} and Q_{US} represent competitive equilibrium in the United States and P_j and Q_j are market equilibrium in Japan. In the absence of trade, the quantities of wheat produced and consumed in each nation represent market equilibrium.

Next, assume that traders from the United States travel to Japan and discover that the price for wheat, P_j, is much higher than at home, P_{US}. Traders realize that they can now buy wheat in the United States and resell it in Japan at a profit, ig-

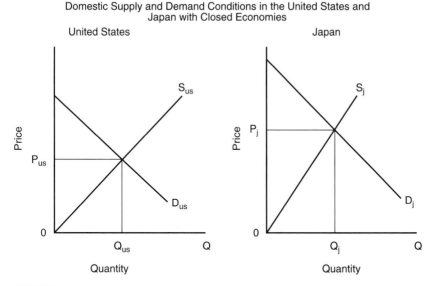

Domestic Supply and Demand Conditions in the United States and Japan with Closed Economies

FIGURE 22.1   Under autarky, or complete self-sufficiency, market equilibrium is determined solely by the interaction of domestic supply and demand conditions. No trade between nations is assumed to occur.

noring transfer costs. When wheat is exported from the United States to Japan, P_j will begin to decline, and P_{US} will begin to rise. This process, known as **arbitrage,** will continue as long as the price in Japan exceeds the U.S. price, indicating a profit from trade. Finally, exports from the United States to Japan will be sufficient to result in the same wheat price in both nations, hence establishing one market for the trade of wheat.

Trade and Partial Equilibrium

To determine the precise price and quantities at which trade will occur, a partial equilibrium model is constructed in Figure 22.2. First, it is necessary to derive the excess supply curve for the United States. **Excess supply** indicates the amount by which quantity supplied exceeds quantity demanded for each price level above P_{US}. Goods will always move from where prices are low to where prices are higher. When S_{US} and D_{US} are equal, competitive equilibrium exists and excess supply is zero. ES_0 can be plotted on the vertical axis of Figure 22.2, *B* to represent the initial point on the excess supply curve for the United States. When prices rise, more wheat will be shipped, so a second point on the excess supply curve can be determined by plotting the horizontal distance between any two quantities. For example, the difference between QS_{US1} and QD_{US2} can be plotted at P_{US2} in Figure 22.2, *B* and labeled ES_1. When ES_0 and ES_1 are connected, *ES* results and indicates the excess supply curve for the United States. This relationship is often referred to as

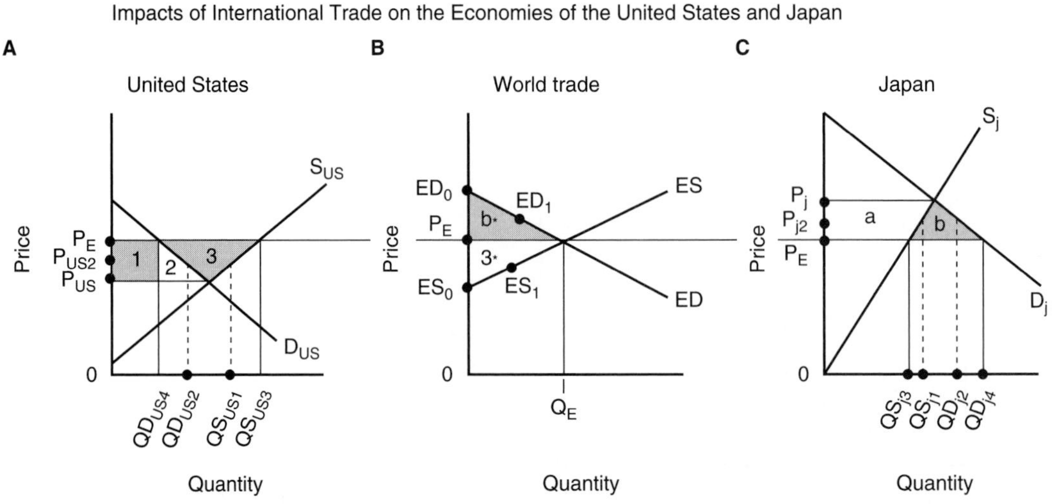

Impacts of International Trade on the Economies of the United States and Japan

FIGURE 22.2 In an open economy, trade is allowed to take place. Prices rise in the United States and decline in Japan. Producers in the United States and consumers in Japan benefit from trade. Consumers in the United States and producers in Japan experience losses in economic welfare.

export supply or exportable surplus. The *ES* function indicates the quantity of wheat supplied to the export market at each price above P_{US}.

Similarly, excess demand can be derived for Japan. **Excess demand** indicates the amount by which quantity demanded exceeds quantity supplied at each price below P_j. At equilibrium price P_j, $S_j = D_j$ and, therefore, excess demand is zero and is plotted as ED_0 in the trade sector. When more wheat is supplied to the Japanese market, prices decline. For any price below P_{j1}, for example, P_{j2}, excess demand, $QD_{j2} - QS_{j1}$, can be determined and plotted as ED_1 in Figure 22.2, *B*. Connecting ED_0 and ED_1 yields *ED*, or the excess demand curve for the Japanese wheat market. *ED* represents the quantity demanded in excess of domestic supply at each price below P_j. Japan will import the quantity $QD_{j4} - QD_{j3}$ from the United States. Excess demand is commonly called import demand.

The point at which *ED* equals *ES* in Figure 22.2, *B* indicates world wheat market equilibrium. One price, P_E, prevails in both countries and indicates the price at which wheat is traded internationally, or the world market price for wheat. In the United States, the higher price, P_E, causes quantity supplied to increase to QS_{US3} and the quantity demanded to decline to QD_{US4}. The quantity exported from the United States, or export supply, would equal $QS_{US3} - QD_{US4}$. In Japan, at the lower price P_E, the quantity demanded increases to QD_{j4} and the quantity supplied declines to QS_{j3}. The difference equals excess demand and is identically equal to excess supply from the United States and the amount entering world trade in Figure 22.2, *B*, quantity Q_E.

Trade impacts are quite different, yet still have important implications for producers and consumers in both countries. Higher wheat prices cause U.S. producers to expand production by bringing additional land, labor, capital, and technology into use. Consumers, however, reduce wheat consumption because its cost to them has risen. The difference between what is produced and consumed generates the excess supplies shipped to Japan.

Lower prices in Japan mean that wheat consumers will increase their purchases, resulting in greater quantity demanded. Producers will reduce output, however, since prices have fallen. Greater excess demand causes imports to rise. The amount of wheat imported from the United States at price P_E will increase to quantity Q_E. It is important to note that with trade, internal market prices are jointly determined by both domestic supply and demand conditions in both countries and the world market.

Welfare Gains from Trade

The welfare effects of free trade mirror the quantitative effects discussed on the preceding page (Figure 22.2). In the exporting nation, consumer surplus declines by areas 1 + 2. Producer surplus, however, increases by areas 1 + 2 + 3. Society in the exporting country is better off by the amount of area 3 as a result of trade. In the importing country, consumers gain areas *a* + *b*, and producer surplus declines by the amount of area *a*. There are net societal gains of area *b* for the importing nation. The net gains to trade can be illustrated in the trade sector. Area 3, net gains to the exporting country, is shown by 3* in Figure 22.2, *B*, and represents the amount by which producer gains exceed consumer losses. Net gains for the importing country are shown by *b** and represent the amount by which consumer gains exceed producer losses. The net gains to trade are summarized in Table 22.1. If gainers compensate losers, then economic agents in both nations will be made better off from free trade.

Several key assumptions are necessary in order to achieve consistent results when applying the partial equilibrium analysis to trade. First, a constant exchange rate is assumed between the currencies in the exporting and importing nations. Second, all input and output prices, per capita incomes, population, production technology, and consumer tastes and preferences are held constant. Third, transfer costs between nations are assumed to be zero. If any of these conditions change, it will be necessary to derive new world market equilibrium conditions. Finally, this partial analysis can handle only one product or commodity at a time.

To summarize, producers in the United States and consumers in Japan gain from free trade. Producers in Japan and consumers in the United States will be forced to adjust to the more open market. If gains are shared, however, both sets of interests can gain from trade. Because trade has potentially very different impacts

TABLE 22.1   Gains to Trade

	United States	Japan
Consumer gains	−(1+2)	+(a+b)
Producer gains	+(1+2+3)	−(a)
Net gains to society	+(3)	+(b)

on special interest groups, it is often not a popular policy option. In this case, producers in Japan would most certainly resist moves to liberalize trade. Although U.S. consumers will be adversely affected, it is unlikely that any strong resistance to freer trade would occur primarily because no one consumer is forced to bear a large share of the burden of adjustment. The prospects for strong consumer reaction are, therefore, diluted.

WHY RESTRICT TRADE?

It has been demonstrated that there are significant gains to freer trade. The **free trade** argument states that resource use will adjust *across* and *within* national boundaries, so that marginal-value products for land, labor, and capital will be equal in all uses. Optimal world welfare will result. Why is trade restricted? Very simply, the economic hardship generated by adjustments to more open borders often leads some sectors or industries to effectively lobby for protection. **Protectionism** occurs when government policy is implemented to remedy domestic economic problems associated with excessive imports. Protectionist policies reduce competitive pressures on producers and allow them to forego the adjustments brought about by freer trade.

Protectionism in Agriculture

Although the URA was important in reducing agricultural trade barriers, protectionism in agriculture still abounds. In 1995, for example, the level of support to producer incomes from government averaged 36% for five OECD countries.[1] During this same period, it was estimated that U.S. farmers received nearly 19% of their income from government programs. Producers in Japan depended upon government for 70% of their income, and farmers in the EU received 48% of income from government sources. The perceived need for a country to be self-sufficient in food production or to have a high degree of food security often transcends the economic benefits of trade, leading to high protective tariffs or the implementation of restrictive import policies.

Although tariffs and quotas are the traditional methods used to restrict trade, many nations employ less overt measures. The use of support prices above the world price, for example, was common during the 1960s and led to overproduction and lower import demand. Surplus disposal policy may employ a trade-distorting subsidy on exports. Taxes on exports may be used to keep products from leaving a country, resulting in greater supplies and lower prices for domestic consumers, but higher prices for foreign consumers. More subtle measures, such as exchange rate controls, may be used to stimulate a country's exports and improve its competitive

---

[1]The Organization for Economic Cooperation and Development (OECD) is a 24-member organization established in 1961 to promote the economic well-being of member countries, while contributing to the overall development of the world economy.

position. Although the intent of these measures may not be to disrupt trade, the effects are trade distorting nonetheless.

Arguments against Trade

Arguments against trade can be classified into the following general categories:

1. to protect a new or infant industry,
2. to counter unfair foreign competition,
3. to improve the balance of payments, and
4. to protect national health, the environment, or food safety.

The argument to protect a new or "infant" industry is often made because, when production of a product first begins, firms are usually less efficient than they will be after economies of scale are achieved. Costs of production tend to be higher and labor may be less skilled and therefore less productive than it will become over the long term. These infant industries call for protection for a period of time so that they do not fail before they have the opportunity to become efficient. Over time, economies of scale will result, allowing the protected firms to become globally competitive. This argument has appeal for nations in the early stages of industrial development. In fact, the United States used the infant industry argument to support tariffs in the 1800s. Most often, however, the argument is used by developing countries that want to protect smaller, less-efficient businesses from large industrial firms in the United States, Japan, and Europe. Although this argument is justified in many cases, it has been used to perpetuate trade barriers and to maintain inefficient industries well beyond their infancy.

Unfair competition by foreign governments and businesses is often cited by special interest groups who seek protection. Much of the recent U.S. trade law has been in response to calls for protection against unfair competition. **Section 301** of the Trade Act of 1974 provides presidential authority to impose duties on products from nations whose trade practices are deemed unfair or found to restrict U.S. commerce.

Another component of this argument is that U.S. wage rates are higher than those in developing countries, and therefore, U.S. businesses face a competitive disadvantage. U.S. Congress was urged by organized labor interests to vote against the establishment of a North American Free Trade Agreement because Mexico has relatively low-cost labor, which might lure some U.S. manufacturing plants to Mexico, taking jobs with them.

In Chapter 20, we learned that a nation's balance of payments position can serve as an indicator of its overall international well-being. When payments to foreigners chronically exceed earnings, confidence in the nation's currency and economic strength may be undermined. Foreign investment may level off, followed by a decline in the value of currency. To stem the outflow of currency, a nation may attempt to reduce its payments to foreigners by limiting imported products. If export earnings remain constant, reduced imports will bring the international payments account more into balance. In reality, however, these actions often invite retaliation by foreign governments and firms, resulting in reduced export earnings. Protectionist policies rarely result in higher, sustained exports over the long run.

Trade in food products is also subject to restrictions related to the protection of human, animal, and plant health; the environment; or food safety. For example, the United States prohibits the import of fresh/chilled meat from nations that have endemic aftosa, or foot-and-mouth disease. U.S. meat imports from Mexico are limited to shipments from those facilities that have been inspected and approved by the USDA. Likewise, U.S. meat exports to the EU must be from facilities approved by the EU. These health provisions are called **sanitary and phytosanitary regulations (SPS).** For U.S. exports of agricultural products, a phytosanitary certificate is issued by the Animal and Plant Health Inspection Service (APHIS), which certifies that a particular shipment is free of harmful disease or pests. Although many of the SPS currently in effect around the world are based upon valid health concerns and scientific evidence, it has become increasingly popular for special interest groups to attempt to convince governments that health or food safety concerns exist when none actually do. The EU has been criticized for implementing a ban on imported meat containing any trace of growth hormones. This action effectively eliminated the EU market for U.S. beef.

Success in the Uruguay Round to reduce tariff and nontariff barriers to trade may result in greater use of technical barriers to trade by special interest groups to influence trade in their favor. Technical barriers to trade (TBT) are a special form of nontariff barrier that includes not only SPS but also restrictive food safety and labeling regulations or other peculiar laws that inhibit trade. As duties and import quotas are reduced and eliminated through trade negotiations, it appears likely that TBT will be used by many countries to restrict imports. During 1996, for example, the United States experienced a large number of complaints from U.S. food exporters claiming that major trading partners were using unfair TBTs to limit imports. About $4.9 billion in processed food trade was affected by these barriers. Japan, South Korea, and Mexico were among the major offenders. The United States has been accused of implementing trade-restricting TBTs as well. Concerns about BSE, or mad-cow disease, prompted the United States to ban the importation of cattle and sheep from the United Kingdom in 1997.

TRADE RESTRICTIONS

Trade restrictions may take many different forms. For discussion purposes, we will classify them into three groups: (1) import policies, (2) domestic food and agriculture policies, and (3) export policies. Import policies relate to those measures designed to restrict imports to protect domestic industries or to raise government revenue. Domestic food and agriculture policies may attempt to influence production and/or consumption and may take the form of support prices, consumption taxes, stocks policies, input use restrictions, or input subsidies. Export policies attempt to stimulate the export of products, often below the world market price.

Import Policies

Border policies can be classified as tariff or nontariff barriers to trade. Tariffs, or duties, were once the dominant restrictions affecting trade. Recent efforts to liberalize trade through the General Agreement on Tariffs and Trade (GATT), however, have resulted in a lower overall level of tariff protection in many developed nations. Nontariff barriers, such as quantitative restrictions limiting imports, increased in importance from 1950 until 1990 because they were often more politically acceptable, and nations have sought ways to restrict trade and still comply with international obligations to reduce tariff protection. Consequently, nontariff barriers have become the most widely used means of restricting agricultural trade.

Tariff Barriers.   A **tariff** is a tax levied on goods when they enter a country. Tariffs may be imposed for protection or to generate revenue. Protective tariffs are implemented to insulate domestic producers from import competition. Protective tariffs are not intended to prohibit imports. In negotiations to liberalize trade, tariffs often go through a process of tariff binding, in which their initial level is agreed upon and can be reduced over time, but not raised without additional consultation. Revenue tariffs represent a tax on imports and are implemented to raise revenue for the government. The general effect of tariffs is to raise the domestic price and reduce the quantity demanded. The U.S. tariff on imported melons, for example, results in higher consumer prices for melons and substitute goods, regardless of whether the melons are from foreign or domestic sources.

Revenue tariffs have declined over time. In 1987, the United States relied on tariff revenue for only 1.6% of government revenue, compared with 41% in 1900 (Carbaugh, 1992). Developed nations, such as Switzerland and the United Kingdom, rely on tariffs for 8.2% and 0.1% of total government revenue, respectively. Developing countries, such as Uganda, Brazil, and Sudan, however, still rely on revenue tariffs for up to 62% of total government revenue.

Tariffs may be classified into three broad categories. An *ad valorem* tariff is assessed as a percentage of the value of imported goods. Cantaloupes imported from Mexico, for example, are assessed a tariff of 35% *ad valorem.* A specific tariff is assessed as a fixed amount of money per unit of imported product. Imports of fresh grapefruit are assessed a tariff of $0.029 per kilogram, for example. A compound tariff is calculated on both an *ad valorem* and a specific basis. Goods may be assessed at the rate of 5% *ad valorem* and $20 per ton. *Ad valorem* and specific tariffs are the most common tariffs applied in the trade of food and fiber products.

A nation's prices, production, and consumption can all be affected by a tariff. Figure 22.3 illustrates these effects. It is assumed that this is a small nation whose imports represent only a small share of world market supply. This nation is therefore a price taker, facing a constant world price for imported beef. Under these assumptions, the nation is not large enough to influence the world price, a fairly realistic case for many countries. If D_d represents domestic demand for beef and S_d is

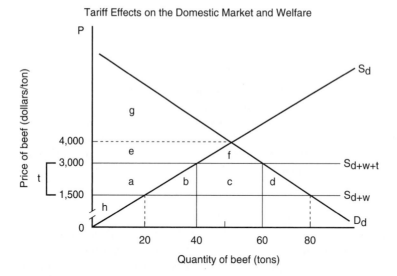

Tariff Effects on the Domestic Market and Welfare

FIGURE 22.3   Free trade equilibrium is denoted by the intersection of S_{d+w} and D_d. Imports equal 60 tons at a price of $1,500/ton. The imposition of a tariff, t, on imports would lower imports to 20 tons (60 tons − 40 tons) and increase domestic production to 40 tons. A tariff has the overall effect of protecting domestic producers from import competition but also causes higher prices for consumers. In this small nation case, the full burden of the $1,500/ton tariff is passed on entirely to consumers. Consumption of domestic output declines to 60 tons.

domestic supply, then in the absence of trade, market price is $4,000/ton, the quantity supplied is 50 tons, and the quantity demanded is 50 tons. With international trade, S_{d+w} is the free trade supply function, indicating the tons of beef available to the small nation from domestic and foreign sources combined. The world price, $1,500/ton, prevails and quantity supplied declines to 20 tons. Quantity demanded rises to 80 tons. Excess demand (80 tons − 20 tons) of 60 tons of beef is supplied by imports. Domestic beef prices have fallen as a result of free trade, making consumers better off because they can buy more at lower prices. Producers, however, are worse off because they sell less at a lower price than before trade.

Under free trade, the domestic beef industry is being damaged by imported products. Sales and profits are declining, producers are going out of business, and workers are losing their jobs. Assume labor and management unite and successfully lobby Congress to impose a specific tariff of $1,500/ton on beef imports. Because this is a small nation and, therefore, a price taker on the world market, world beef supplies remain constant. This new policy raises the domestic beef price by the full amount of the tariff, placing the burden of adjustment entirely on beef consumers. Available supply increases by the amount of the tariff to S_{d+w+t}.

With the protective tariff, a new world market equilibrium price, $3,000/ton, is established. Due to higher prices, domestic consumption declines by 20 tons, from 80 tons before the tariff to 60 tons after the tariff. Production increases by 20 tons, from 20 tons to 40 tons. Because domestic production increases and domestic consumption declines, imports fall to 20 tons. The overall effects of the tariff are to restrict imports, raise domestic prices, and increase production. Domestic beef producers are protected from import competition because prices are raised and imports are reduced. Note that if the tariff were high enough to raise the price to $4,000/ton, beef imports would cease. In such a case, the policy is called a **prohibitive tariff,** because imports are completely eliminated and the market is supplied only from domestic sources.

Consumer and producer surplus can be used to evaluate the welfare effects of tariff policy. Before the tariff, consumer surplus equaled the area of the large triangle, areas $a+b+c+d+e+f+g$. After the tariff, consumer surplus falls to areas $e+f+g$, or an overall loss of areas $a+b+c+d$. Producer surplus increases from area h before the tariff to areas $a+h$ after the tariff for a net gain of area a.

Tariff revenue accruing to the government equals area c. Tariff revenue can be calculated from this example also. The number of imported tons of beef, 20, times the tariff, $1,500/ton, yields government tariff revenue of $30,000.

Areas $b+d$ together represent the dead-weight loss to society from imposing the tariff. Area b is the protective effect of the tariff. It represents the loss to the domestic economy from producing additional beef at higher per unit costs. Resources less suited to beef production are employed when tariff-induced output expands. The protective effect occurs because less-efficient domestic beef production is substituted for more-efficient imported beef. Area d is called the consumption effect and results from the tariff artificially increasing the price of beef. Domestic consumers are denied the opportunity to purchase beef at the lower world price of $1,500/ton and must therefore pay the higher cost of $3,000/ton. This loss in welfare represents a real net cost to society, which is not captured by any other economic agent.

The net welfare effects of a tariff imposed by a small nation are to:

1. reduce consumer welfare,
2. increase producer welfare,
3. raise government revenue, and
4. cause a dead-weight loss to society.

A tariff imposed by a small nation reduces net social welfare.

Nontariff Barriers.   In recent years, barriers to trade other than tariffs have become much more important. **Nontariff barriers (NTB)** include any government policy, other than tariffs, that reduces imports but does not limit the domestic production of goods that substitute for imports. Governments have sought ways to limit imports without violating their international obligations to reduce tariffs. Politicians are often reluctant to levy a tariff because it acts as a tax not only on imports, but on domestic consumers. NTBs may, therefore, offer a more politically acceptable alternative. NTBs include restrictions at national borders, laws and regulations that discriminate against imports, export subsidies, price supports, favorable exchange rates and credit terms for exporters, and other programs or production assistance that substitutes domestic output for imports or stimulates exports.

Import quotas restrict the physical amount of a product that may be imported during a specified period of time. Presently, the single most important NTB is the import quota. Quotas may be global, meaning that they do not discriminate among nations that supply imported goods. Quotas may also be allocated, meaning that each exporting nation is allocated a maximum amount of product that can be shipped.

The U.S. Meat Import Law of 1964 required the president to consider the imposition of import quotas on frozen, chilled, or fresh veal, mutton, beef, and goat meat when it was estimated that imports would reach the level of an adjusted base quota, near 7% of domestic production. The law was amended in 1979 to allow more imports in years when U.S. domestic meat production is low and less imports in years when U.S. meat production is high. This feature is referred to as a countercyclical measure, which attempts to account for the biological nature of the cattle production cycle and weather variability. The law was replaced by a tariff-rate quota under the URA in 1995.

Although special laws, such as that for meat, may be passed by Congress, U.S. agriculture received import protection through basic legislation passed during the Great Depression. Section 22 of the Agricultural Adjustment Act of 1933 authorized the president to impose import quotas or fees if it were determined by the Secretary of Agriculture that imports would interfere with federal price support programs or substantially reduce U.S. production of goods processed from farm commodities. Section 22 was designed to protect the integrity and operation of domestic farm programs. Imports of cotton, peanuts, some dairy products, and some sugars and sugar-containing products have been limited by Section 22 authority in recent years. Since 1935, Section 22 has been used to restrict the import of several important commodities, including wheat and flour, rye, rye flour and meal, almonds, filberts, peanuts and peanut oil, tung nuts and oil, flaxseed and linseed oil, sugars, and syrups.

The effects of a quota are similar to those of a tariff. In Figure 22.3, for example, if a quota of 20 tons were imposed, the domestic price would rise to $3,000/ton. By restricting available supplies on the market, domestic prices for beef could not fall below $3,000/ton, assuming supply and demand conditions remain unchanged. When domestic prices rise due to limited supplies of beef, consumer surplus declines and producer surplus increases as before. An import quota has been referred to as one of the most insidious trade restrictions employed because it satisfies special interest groups by affording higher prices, but limits the physical quantity available on the market, often causing shortages of critical goods and exorbitantly high prices.

Revenue effects of a quota are quite different from the tariff case, however. Area *c* represented tariff revenue to the government imposing the tariff. With a quota, however, this revenue may accrue to importers, exporters, or government. Beef consumers must now pay an additional $1,500 for each of the 20 tons imported under the quota. If U.S. importers organize collectively and bargain with foreign exporters to buy beef on the world market at \$1,500/ton and resell at \$3,000/ton, quota revenue would accrue to the import firms in the form of monopoly, or windfall profits.

Alternatively, foreign exporters could organize as sellers and limit market supplies to 20 tons. This action would drive up the price to \$3,000 and allow the exporters to capture and share the profits. However, as with any cartel scheme, the incentive for each individual exporter to cheat and attempt to undersell the market at a lower price would be great and might lead the collective action to disintegrate.

Finally, the government of the importing nation could auction off import licenses to domestic importers according to the highest bidder. The government could then capture the quota revenue that would have accrued to the importers.

Under a quota, the distribution of revenue among the economic agents—domestic importer, foreign exporter, government—is indeterminate. Windfall profits from the quota are shared among the agents depending upon the degree of bargaining power each possesses. These results lead to the often surprising conclusion that in some instances, exporters may actually prefer to limit their sales to a particular country.

Voluntary export restraints (VERs) are agreements between the government of an importing country and exporters that limit exports to a specified amount. Authority to negotiate VERs is provided to the president in Section 203 of the Trade Act of 1974. The U.S. Trade Representative conducts the actual negotiation. VERs have been used most recently to limit U.S. imports of soft wood lumber from Canada, steel, automobiles, and textiles. Under a VER, the importing nation does not establish a limit on imports, but exporters agree to voluntary limits on the amount shipped during a specified time period. Exporting nations often agree to participate in VER schemes out of fear that the importing country may adopt more severe import restrictions. In practice, VERs are often negotiated between an exporting nation and the U.S. government as Congress is considering the imposition of an import quota. VERs can also benefit exporters.

By limiting the available supplies on the market, a VER leads to higher prices in the importing nation. Higher prices cause a decline in consumer surplus within the importing country. Because the exporting nation limits the amount shipped, it can allocate the rights and determine who receives the windfall profits. Exporters receive those rights in order to not retaliate against the import country. It is not so surprising to see that some nations would agree to limit their own exports to another country. Together, these exporters behave in much the same way as a monopoly.

The variable levy is a key nontariff barrier that was used by the EC to protect domestic agriculture, while transforming the Community from one of the largest net grain importers to one of the largest net grain exporters. A **variable levy** is a variable import tariff equal to the difference between a designated domestic price and the lowest landed import price. The variable levy in the EC was adjusted daily for grains and sugar; weekly for dairy, beef and live cattle, and rice; monthly for olive oil; and quarterly for pork, poultry, and eggs. The common agricultural policy (CAP) of the EC set the minimum price at which any import could enter the market. If the import price fell below this level, a levy was imposed that equaled the difference between the minimum price and the import price. Using this mechanism, the EC has effectively regulated the import of food products into the Community and, therefore, protected agricultural producers from lower-cost, more-efficient foreign competition. The EC has adopted major reform of the CAP by agreeing to reduce support prices, institute direct payments to producers to compensate them for lower prices, and implement supply controls.

Another NTB used by the United States and many other countries to protect agriculture is the **tariff-rate quota (TRQ).** This measure is used extensively by WTO members in order to comply with the URA provisions requiring the conversion of import quotas to their tariff equivalent. The TRQ allows a specified amount or number of products to enter an importing country at one tariff rate, often zero (the within-quota rate), whereas imports above this level are assessed a higher rate of duty (the over-quota rate). Although not often used to restrict imports of manufactured goods, it has been used to limit the import of milk, cattle, fish, brooms, tobacco products, coconut oil, and, more recently, sugar to the United States. In the early 1970s, a TRQ on fluid milk was implemented that set a nonbinding import quota of 3.0 million gallons annually. The within-quota tariff was set at $0.02/gallon, and the over-quota tariff was $0.065/gallon.

The TRQ is viewed as a compromise between consumers, who desire low-cost imports, and producers, who want higher prices and protection from foreign competition. Compromise between these two sets of diverse interests results because the extremely adverse impacts and higher prices from quota are averted, and at the same time producers are offered some insulation from the impacts of severe import competition by the higher over-quota tariff.

To liberalize trade in the future, it is likely that the WTO will rely upon the use of tariff-rate quotas for the tariffication of nontariff barriers. **Tariffication** is the process whereby quotas, licenses, variable levies, and other nontariff barriers to trade are converted to their tariff equivalents. These tariffs are then bound at negotiated tariff rates and reduced over a specified period of time.

Figure 22.4 illustrates the effects of a tariff-rate quota on trade and economic welfare. Domestic supply and demand for a small-nation importer are denoted by S_d and D_d, with an autarkic equilibrium price of $300. With free trade, this nation produces 10 tons of corn, consumes 50 tons, and imports the balance, 40 tons, at a price of $100/ton.

Now assume that a tariff-rate quota of 5 tons is imposed. The within-quota tariff is reflected by S_{d+w+t1}, and the over-quota tariff is higher and denoted by S_{d+w+t2}. Because imports initially exceed the quota amount, both the within-quota and over-quota rates apply. The TRQ results in an increase in corn prices from $100/ton to $200/ton. Domestic production increases to 20 tons, domestic consumption falls to 40 tons, and imports decline to 20 tons. Higher domestic prices and increased production result in an increase in producer surplus equal to area e, and dead-weight losses equal to areas $(f+g)$.

Revenue generated by the TRQ, however, is apportioned between the government as tariff revenue and businesses as windfall profits. In this case, 20 tons of

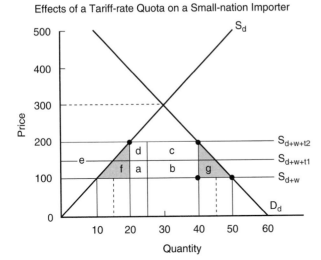

Effects of a Tariff-rate Quota on a Small-nation Importer

FIGURE 22.4   The tariff-rate quota combines both a tariff and a quota. Imports are allowed to enter the market at a low duty rate up to the quota, and imports over-quota are assessed a higher duty rate. The TRQ represents a compromise between consumers who desire low prices, and producers who want high prices.

corn are imported after the TRQ is imposed. The government collects area *a* when tariff revenue is equal to the within-quota tariff, t1 = $50/ton, times 5 tons. Areas *b* + *c* also accrue to government and are found by multiplying the over-quota tariff, t2 = $100/ton, times 15 tons.

Area *d* represents a windfall profit to domestic or foreign businesses. Following the imposition of the TRQ, the domestic price for the first 5 tons of corn increased to $150, reflecting the world price of $100/ton plus the tariff of $50/ton. If importers can obtain foreign corn at $150/ton and resell on the domestic market for $200/ton, or the over-quota price, area *a* would be captured as windfall profit by the importers. However, if foreign exporters can restrict corn shipments, they can force up the price of corn and extract profits from importers, and capture area *a* by raising their supply price to $200/ton. Any portion of area *a* captured by foreign exporters represents a welfare loss to the importing nation.

Domestic Agriculture and Food Policies

Trade is heavily influenced by the type and degree of domestic agriculture and food policies adopted by countries. To achieve domestic policy objectives related to food security, market stability, or economic and social development, many nations, either explicitly or implicitly, subsidize domestic agricultural production or food consumption. Policies to achieve these often-conflicting objectives usually are

1. those that alter market price, thereby influencing production and consumption,
2. those that only affect the production and/or processing of foods, and
3. those that affect consumption or utilization of raw commodities, semi-processed products, or consumer-ready products.

Direct payments to producers, deficiency payments, payments-in-kind, marketing loans, and cash bonuses all directly affect the planting and production decisions of producers. Policies that subsidize production inputs, such as fertilizer, credit, transportation, irrigation, and the development of infrastructure, such as roads, bridges, and storage facilities, affect agricultural production either directly or indirectly. Consequently, these policies often have the same trade-distorting impacts as tariffs and nontariff barriers. Many countries decouple farm program payments to comply with URA provisions. These are less distorting but can still distort resource use in agriculture.

Consumption policies also impact trade. Many developing nations subsidize consumption of food to achieve political objectives and influence large shares of the population who have migrated into urban areas and cities searching for higher paying jobs and a better standard of living. Food consumption policies may include providing staples such as corn, bread, meat, or milk to consumers at below the free market price. Other policies may provide low cost or nearly free food to some parts of the nation, but charge consumers in more affluent areas the full market price. Whichever form or mix of policies are chosen, they almost always have major impacts on trade.

Export Policies

Trade barriers need not apply only to policies that restrict imports. Government control over exports, particularly sales of food, have increased during the past two decades. Fears associated with the prospects of food shortages and high consumer prices resulted in the widespread use of measures to limit the export of food in the 1970s. Because of large crop surpluses during the late 1970s and early 1980s, policies designed to expand or promote exports became popular tools for increasing food shipments.

Policies designed to either restrict or promote agricultural exports are undertaken to:

1. dispose of surplus production or stored commodities,
2. reduce or halt the export of domestic commodities, thereby limiting the potential for consumer price increases,
3. develop or maintain domestic processing industries and increase employment rather than export raw materials,
4. limit the economic or military capability of another nation, and
5. encourage another country to undertake policy reform by denying critical food or fiber supplies.

The use of food as a diplomatic tool became important to the United States during the 1970s and 1980s. The export of U.S. soybeans was embargoed in 1973 to avoid increases in domestic prices. Grain exports to the former Soviet Union were limited in 1974 and again in 1975 due to public concern regarding the prospects of food shortage and the resultant impacts on consumer prices. In 1980, the United States suspended the sale of wheat and corn to the Soviet Union because of its invasion of Afghanistan. Although the ban was lifted in 1981, many observers claim that the reputation of the United States as a reliable supplier of food products came into serious question. As a result, legislation has been passed by Congress that limits the president's authority to single out agricultural goods for export embargo.

Because the EC sets minimum internal producer prices well above the world price, surplus production results. This surplus is disposed of on the world market using another nontariff barrier, the export subsidy, or restitution. An **export subsidy** is a direct cash or in-kind payment provided by a government to encourage export sales. The **export restitution** used by the EU is normally a cash payment to exporters of a specified commodity and equals the difference between the higher, domestic EU market price and the lower world price. Surplus disposal by the EU, and the associated decline in world prices, has often resulted in threats or actual trade retaliation by other major exporting nations, particularly the United States, Australia, and Canada.

To counter EU export restitutions, the 1985 Farm Bill contained provisions for the **export enhancement program (EEP).** EEP was designed as a tool to assist U.S. producers in regaining markets lost to subsidized EC wheat and flour exports. EEP is an export incentive program that uses surplus commodities or cash payments as export bonuses to make U.S. products more competitive on the world market. Although EEP has accomplished its primary objective of regaining lost markets, it has invited complaints from trading partners and allies. EEP also displaces commercial sales, lowers world prices, and results in a welfare loss for U.S. consumers. EEP must be reduced 21% in tonnage and 36% in value to comply with URA provisions.

The Market Access Program (MAP) was authorized by the 1996 FAIR Act to stimulate the export of high-value food and agricultural products. The MAP provides assistance in cash or commodities to trade promotion associations to partially fund foreign market development, advertising, and promotional activities. Although MAP was designed to create commercial markets for U.S. products, it has been criticized because a few, large multinational firms have received substantial government assistance, perceived as a subsidy under this program. Funding for MAP was subsequently reduced by nearly one-half. MAP was referred to as the Market Promotion Program (MPP) in the 1990 Farm Bill and as the Targeted Export Assistance (TEA) program in the 1985 Farm Bill.

Export policies can have major impacts on the trade of food and agricultural products. Policies designed to restrict exports obviously affect the perception of the exporter as a reliable supplier, leading to greater uncertainty, reduced purchases by other nations, and possible retaliation by importing nations. Market disruption occurs, which leads to greater variability in trade volumes, increased uncertainty by private and public decision makers, and resource misallocation and inefficiencies in production and marketing. Domestic consumers and foreign producers may experience short-term gains from these policies because domestic prices fall and world prices rise; however, domestic producers and foreign consumers generally lose. Export embargoes can be especially disruptive and painful for foreign consumers if imposed during times of tight world food supplies. These measures may also damage the reputation of the implementing nation as a serious advocate of freer trade.

Policies that subsidize exports can be equally disruptive to world markets. EU export restitutions and the U.S. EEP have caused great concern among those nations that claim that they do not subsidize agricultural trade. Direct export subsidies lower the world price and consequently invite criticism and retaliation from nations that may be competitors but are also important political allies, such as Australia and Canada. Foreign consumers and domestic producers gain from export subsidies, and domestic consumers and foreign producers lose. Although export subsidies may be effective in achieving short-term policy goals, they are almost always costly in terms of both money and image.

AGRICULTURAL TRADE POLICY MAKING

The United States faces critical policy choices as it adjusts to changes emanating from the international marketplace. Agricultural trade policy has taken on added importance as trade has become crucial to the future prosperity of U.S. agriculture. Policy formulation related to trade, domestic programs, and economic growth and development occurs in a **multilateral** setting. Major international institutions and policy-making bodies now provide an important linkage between efforts to reduce distortions to trade and accomplish social goals to support sustainable domestic agricultural production, and the need to ensure consumers of a low-cost, safe food supply. The industrial nations, particularly the United States, Japan, and the EU countries, have received harsh criticism from trading partners for failing to eliminate agricultural subsidies that distort trade.

The General Agreement on Tariffs and Trade

The General Agreement on Tariffs and Trade (GATT) resulted from the Bretton Woods Agreement negotiated at Bretton Woods, New Hampshire, in 1947. The GATT was one of the three global institutions created after World War II to assist in rebuilding nations devastated by war and fostering economic recovery worldwide. Although GATT was replaced by the **World Trade Organization (WTO)** in 1995, its principles were incorporated into WTO. The International Monetary Fund (IMF) and the International Bank for Reconstruction and Development (IBRD), or the World Bank, were the other two. The GATT emerged as the multilateral body governing trade after the failure of many of the 22 member nations and the United States to ratify the International Trade Organization. GATT's primary objective was to encourage economic growth and development by liberalizing world trade. Although GATT was a multilateral agreement establishing the rules governing international trade, it also served as an institutional forum for negotiating reductions to trade barriers and for settling trade disputes. GATT was replaced by the World

Trade Organization and has 135 member nations, which account for more than 90% of world merchandise trade.

Several key principles governed the operation of GATT, and subsequently the WTO. First, and most important, is the **most-favored-nation (MFN)** clause, which states that trade must be nondiscriminatory. Concessions granted to one nation automatically apply to all other WTO members. Second, tariffs, rather than nontariff barriers, should be used to protect domestic industries. Third, agreed-upon tariff levels are binding. If duties are raised by one country, compensation may be required by its trading partners. Finally, imported goods should be treated no less favorably than domestically produced goods.

In seven previous rounds of negotiations, GATT was effective at reducing tariff protection in the manufacturing sector from an average of 50% in 1947 to 5% in the mid-1980s. However, protection for agriculture in industrial countries rose from 21% in 1965 to more than 40% during the same period (OECD).

Despite the overall increase in protection of agriculture, some important concessions were granted in agricultural trade. Most notable was during the Dillon round when the EEC granted the United States zero tariff binding on imported soybeans and corn gluten feed. Next was the Tokyo Round, which led to Japanese concessions to liberalize its market for imported beef and citrus. Major attempts to bring agriculture under GATT discipline were unsuccessful prior to 1986.

Important exceptions to GATT rules exist. Many of these exceptions relate to agricultural trade. The United States requested and received a waiver from GATT to employ quotas and other import restrictions on imported agricultural products that interfere with the operation of domestic farm programs. Other nations have also used this exemption, most notably the EC, to impose a variable levy that blocks many agricultural imports. Although GATT prohibited the use of export subsidies, agricultural products are exempt if such subsidies do not distort "historical" market shares. These exemptions led to a proliferation of nontariff barriers to trade and the current level of distortion in international trade of food and agricultural products.

The Punta del Este Declaration of September 1986 launched the eighth and final GATT round, referred to as the Uruguay Round. All GATT contracting parties affirmed their commitment to liberalize agricultural trade. It was anticipated that the Uruguay Round would allow a multilateral reformulation of agricultural policies, leading to lower taxpayer costs and less market distortion. The Uruguay Round represented the first serious attempt to liberalize agricultural trade. Previous piecemeal approaches to reform domestic and agricultural trade policies had failed, and by the mid-1980s, the United States and other world powers were concerned that the cost of protecting agriculture had become excessive.

The initial U.S. proposal in the Uruguay Round was to press for the comprehensive reform of agricultural trade. The U.S. proposal called for a 10-year phaseout of all agricultural subsidies that directly or indirectly distort trade, a 10-year phaseout of all import barriers, and the harmonization of health and sanitary regulations based on internationally agreed-upon standards. The United States insisted that agriculture should respond to market signals rather than government policies and that it was in the best interests for world economic development if nations adjusted to specialize in those areas where they possess a long-run comparative advantage. This sweeping proposal for reform met with strong opposition from members of the EC and Japan.

After three years of negotiation, the parties issued a mid-term review reflecting the status of progress in the Uruguay Round. It was significant that all parties could agree to "substantial, progressive reduction" in those policies that distort trade. Differences among the United States, the EC, and Japan again resulted in little real progress to reform trade. A special session was held at Blair House, Virginia, in November 1992 to settle the differences between the United States and EC. During this session, the United States and EC agreed to adopt the following measures to liberalize agricultural trade:

1. reduce domestic support by 20%—U.S. and EC deficiency payments would be exempt.
2. reduce subsidized export tonnage by 21% and expenditures by 36%.
3. implement tariffication of nontariff barriers and bindings under GATT.

This dispute became so heated that French farmers organized blockades of major roadways to stall traffic, attract public attention, and develop public support for their position. France threatened to exercise its right to veto the agricultural trade reforms proposed by the EC. The Uruguay Round was three years beyond its scheduled completion date when finalized. Although the final outcome fell short of expectations, it represents an important first step to agricultural trade liberalization. The stage is now set to further reduce barriers to agricultural trade in further future negotiations.

The United Nations Conference on Trade and Development

The United Nations Conference on Trade and Development (UNCTAD) was formed in 1964 as an agency of the General Assembly of the United Nations to stimulate the economies of developing countries through trade. These nations, concerned about domination of GATT and IMF by industrial countries, believed that their needs were not being met (Tweeten, 1992). Many developing countries were also convinced that a conspiracy existed because prices for agricultural products had fallen relative to prices for manufactured products. With a majority of votes in the United Nations, developing countries successfully established the UNCTAD in 1964.

UNCTAD has the goal to eliminate or reduce tariff and nontariff barriers to trade levied by industrial countries. In recent years, UNCTAD has served as a forum for industrial and developing countries to discuss trade issues. Any member of the United Nations may join UNCTAD.

The most significant policy development occurring under the auspices of UNCTAD was the granting of nonreciprocal trade concessions to developing countries by industrial countries. The **generalized system of preferences (GSP)** permits duty-free entry into industrial countries of many manufactured and agricultural products produced and exported by developing countries. The objective of GSP is to expand exports by developing countries to bring about economic development. The United States grants duty-free entry to developing countries for about one-half of the items classified in the tariff schedule of the United States. The basic U.S. legislation granting these tariff concessions was the Trade Act of 1974. Agricultural products entering the U.S. market under GSP include okra, citrus, melons, nuts, cucumbers, and sugar. More than 100 developing nations now receive preferential tariffs under the GSP for products shipped to the United States.

The GSP was legal under the GATT because a waiver of most-favored-nation status was granted in 1971. Under this waiver, industrial countries may grant special, low tariffs to developing countries without extending those same tariff preferences to other industrial countries.

U.S. Agricultural Trade Policy Formulation

Although the president and Congress are both responsible for U.S. agricultural trade policy, Congress has delegated much of its authority to the president. The Executive Branch has the overall responsibility to formulate and administer the everyday operational details of trade policy. The Office of the U.S. Trade Representative (USTR), the U.S. International Trade Commission (ITC), and the U.S. Department of Agriculture have specific roles in implementing various aspects of U.S. trade policy related to agriculture.

The U.S. Trade Representative.   The USTR is the primary agency responsible for all U.S. trade policy matters. The USTR is a cabinet-level position created by executive order in 1980 when the president moved the office from the Department of State to the cabinet. The USTR holds the rank of ambassador. The USTR is responsible for all trade activities related to the WTO, bilateral and multilateral trade negotiations, direct investment incentives and disincentives, and negotiations before the United Nations that relate to trade and before any other organization where trade is the primary issue. Public input from business, individuals, and members of Congress is received by the USTR on matters related to trade and investment.

The Economic Policy Council

Interagency coordination is critical to the effective formulation and implementation of U.S. trade policy. The Economic Policy Council (EPC) is the primary unit of the Executive Branch for coordinating trade policy (Tweeter, 1992). The EPC is composed of the Secretaries of Agriculture, Commerce, Defense, Labor, Transportation, Treasury, and State. The Secretary of Treasury serves as the President *Pro Tempore*. The Director of the Office of Management and Budget and the Chairman of the Council of Economic Advisers also serve on this interagency committee.

The president also has more than 40 other committees that serve in an advisory role in evaluating and formulating trade policy. The Advisory Committee for Trade Policy Negotiations is the lead committee charged with advising the president on trade policy negotiations. The ACTP is composed of private sector business interests and is usually administered by a departmental staff assistant. The Agricultural Policy Advisory Committee (APAC) represents the interests of Agriculture. It is composed of private sector representatives from producer, processor, cooperative, trade, and other organizations. The Agricultural Technical Advisory Committee is composed of ten sector committees related to cotton, dairy, grain and feed, livestock and products, oilseeds and products, poultry, processed foods, sweeteners, and tobacco.

The U.S. International Trade Commission.   The U.S. International Trade Commission (ITC) is an independent government agency charged with investigating U.S. trade law violations. A chairman, vice-chairman, and four commissioners appointed by the president make up the ITC. Commissioners serve nine-year terms and may not be reappointed. The ITC conducts inquiries into the laws of the United States and other countries, examines complaints related to foreign competition, and compares imports with U.S. consumption levels. Injuries to domestic industries are investigated, and when appropriate, the president may impose import protection.

Investigations of possible agricultural trade law violations usually involve complaints related to Sections 201 and 301 of the Trade Act of 1974 and the Tariff Act of 1930. Section 201 allows the president to impose duties or other import restrictions on imports that threaten U.S. industries producing like goods. This provision is intended to provide temporary relief from injurious competition. The Tariff Act of 1930 authorizes investigation of cases in which imports may be subsidized and dumped at less than fair value on the U.S. market. This act, commonly referred to as the Smoot-Hawley Act, established the highest tariff rates ever imposed by the United States, largely in response to high domestic unemployment during the Great Depression. It remains the cornerstone of U.S. import legislation (Miller, 1985).

SUMMARY

The purpose of this chapter was to compare free trade with more restrictive trade policies such as tariffs and quotas. Social welfare resulting from these policies was examined, emphasizing the net gains to both exporting and importing nations. Protection of agriculture was discussed, focusing on reasons why government intervention in agricultural trade is so prevalent. The role and importance of the General Agreement on Tariffs and Trade (GATT) was examined, along with a discussion of waivers granted for agriculture, which have led to the proliferation of nontariff barriers to trade. Significant reductions in agricultural trade barriers in each of the seven previous GATT rounds were noted. The major points made in this chapter may be summarized as follows:

1. Free trade affects exporting and importing nations differently. With freer trade, prices in the exporting country rise, production increases, consumption declines, and exports rise. Prices in the importing nation decline, resulting in lower production, higher consumption, and increased imports. In summary, producers in the exporting nation and consumers in the importing country gain from free trade, and consumers in the exporting country and producers in the importing country lose. There is a net gain in social welfare associated with free trade. If gainers are willing to compensate losers, both sets of economic agents will become better off from free trade.

2. Although there are gains to trade, protectionism in agriculture abounds. Domestic policy goals related to food security and self-sufficiency often transcend the economic benefits of free trade. Protective nontariff barriers now dominate much of the trade in food and agricultural products. Arguments against free trade are often based on the need to protect a new industry, counter unfair foreign competition, improve the balance of payments, and protect national health, the environment, or food safety.

3. Trade restrictions can take the form of tariff or nontariff barriers. A tariff is a tax levied on goods when they enter a country. When a tariff is imposed by a small nation, consumer welfare declines, producer welfare and government revenue both increase, but net social welfare is reduced when compared with free trade. Nontariff barriers are government policies, other than tariffs, that reduce imports but do not limit domestic production of goods that substitute for imports. Import quotas, variable levies, voluntary export restraints, tariff-rate quotas, and other domestic policies are included. Nontariff barriers are now the most important set of barriers limiting agricultural trade. Revenue generated by a quota may accrue to government, importers, or exporters as windfall profits depending upon the bargaining power of each economic agent.

4. Agricultural trade policy making has taken on added importance as trade has become more crucial to the future prosperity of U.S. agriculture. The General Agreement on Tariffs and Trade (GATT), now the World Trade Organization (WTO), is a specialized agency of the United Nations designed to encourage economic growth and development by liberalizing international trade. WTO's 125 member nations account for more than 90% of total world trade. Seven rounds of GATT negotiations failed to make substantial progress to liberalize trade in agriculture, because many trade barriers are linked to domestic policies supporting farm income or prices. Previous rounds have also focused on the reduction of tariffs on industrial products. The Uruguay Round Agreements (URA) represent the first serious attempt to liberalize agricultural trade. Progress to substantially reduce trade-distorting domestic agricultural policies led to a major confrontation between the United States—a proponent of trade liberalization—and some of its major trading partners and allies, such as Japan and France, which oppose reductions in farm support measures. Divisiveness characterized the Uruguay Round and limited its ability to completely eliminate trade barriers affecting agriculture.

5. U.S. agricultural trade policy formulation and implementation is the joint responsibility of the Executive Branch and Congress. Day-to-day operations are delegated to the president and operationally handled by the Office of the U.S. Trade Representative, the U.S. International Trade Commission, and selected Secretaries and interagency advisory committees. Agriculture has specially appointed committees that interact with the USTR and Congressional committees to provide input on agricultural trade policy issues and technical matters. Coordination of trade policy is necessary to ensure effective and timely implementation of U.S. laws and regulations designed to influence trade in food and agricultural products.

DEFINITION OF KEY TERMS

Arbitrage: the process of purchasing commodities in one market at a low price and rapidly selling them in another market at a higher price.

Autarky: each nation is self-sufficient and no trade occurs between it and any other nation.

Excess demand: the amount by which quantity demanded exceeds quantity supplied at each price below equilibrium.

Excess supply: the amount by which quantity supplied exceeds quantity demanded at each price level above equilibrium.

Export enhancement program (EEP): an export incentive program that uses surplus commodities or cash payments as export bonuses to make U.S. products more competitive on the world market.

Export restitution: measure used by the European Union; normally a cash payment to exporters of a specified commodity that equals the difference between the higher, domestic EU market price and the lower world price.

Export subsidy: a direct cash or in-kind payment provided by a government to encourage export sales.

Free trade: the absence of direct or indirect government intervention to alter market prices and quantities.

General Agreement on Tariffs and Trade (GATT): the multilateral organization that resulted from the Bretton Woods Agreement in 1947 with the primary objective of encouraging economic growth and development through the reduction of tariff and nontariff barriers to trade.

Generalized system of preferences (GSP): permits duty-free entry of many manufactured and agricultural products into industrial countries.

Import quotas: a policy that restricts the physical amount of a product that may be imported during a specified period of time.

Most-favored-nation (MFN): GATT principle that states that trade among member nations must be nondiscriminatory. Concessions granted to one nation automatically apply to all other WTO members.

Multilateral: refers to negotiations among many nations, as opposed to two nations, which is bilateral; working together for a common purpose such as reducing trade barriers. The WTO is an example of a multilateral institution.

Nontariff barriers (NTBs): government policies, other than tariffs, that reduce imports but do not limit domestic production of goods imported as substitutes.

Prohibitive tariff: a tariff so high that imports are completely eliminated and the market is supplied only from domestic sources.

Protectionism: any government policy implemented to remedy domestic economic problems associated with trade.

Sanitary and phytosanitary regulations (SPS): domestic regulations that require that food, animal, and plant products are free of harmful diseases or pests.

Section 301: provision of the Trade Act of 1974 and its counterpart in the Trade and Competitiveness Act of 1988 that was designed to provide presidential authority to impose duties on products from nations whose trade practices were deemed "unfair" or that restrict U.S. commerce.

Tariff: a tax levied on goods when they enter a country. Tariffs may be imposed for protection or to generate revenue (also referred to as a duty).

Tariffication: the process whereby quotas, licenses, variable levies and other nontariff barriers to trade are converted to their tariff equivalents.

Tariff-rate quota (TRQ): allows a specified amount or number of products to enter an importing country at one tariff rate (the within-quota rate), whereas imports above this level are assessed a higher rate of duty (the over-quota rate).

Trade liberalization: the removal of government policies that restrict a nation's imports or exports.

Variable levy: a variable import tariff (duty) imposed by the European Union equal to the difference between a predesignated domestic price and the lowest landed import price.

Voluntary export restraints (VERs): an agreement between the government of an importing country and exporters limiting exports to a specified amount.

World Trade Organization (WTO): created in 1995 to provide a single institutional framework encompassing all articles of the GATT and all agreements and arrangements concluded under the Uruguay Round of GATT. Timely dispute settlement and establishment of a trade policy review mechanism are two important functions of the WTO.

REFERENCES

Carbaugh RJ: *International economics,* 4th ed., Belmont, Calif, 1992, Wadsworth.

Houck J: *Elements of agricultural trade policies,* New York, 1986, Macmillan.

Ingram JC, Dunn RM: *International economics,* 3rd ed., New York, 1993, John Wiley & Sons.

Miller WJ: *Encyclopedia of international commerce,* Centerville, Md, 1985, Cornell Maritime Press.

Runge CF: The assault on agricultural protectionism, *Foreign Affairs,* Fall 1988.

Secretariat, Organization for Economic Cooperation and Development: *Agricultural policies, markets, and trade: monitoring and outlook 1993,* Paris, 1993.

Tweeten L: *Agricultural trade: principles and policies,* Boulder, Colo, 1992, Westview Press.

EXERCISES

1. Under autarky, each nation operates as a closed economy and market equilibrium is determined solely by the interaction of domestic supply and demand conditions. T F

2. Excess supply is the amount by which quantity demanded exceeds quantity supplied for each price level above equilibrium. T F

3. Protectionism occurs when government policy is implemented to remedy domestic economic problems associated with excessive exports. T F

4. The infant industry argument says that new firms need to be protected until their products become widely known in the market. T F

5. Nontariff barriers reduce imports and limit the domestic production of goods that substitute for imports. T F

6. Tarrification is the process whereby quotas, licenses, variable levies, and other nontariff barriers to trade converted to their tariff equivalents. T F

7. Some of the net welfare effects of a tariff imposed by a small nation are:

   a. to reduce consumer welfare, increase producer welfare, and raise government revenue.

   b. to increase consumer welfare, decrease producer welfare, and raise government revenue.

   c. to increase consumer and producer welfare, and raise government revenue.

   d. none of the above.

8. A quota represents:

   a. a quantitative restriction on the amount of a good imported.

   b. a combination of *ad valorem* and specific duties.

   c. a compromise between taxpayers and consumers.

   d. none of the above.

9. Nontariff barriers (NTB) include which of the following?

    a. export subsidies, specific tariffs, restrictions at national borders.

    b. export subsidies, *ad valorem* tariffs, price supports.

    c. price supports, restrictions at national borders, favorable credit terms for exporters.

    d. price supports, favorable credit terms for exporters, specific and *ad valorem* tariffs.

10. The three global institutions created after World War II to assist in rebuilding nations devastated by war and fostering economic recovery worldwide are:

    a. GATT, World Bank, International Bank for Reconstruction and Development.

    b. United Nations Conference on Trade, GATT, International Monetary Fund.

    c. GATT, International Monetary Fund, International Bank for Reconstruction and Development.

    d. International Monetary Fund, International Bank for Reconstruction and Development, United Nations Conference on Trade.

11. The World Trade Organization:

    a. is a multilateral agreement establishing the rules for governing international trade.

    b. serves as an institutional forum for negotiating reductions to trade barriers and trade disputes.

    c. has the main objective of encouraging economic growth and development by liberalizing world trade.

    d. all of the above.

12. Identify and discuss the four general categories of arguments against trade. Discuss at least four reasons for undertaking policies designed to either restrict or promote agricultural exports.

13. Discuss the Uruguay Round of GATT negotiations and the measures taken to liberalize agricultural trade.

14. Discuss the role of the U.S. Trade Representative, the Economic Policy Council, and the U.S. International Trade Commission in implementing various aspects of U.S. trade policy related to agriculture.

23

EMERGING ISSUES IN AGRICULTURAL TRADE: THE FORMATION OF PREFERENTIAL TRADING ARRANGEMENTS

The greatest meliorator of the world is selfish, huckstering trade.

Ralph Waldo Emerson

TOPICS OF DISCUSSION

Before the 1980s, the United States depended upon GATT for increased market access and resolution of trade disputes. During the second half of the 1980s, however, the United States began to move toward preferential bilateral or regional trading arrangements known as **preferential trading arrangements** (PTAs). PTAs provide for the elimination of tariff and nontariff barriers (import quotas and licenses) to trade. Investment laws, transportation regulations, and mechanisms for resolving trade disputes are often included.

THE IMPORTANCE OF PREFERENTIAL TRADING ARRANGEMENTS

In many cases, PTAs are used to strengthen existing trade ties among nations, creating opportunities for additional employment and stimulating investment. More recently, concerns about air and water pollution, worker safety, and unfair labor regulations have led to the inclusion of environmental and labor issues in the negotiation of PTAs.

The **Reciprocal Trade Agreements Act of 1934 (RTA)** authorizes the president to fix tariff rates, and has resulted in a liberal tariff stance for the United States. From 1934 to 1947, in fact, the United States negotiated bilateral tariff reduction agreements with 29 nations. However, when the GATT emerged as the primary trade forum, the RTA declined in importance as a mechanism for trade liberalization.

Although PTAs are relatively new in U.S. trade policy, many nations have used and participated in various forms of preferential trade for decades. Currently there are more than 23 forms of preferential trading arrangements among the 119 countries that account for 82% of world trade (Fieleke, 1992). The use of PTAs to achieve both domestic and international trade policy objectives is clearly increasing.

Since 1985, the United States has negotiated bilateral PTAs with Israel and Canada, and a regional agreement with Mexico and Canada. Public concerns regarding the ability of the United States to compete with low-wage economies that produce low-cost goods for export have been raised. The loss of U.S. jobs to countries such as Mexico, the environmental consequences of more-open common borders and increased industrialization in Mexico, and the prospects of the world economy fracturing into openly hostile trading groups are major public policy issues related to the use of PTAs as trade policy tools. Some writers doubt that free trade is achievable, particularly in highly protected sectors such as agriculture.

In September 1985, the United States and Israel entered into a bilateral agreement to reduce tariff and nontariff trade barriers in merchandise trade. Services trade was liberalized and provisions were made to protect intellectual property rights.

The Canada–United States Trade Agreement (CUSTA) went into effect on January 1, 1989. It provides for the elimination of both tariff and nontariff barriers to trade between the two countries, specifies rules governing trade in services, and reduces restrictions on investments by both nations over a 10-year transition period.

The North American Free Trade Agreement (NAFTA), submitted to the U.S. Congress on November 3, 1993 was implemented on January 1, 1994. NAFTA will eliminate tariff and nontariff barriers to trade among the United States, Mexico, and Canada over a 5-, 10-, or 15-year period. NAFTA substantially reduces or eliminates most agricultural trade barriers between the United States and Mexico, but does not liberalize U.S.–Canada agricultural trade.

FORMS OF ECONOMIC INTEGRATION

For the most part, PTAs entered into by the United States have taken the form of free trade areas. A **free trade area** consists of provisions to remove both tariff and nontariff barriers to trade with members, while retaining individual trade barriers with nonmember nations. Trade barriers with the rest of the world differ among member nations and are determined by each member's policymakers. The **North American Free Trade Agreement (NAFTA)** is an example of a free trade area. However, free trade areas only represent one form of economic integration.

A **customs union** requires that members eliminate trade barriers with each other and takes economic integration a step further by establishing identical barriers to trade against nonmembers, most often in the form of a common external tariff. The European Community was probably the world's most well-known customs union. It includes a common agricultural policy (CAP) for all 15 member states. CAP is characterized by a variable levy system that limits imports from nonmembers by applying a common variable duty.

A **common market** includes all the aspects of a free trade area and customs union, but takes integration even further by permitting the free movement of goods and services, labor, and capital among member nations. In January 1993, the EC moved closer to full common market status by implementing the Single European Act (SEA). Further unification of the EC by creating a single, barrier-free market for most goods and services will be attempted. The SEA calls for elimination of barriers to trade in goods and services, labor, and capital.

Finally, an **economic union** represents the most complete form of economic integration. Member nation social, taxation, fiscal, and monetary policies are harmonized or unified. Belgium and Luxembourg formed an economic union in the 1920s. Customs documents and procedures, value added taxation, and nontariff barriers to trade have been harmonized to facilitate the movement of goods among EC member nations. However, more contentious issues, such as implementing a common currency, delayed the ratification of the Treaty of Maastricht, the document that completes the economic integration of the EU. Finally in late 1993, plans to move the EC to an economic union were approved, calling for common banking laws, coordinated macroeconomic policy, and a common EU currency by the year 1999.

Various forms of economic integration have occurred within the framework of the General Agreement on Tariffs and Trade (GATT), and now the WTO. Although Article I of GATT prohibits the use of preferential tariff rates, an important exception is allowed by Article XXIV. This latter article specifies the conditions under which signatories may form and legally operate PTAs. These conditions are: (1) the elimination of trade barriers on substantially all trade among members, (2) remaining trade barriers against nonmembers are not more severe or restrictive than those previously in effect, and (3) interim measures leading to the formation of the agreement are employed for only a reasonable period of time. When these three conditions are met, international trade agreements do not violate any of the GATT articles.

Vague phrases such as "substantially all trade" and "reasonable time period" have allowed much latitude for interpretation of the articles. Since its inception, GATT has been notified of more than 70 PTAs, some establishing interim agreements with no final date for completing the PTA being specified. None have been formally disapproved (Jackson, 1989). PTAs were allowed to proliferate under GATT and it remains to be determined if the WTO will take a less conciliatory stance toward PTA formation.

REASONS FOR PREFERENTIAL TRADING ARRANGEMENTS

The emergence of the United States as a major player in the formation of regional trading blocs has raised questions about why PTAs have become such a popular policy tool and why the United States is seeking to participate. Reasons might include the following:

1. to provide for special trading arrangements with countries that are economically or politically important to the strategic interests of the United States,
2. to achieve timely, substantial reduction in barriers to trade, particularly agriculture, intellectual property rights, services trade, nontariff barriers, and dispute settlement procedures,
3. to counter the economic and political power created in Europe by further integration of the EU and the prospects for trade and economic cooperation with former Eastern bloc nations, Central and South America, and Asia,
4. to reduce the flow of illegal immigration and other side effects of trade barriers by stimulating economic growth and development, and
5. to foster political stability and economic prosperity, thereby supporting the continuation of the democratic process and reducing the likelihood of political and social disruption (Rosson, Runge, Hathaway, 1994).

Preferential Trading Arrangements and the GATT

The GATT, along with the World Bank and the International Monetary Fund, was created to prevent the reemergence of trade protectionism and the slow economic growth that preceded World War II. Although the GATT was successful at reducing tariffs in manufacturing from 50% in 1947 to about 5% in the mid-1980s, trade in agriculture, services, intellectual property rights, and trade-related investment were not effectively dealt with during previous GATT rounds (Runge, 1992). The Uruguay Round was convened as an attempt to liberalize trade in these more sensitive sectors. Rising farm program costs, the inability to separate international agriculture from domestic farm programs, and trade distortions caused by government intervention in agriculture all contributed to the complex set of policy issues facing trade negotiators as the Uruguay Round opened. As Runge (1988) points out, "If farm policy is to be liberal, rather than protectionist, it must reflect greater market orientation both at home and abroad, and reduce the distortions that now separate the domestic from the global markets." PTAs, therefore, may provide an effective alternative to multilateral trade negotiations, because it is often easier to liberalize trade among a few, rather than many, countries.

Counter Economic and Political Power in Other Parts of the World

NAFTA created one of the world's largest free trade areas, consisting of 385 million people, $8.0 trillion in economic output, and more than $2.0 trillion in trade. The EU has 370 million people, $8.0 trillion in economic output, and trade of $3.8 trillion. However, the EU already has extended preferential market access to Poland, Hungary, the Czech Republic, Estonia, and Slovenia and has expanded to 15 members with the inclusion of Austria, Finland, and Sweden on January 1, 1995. As economic and political integration in Europe continues, it is likely that the United States will experience more pressure to form larger and economically stronger PTAs in the Western Hemisphere.

In 1991, Argentina, Brazil, Paraguay, and Uruguay initiated the Southern Cone Common Market, MERCOSUR, to foster economic growth and mutual interests among the member nations. Although smaller in economic strength than NAFTA, MERCOSUR has stimulated interest and concern among many publics. Some analysts speculate that the U.S. Enterprise for the America's Initiative, designed to stimulate investment and reduce debt in the Caribbean and Latin America, may be a key first step to linking the economies of the Americas through trade.

Though less well organized than the EU or NAFTA, five Asian nations are attempting to strengthen trade ties. Japan, Singapore, Hong Kong, South Korea, and Taiwan have developed plans to extend preferential trading terms to each other. Although this does not constitute a formal trading relationship, it may create the economic incentive for the formation of a free trade area or other trading arrangement in the near future.

Economic and political interests of the United States may dictate that, as economic blocs emerge around the world, steps must be taken to ensure that existing and future trading relationships are maintained, and even strengthened, by participation in one or several forms of economic integration. As Salvatore (1993) points out, any free trade area or customs union acting as a single unit in international trade negotiations will likely possess much more bargaining power than the total of its members acting separately. Further, reaching a compromise with a unified group of nations may sometimes be easier than dealing with each nation individually.

Reduce Side Effects

PTAs may be used to encourage member nations to undertake economic, social, or political reform. For instance, over the long-run, NAFTA will create an economic incentive for Mexican workers to seek employment in Mexico, rather than in the United States. It is doubtful that any other policy action taken by Mexico could or would have been as instrumental in instituting policy reform as NAFTA. This could have other consequences for the United States, such as reducing the poverty in border colonies and eliminating the strain on border health care, water, sewage, and other facilities required to accommodate a near doubling of population over the last decade. It will also reduce the flow of low-wage labor to the United States and increase costs in some industries.

Other reforms include an increased effort on the part of the Mexican government to control air and water pollution in its major cities and along its border with the United States. There is evidence that the government shut down the state-owned and operated petroleum refinery in Mexico City to indicate to the United States that Mexico was serious about environmental degradation and its consequences. Sewage treatment facilities have been constructed in border cities to stop the dumping of untreated waste into the Rio Grande and its tributaries. Meat packing facilities have been relocated outside of cities to improve air quality and reduce problems with disposal of animal waste.

Foster Political Stability and Economic Prosperity

Probably an overriding goal of the United States in liberalizing trade with Mexico was to foster the continuation of democratic government and political stability of the past few decades. More open markets and the ensuing economic growth may be one of the most effective means of improving the standard of living in Mexico and ensuring that economic prosperity leads to a stable political environment.

Although various reasons may cause the United States to participate in PTAs, it is likely that strong economic and political interests will be the driving force behind such efforts. In the future, leverage may be needed to negotiate continued or additional access to major markets. Negotiating from the strength of an economic bloc may prove to be one effective alternative.

DO PREFERENTIAL TRADING ARRANGEMENTS CREATE OR DIVERT TRADE?

Economic integration that leads to the formation of regional trading blocs can have both trade-creating and trade-diverting effects. One argument against PTAs is that they may lead to increased trade within the bloc, but less trade outside the bloc. Arguments in support of PTAs state that they may lead to the merger of regional blocs and, therefore, multilateral trade liberalization more quickly than other forms, such as the WTO. Although both arguments have merit, theory indicates that the effects of PTAs are largely empirical and depend upon several important conditions, which will be discussed in the next section.

With the prevalence of PTAs, a question of critical importance is whether or not the United States will become better off economically as the economies of North America become more closely integrated. Each case of PTA creation must be examined empirically and on its own merits. It is impossible to assert unequivocally that PTAs may be more efficient than free trade among all nations. Given that trade is less than free, the creation of a PTA may or may not result in welfare gains to society.

Recent advances in international trade theory, however, allow some conclusions to be made about the likely outcomes of PTA creation. Viner (1950) and Johnson (1965) were among the first to explain the costs and benefits of PTAs through the example of a customs union. These writers focused primarily on trade creation and trade diversion, along with total welfare gains. Later efforts by Salvatore (1993) focused on the dynamic effects of PTA creation.

Static Effects

The static effects of creating a PTA are measured by trade creation and trade diversion. Trade creation occurs when some domestic production of a member nation is replaced by lower-cost imports from another member nation. Assuming full employment of domestic resources, trade creation increases the economic welfare of member nations, because it leads to greater specialization in production and trade, based on comparative advantage. PTA members will import from one another certain goods not previously imported at all due to high tariffs. The trade-creation effect results in efficiency gains for member nations because some members shift from a higher-cost domestic source of supply to a lower-cost foreign source. A trade-creating PTA may also increase the welfare gains of nonmembers, because some of the increase in its economic growth will produce real increases in income that will in turn translate into increased imports from the rest of the world.

The impacts of PTAs can be determined using the partial equilibrium analysis (Tweeten, 1992). Assume that the case to be examined is that of Mexico and the United States, in which Mexico is a small-nation importer. In Figure 23.1, it is assumed that the United States is the low-cost supplier of a product imported by Mexico. Domestic supply and demand are S_M and D_M, respectively. Before the free trade area is created, the price in Mexico is P_{M+t} and the supply of U.S. imports with the tariff is M_{US+t}. Domestic production and consumption are Q_{S1} and Q_{D1}, respectively. Quantity imported with the tariff is $Q_{D1}-Q_{S1}$. Creating a free trade area removes the tariff and increases imports from the United States to M_{US} and lowers the price to P_M. When the price in Mexico declines, domestic production falls and domestic consumption increases. Imports from the United States rise to $Q_{D2}-Q_{S2}$, which exceeds the quantity imported with the tariff.

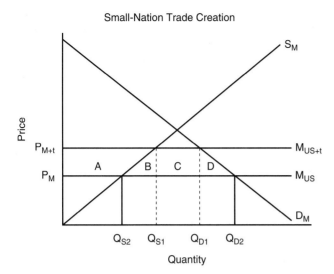

Small-Nation Trade Creation

FIGURE 23.1 When tariffs imposed by a small-nation importer are removed, consumers experience welfare gains, and producers and government experience welfare and revenue losses, respectively. The nation as a whole experiences welfare gains.

Trade Creation. Welfare impacts of forming a trade-creating free trade area can be determined by examining changes in producer and consumer surplus. Mexican consumers gain areas $A + B + C + D$ after the free trade area is implemented. Mexican producers, on the other hand, lose area A, and the government loses tariff revenue equal to area C. Total welfare gains to Mexico are the sum of areas $B + D$. The free trade area allows the importing nation to regain the deadweight loss from a tariff, which was discussed in Chapter 22.

A summary of the welfare effects of a small-nation trade creating free trade area are:

Consumer gains	$A + B + C + D$
Producer gains	$-A$
Government revenue	$-C$
Net gains to society	$= B + D$

Gains from a free trade area are expected to be large if the tariff that is to be removed is large and as domestic supply and demand become more elastic over the long run. Finally, consumer income gains within a free trade area can be expected to create trade with nonmember countries.

Trade Diversion. Trade diversion occurs when lower-cost imports from a non-member nation are replaced, or diverted, by higher-cost imports from a member nation of a free trade area. This occurs naturally with a PTA, due to the preferential trade treatment provided by member nations. Trade diversion reduces global

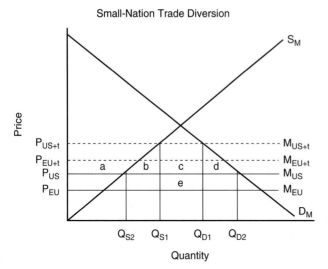

FIGURE 23.2   Trade diversion reduces global welfare because production is shifted from more-efficient nonmembers to less-efficient member producers. The small-nation importer gains from trade diversion only when the recovery of dead-weight losses due to the tariff exceeds losses of government revenue from the tariff.

welfare because it shifts production from more-efficient producers outside the PTA to less-efficient producers within the PTA. The international allocation of resources becomes less efficient and production shifts away from comparative advantage. Member nations may gain or lose from a PTA, as shown in this section.

The trade and welfare effects of trade diversion are demonstrated in Figure 23.2. Assume that both the United States and the EU compete for the Mexican market. The EU is initially the largest supplier to this market, noted by M_{EU}, which exceeds U.S. imports denoted by M_{US}. With an equal tariff of amount t applied to imports from both supplying countries, the EU is the sole source of import supply because P_{EU+t} is below P_{US+t}. In the absence of a free trade area, the price in Mexico is P_{EU+v} domestic production and consumption are Q_{S1} and Q_{D1}, respectively. Imports from the EU are equal to $Q_{D1}-Q_{S1}$.

If the United States and Mexico enter a free trade area, the tariff, t, is removed from imports supplied by the United States, but the tariff is not removed from imports supplied by the EU. P_{US} now prevails in the Mexican market, giving the United States a competitive advantage relative to the EU price, P_{EU+t}. Mexican consumption increases to Q_{D2}, and domestic production falls to Q_{S2}. Imports from the United States increase to $Q_{D2}-Q_{S2}$. Imports from the EU are no longer competitively priced and it no longer supplies the Mexican market. The result of the free trade area was to replace EU imports with those from the United States.

Welfare impacts of the trade-diverting free trade area reveal that consumers gain at the expense of producers and government. Because market price declines from P_{EU+t} to P_{US}, consumers gain areas $a+b+c+d$. Producers lose area a. Government tariff revenue declines by the sum of areas $c+e$. Therefore, net gains to Mexico from entering a free trade area with the United States occur only if the sum of areas $b+d$ exceeds area e.

A summary of the welfare effects of a small-nation, trade-diverting free trade area are:

Consumer gains	$a + b + c + d$
Producer gains	$- a$
Government revenue	$- (c + e)$
Net gains to society	$= b + d - e$

Most attempts to create PTAs contain both trade-creation and trade-diversion effects and can increase or decrease member welfare depending on the relative strength of the two opposing forces. PTAs will most likely lead to trade creation and increased welfare of member nations under the following conditions:

1. High pre-PTA trade barriers increase the probability that trade will be created among members, rather than diverted from nonmembers to members.
2. The more countries included in the PTA and the larger their size, the more likely that low-cost producers will be found among its members.
3. A PTA formed by competitive, rather than complementary, economies is more likely to produce opportunities for specialization in production and trade creation.
4. When member nations are in proximity to one another, transportation costs become less of an obstacle to trade creation.
5. If the free trade area contains countries with the lowest-cost source of goods and services consumed by member nations.

Dynamic Effects

The formation of a PTA can be expected to have major dynamic benefits that should be considered important to the participating nations. In fact, it has been recently estimated that the dynamic gains from forming a PTA often exceed the static or welfare gains by a factor of five or six (Salvatore, 1993). The more-important dynamic gains include increased competition, economies of scale, stimulus to investment, and more-efficient use of economic resources.

Increased Competition.   Possibly the most important single gain from a PTA is the potential for increased competition. Producers, especially those in monopolistic and oligopolistic markets, may become sluggish and complacent behind barriers to trade. With the formation of a PTA, trade barriers among members are eliminated, and producers must become more efficient to effectively compete. Some may merge with other firms; others will go out of business. The higher level of competition is likely to stimulate the development and adoption of new technology. These forces combined will likely reduce costs of production and, in turn, consumer prices for goods and services. Importantly though, the PTA must ensure that collusion and market-sharing arrangements are minimized if competitive forces are to operate efficiently.

Economies of Scale. Another major benefit of PTAs is that substantial economies of scale may become possible with the expanded market. If firms were serving only the domestic market, the expanded market with the PTA will likely create substantial export opportunities, resulting in more output, lower costs per unit, and greater economies of scale. For instance, it has been determined that before joining the EEC, many firms in small nations such as Belgium and the Netherlands, were comparable in size to U.S. plants, and thus enjoyed economies of scale by producing for the domestic market and for export. However, after becoming members of the EEC, significant economies of scale were gained by reducing the range of differentiated products manufactured in each plant, thereby gaining from increased specialization and greater reliance on comparative advantage.

Stimulus to Investment.   The formation of a PTA is likely to stimulate outside investment in production and marketing facilities to avoid the discriminatory barriers imposed on nonmember products. Further, in order for firms to meet the increased competition and take advantage of the enlarged market, investment is likely to increase. In most cases, investment is an alternative to the export of goods—a benefit provided to PTA members. The large investments made by U.S. firms in Europe after the mid-1950s and in the years and months leading up to the Single European Act were fostered by their desire not to be excluded from this large potential market; investing ensured that their products would not be restricted by tariff and nontariff barriers.

Efficient Resource Use.   Finally, if the PTA is a common market, the free movement of labor and capital is likely to stimulate more efficient use of the economic resources of the entire community. Efficiency of industries and individual firms will likely increase with increased access to lower-cost capital and additional labor. Lower consumer costs and higher real incomes should follow.

PROVISIONS OF THE NORTH AMERICAN FREE TRADE AGREEMENT

The North American Free Trade Agreement (NAFTA), involving the United States, Mexico, and Canada, has separate bilateral agreements in agricultural trade, one between the United States and Mexico and the other between Canada and Mexico. Because of the low level of Canada-Mexico agricultural trade, this discussion focuses primarily on the United States–Mexico bilateral agreement in agriculture. NAFTA became effective on January 1, 1994. U.S. agricultural trade within NAFTA reported $26 billion in 1999, growing by 50% since the early 1990s. Its general provisions include the following:

1. Immediate elimination of all import tariffs on a broad range of agricultural products that faced low or negligible duties. About one-half of United States and Mexico bilateral agricultural trade was duty free when NAFTA took ef-

fect. These commodities represent more than $1.5 billion in U.S. exports to Mexico and $1.6 billion in Mexican exports to the United States. Duties on beef, live cattle, and grain sorghum were eliminated under this provision.

2. Systematic reduction of all remaining tariffs on agricultural trade between the United States and Mexico. Some commodities will be covered by special safeguard provisions. A small share of trade (about 10%) will be liberalized over a five-year period. These products were deemed too sensitive for immediate liberalization but not sensitive enough to require more than five years for transition to freer trade.

3. Elimination of tariffs for most sensitive products over a 10- or 15-year transition period. Some of these products will be eligible for special safeguards in the form of tariff-rate quotas (TRQs) during the transition period. TRQs impose a low or zero duty on a specified import quantity. Imports in excess of the specified TRQ quantity are assessed a higher tariff. Both the within-quota and the over-quota tariffs will decline to zero during a specified time period. Initial TRQ quantities will be determined by recent average trade levels and will expand at a 3% annual compounded rate over the transition period. The United States will use 10-year TRQs for selected fruit and vegetable imports from Mexico valued at $330 million. Mexico will apply the 10-year TRQs on $155 million in imports from the United States, mainly hogs, pork, potatoes, and apples.

4. Elimination of tariffs on a few selected fruits and vegetables over a 15-year period. A 15-year period with TRQs is provided for products that are economically and politically sensitive, including U.S. imports of sugar, peanuts, and frozen/concentrated orange juice. Mexico will employ a 15-year transition with TRQs for corn, dry beans, and nonfat dry milk.

5. Elimination of nontariff barriers over specified transition periods. Mexico will eliminate its import licensing requirements on U.S. products. The United States will exempt Mexico from its Meat Import Act. The United States will also replace Section 22 (Agricultural Adjustment Act of 1933) quotas on imports from Mexico with TRQs during specified transition periods.

6. Provisions provide that complaining parties in disputes involving environmental measures or sanitary/phytosanitary regulations bear the burden of proof. This is in contrast to the WTO, where the burden of proof rests with the defending party. Dispute settlement must rely on scientific proof, and not on speculation as to the particular merits of a regulation.

7. A working group that will study ways to avoid using export subsidies. However, the United States may use export subsidies on products shipped to Mexico to counter subsidized competition from non-NAFTA countries such as the EU.

8. Two primary differences between the Uruguay Round Agreements (URA) and NAFTA are that: (1) NAFTA is a trilateral agreement calling for the phased reduction and complete elimination of tariff and nontariff barriers to trade, and (2) the URA is a multilateral agreement that reduces tariff and nontariff barriers to trade but does not eliminate them altogether. NAFTA is much more thorough at liberalizing trade, while the URA only begins the process of market opening and reducing trade barriers.

A historic first in trade agreements led to the negotiation of "side agreements" for the enforcement of environmental laws, labor regulations, and protection against import surges. NAFTA secretariats have been appointed to implement provisions of these important side agreements.

SUMMARY

The purpose of this chapter was to discuss the formation of preferential trading arrangements and their impacts on producers, consumers, taxpayers, and trade. Preferential trading arrangements have been growing in importance in recent years. Presently, about two-thirds of world trade is accounted for by "trading blocs." Trade among members has grown at the expense of the rest of the world. By their very nature, PTAs discriminate against nonmembers, favoring trade with member nations. The major points made in this chapter may be summarized as follows:

1. Preferential trading arrangements (PTAs) function legally within the framework of existing GATT and WTO rules. Because of the vague nature of these rules, nations have been able to discriminate rather freely against nonmember nations with little fear of retaliation.

2. PTAs may take the form of a free trade area, the least complete type of economic integration, a customs union, a common market or an economic union, the most complete form of economic integration.

3. A trade-creating, free trade area results in welfare gains for consumers in a small-nation importing country. Losses in economic welfare accrue to producers and government. Net gains to society occur because economic gains offset losses. In a trade-diverting trade area, net gains to society are realized only if consumer gains offset losses in government revenue from tariff elimination.

4. Trade theory provides only ambiguous conclusions regarding the consequences of PTA formation. Free trade is certainly more efficient than discriminatory trade, but in a world of less-than-free trade PTAs may permit major economic gains under certain conditions. It is likely that dynamic gains, such as increased competition, economies of scale, and greater investment will far exceed static gains from trade creation or trade diversion.

5. PTAs will most likely lead to trade creation and increased welfare of member nations under the following conditions:

   - High pre-PTA trade barriers increase the probability that trade will be created among members, rather than diverted from nonmembers to members.

   - The more countries included in the PTA and the larger their size, the more likely that low-cost producers will be found among its members.

   - A PTA formed by competitive, rather than complementary, economies is more likely to produce opportunities for specialization in production and trade creation.

   - When member nations are in proximity to one another, transportation costs become less of an obstacle to trade creation.

- If the free trade area contains countries with the lowest cost source of goods and services consumed by member nations, gains can be expected.

6. It appears likely that PTAs will take on added importance as policy tools for negotiating fewer barriers to trade among groups of nations. Although the United States may participate in the continued formation of PTAs for many reasons, it is likely that strong economic and political interests will be the driving force behind these actions. In the future, it will likely become more important to negotiate continued or additional access to major markets using the additional leverage provided by nations acting as a single unit.

DEFINITION OF KEY TERMS

Common market: includes all the aspects of a free trade area and customs union, but takes integration even further by permitting the free movement of goods and services, labor, and capital among member nations. In January 1993, the EC moved closer to full common market status by implementing the Single European Act (SEA).

Customs union: requires that members eliminate trade barriers with each other and takes economic integration a step further by establishing identical barriers to trade against nonmembers, most often in the form of a common external tariff. The European Economic Community was probably the world's most well-known customs union.

Economic union: represents the most complete form of economic integration. Member nation social, taxation, fiscal, and monetary policies are harmonized or unified. Belgium and Luxembourg formed an economic union in the 1920s. Customs documents and procedures, value-added taxation, and nontariff barriers to trade have been harmonized to facilitate the movement of goods among EEC member nations.

Free trade area: consists of provisions to remove both tariff and nontariff barriers to trade with members, while retaining individual trade barriers with nonmember nations. Trade barriers with the rest of the world differ among member nations and are determined by each member's policymakers. The North American Free Trade Agreement is an example of a free trade area.

North American Free Trade Agreement (NAFTA): free trade area among the United States, Mexico, and Canada.

Preferential trading arrangements: provide for the elimination or phased reduction of tariff and nontariff barriers among nations. PTAs may be bilateral or multilateral and often include the liberalization of investment laws, transportation regulations, protection of intellectual property, and provisions for resolving trade disputes.

Reciprocal Trade Agreements Act of 1934 (RTA): authorizes the president to fix tariff rates and has resulted in a liberal tariff stance for the United States. From 1934 to 1947, in fact, the United States negotiated bilateral trade agreements with 29 nations. However, when the GATT emerged as the primary trade forum, the RTA declined in importance as a mechanism for trade liberalization.

REFERENCES

Fieleke NS: One trading world or many: the issue of regional trading blocs, *New England Economic Review,* May/June 1992.

Jackson JH: *The world trading system: law and policy of international economic relations,* Cambridge, Mass, 1989, MIT Press.

Johnson HG: 1965. An economic theory of protectionism, tariff bargaining, and the formation of customs unions, *Journal of Political Economy,* September 1965.

Rosson CP, Runge CF, Hathaway D: *Food and agricultural policy issues and choices for 1995,* Boulder, Colo, 1994, Westview Press.

Runge CF: The assault on agricultural protectionism, *Foreign Affairs,* Fall 1988.

Runge CF: Economic stability rides on success of GATT negotiations, *Feedstuffs* 64:28, 1992.

Salvatore D: *International economics,* 4th ed., New York, 1993, Endowment of International Peace.

Tweeten L: *Agricultural trade: principles and policies,* Boulder, Colo, 1992, Westview Press.

Viner J: *The customs union issue,* New York, 1950, Carnegie Endowment for International Peace.

EXERCISES

1. Preferential trading arrangements (PTAs) provide for the elimination of tariff and nontariff barriers to trade. T F

2. The North American Free Trade Agreement creates the world's largest free trade area. T F

3. Trade creation occurs as some domestic production of a member nation is replaced by lower-cost imports from another member nation. T F

4. A trade-creating preferential trading agreement can increase the welfare gains of nonmember nations as well as member nations. T F

5. Under Article I of GATT, now the WTO, preferential tariff rates are prohibited, but an exception is allowed by Article XXVI if which of the following conditions are met?

   a. trade barriers are eliminated on substantially all trade among members.

   b. trade barriers remaining against nonmembers are not more severe or restrictive than those previously in effect.

    c. interim measures leading to the formation of the agreement are employed for only a reasonable period of time.

    d. all of the above.

6. The Canada–U.S. Trade Agreement (CUSTA) provides for:

    a. the elimination of tariff and nontariff barriers to trade between the two countries.

    b. specific rules governing trade in services.

    c. reduced restrictions on investments by both nations over a 10-year period.

    d. all of the above.

7. The Southern Cone Common Market, MERCOSUR, was created to foster economic growth and mutual interests among the member nations of:

    a. Brazil, Guatemala, Paraguay, Argentina.

    b. Paraguay, Uruguay, Argentina, Brazil.

    c. Argentina, Guatemala, Panama, Brazil.

    d. Brazil, Paraguay, Argentina, Panama.

8. Important dynamic gains from the formation of a preferential trading arrangement include:

    a. economies of scale.

    b. increased competition.

    c. stimulus to investment.

    d. more efficient use of economic resources.

9. List the five reasons discussed for preferential trading arrangements.

10. Discuss the five conditions under which preferential trading arrangements will most likely lead to trade creation and increased welfare of member nations.

11. Discuss at least four of the general provisions included in the North American Free Trade Agreement.

12. Draw two graphs, one to illustrate small nation trade diversion and one to illustrate small nation trade creation. Label and explain each one.

INDEX